Communications in Computer and Information Science 1780

Rationale

The CCIS series is devoted to the publication of proceedings of computer science conferences. Its aim is to efficiently disseminate original research results in informatics in printed and electronic form. While the focus is on publication of peer-reviewed full papers presenting mature work, inclusion of reviewed short papers reporting on work in progress is welcome, too. Besides globally relevant meetings with internationally representative program committees guaranteeing a strict peer-reviewing and paper selection process, conferences run by societies or of high regional or national relevance are also considered for publication.

Topics

The topical scope of CCIS spans the entire spectrum of informatics ranging from foundational topics in the theory of computing to information and communications science and technology and a broad variety of interdisciplinary application fields.

Information for Volume Editors and Authors

Publication in CCIS is free of charge. No royalties are paid, however, we offer registered conference participants temporary free access to the online version of the conference proceedings on SpringerLink (http://link.springer.com) by means of an http referrer from the conference website and/or a number of complimentary printed copies, as specified in the official acceptance email of the event.

CCIS proceedings can be published in time for distribution at conferences or as post-proceedings, and delivered in the form of printed books and/or electronically as USBs and/or e-content licenses for accessing proceedings at SpringerLink. Furthermore, CCIS proceedings are included in the CCIS electronic book series hosted in the SpringerLink digital library at http://link.springer.com/bookseries/7899. Conferences publishing in CCIS are allowed to use Online Conference Service (OCS) for managing the whole proceedings lifecycle (from submission and reviewing to preparing for publication) free of charge.

Publication process

The language of publication is exclusively English. Authors publishing in CCIS have to sign the Springer CCIS copyright transfer form, however, they are free to use their material published in CCIS for substantially changed, more elaborate subsequent publications elsewhere. For the preparation of the camera-ready papers/files, authors have to strictly adhere to the Springer CCIS Authors' Instructions and are strongly encouraged to use the CCIS LaTeX style files or templates.

Abstracting/Indexing

CCIS is abstracted/indexed in DBLP, Google Scholar, EI-Compendex, Mathematical Reviews, SCImago, Scopus. CCIS volumes are also submitted for the inclusion in ISI Proceedings.

How to start

To start the evaluation of your proposal for inclusion in the CCIS series, please send an e-mail to ccis@springer.com.

Claudio De Stefano · Francesco Fontanella ·
Leonardo Vanneschi
Editors

Artificial Life and Evolutionary Computation

16th Italian Workshop, WIVACE 2022
Gaeta, Italy, September 14–16, 2022
Revised Selected Papers

Editors
Claudio De Stefano
University of Cassino and Southern Lazio
Cassino, Italy

Francesco Fontanella
University of Cassino and Southern Lazio
Gaeta, Italy

Leonardo Vanneschi
Universidade Nova de Lisboa
Lisbon, Portugal

ISSN 1865-0929 ISSN 1865-0937 (electronic)
Communications in Computer and Information Science
ISBN 978-3-031-31182-6 ISBN 978-3-031-31183-3 (eBook)
https://doi.org/10.1007/978-3-031-31183-3

This Springer imprint is published by the registered company Springer Nature Switzerland AG
The registered company address is: Gewerbestrasse 11, 6330 Cham, Switzerland

Preface

This volume of *Communications in Computer and Information Science* contains the proceedings of WIVACE 2022, the XVI Workshop on Artificial Life and Evolutionary Computation. The event was organized by the University of Cassino and Southern Lazio and successfully held at the Castle of Gaeta, Italy, on September 14–16, 2022. WIVACE aims to bring together researchers working in the field of artificial life and evolutionary computation to present and share their research in a multidisciplinary context. The workshop provided a forum for discussing new research directions and applications in different fields, including Bioinformatics, Bioinspired Algorithms and Robotics, Complex Systems, Synthetic and Systems Biology and Chemistry, and theories and applications of Artificial Life, among others.

Overall, WIVACE 2022 hosted 31 invited or contributed talks. In total, 24 high-quality papers were accepted for this proceedings volume from 45 initial submissions after a single-blind review round performed by at least three Program Committee members. Submissions and participants in WIVACE 2022 came from 14 different countries, making WIVACE an increasingly international event despite its origins as an Italian workshop. Following this ever-increasing international spirit, the previous edition was held in Switzerland, and it is expected that future WIVACE editions will also be held outside Italy.

Many people contributed to this successful edition. We would like to thank the authors for submitting their works, the members of the Program Committee, who devoted so much time to reviewing papers despite a tight schedule, the session chairs, who stimulated interesting discussions during the workshop, and finally, the two invited speakers.

Our thanks also go to the University of Cassino and Southern Lazio for providing the venue for the event, the Angevin-Aragonese Castle of Gaeta (a venue rich in history and surrounded by wonderful coastal scenery, creating a peaceful atmosphere for the workshop), and the Municipality of Gaeta, in particular the Deputy Mayor, Mrs. Teodolinda Morini, for affording the WIVACE 2022 participants such a pleasant stay in Gaeta.

We would also like to mention Stefano Cagnoni, Roberto Serra, and Marco Villani for being a constant source of fruitful inspiration and ideas for WIVACE.

Lastly, we would also like to thank Tiziana D'Alessandro, Davide De Giuli, Alessandro De Rosa, and Andrea Omodei for their precious help before and during the workshop.

September 2022

Claudio De Stefano
Francesco Fontanella
Leonardo Vanneschi

Organization

General Chairs

Claudio De Stefano	University of Cassino and Southern Lazio, Italy
Francesco Fontanella	University of Cassino and Southern Lazio, Italy

Program Committee

Marco Antoniotti	University of Milan "Bicocca", Italy
Marco Baioletti	University of Perugia, Italy
Vito Antonio Bevilacqua	Politecnico di Bari, Italy
Leonardo Bich	University of the Basque Country (UPV/EHU), Spain
Leonardo Bocchi	University of Firenze, Italy
Eleonora Bilotta	University of Calabria, Italy
Andrea Bracciali	University of Stirling, UK
Michele Braccini	University of Bologna, Italy
Angelo Cangelosi	University of Manchester, UK
Timoteo Carletti	University of Namur, Belgium
Mauro Castelli	Universitade Nova de Lisboa, Portugal
Antonio Chella	University of Palermo, Italy
Franco Cicirelli	ICAR-CNR, Italy
Nicole Dalia Cilia	Kore University of Enna, Italy
Chiara Damiani	University of Milan "Bicocca", Italy
Luca Di Gaspero	University of Udine, Italy
Alessia Faggian	University of Trento, Italy
Harold Fellermann	Newcastle University, UK
Rudolf Füchslin	Zurich University of Applied Sciences, Switzerland
Mario Giacobini	University of Turin, Italy
Alex Graudenzi	IBFM-CNR, Italy
Antonio Guerrieri	ICAR-CNR, Italy
Giovanni Iacca	University of Trento, Italy
Hans-Georg Matuttis	University of Electro-Communications, Japan
Giancarlo Mauri	University of Milan "Bicocca", Italy
Sara Montagna	University of Bologna, Italy
Marco Mirolli	ISTC-CNR, Italy

Antonella Nardin	Università degli Studi Roma Tre, Italy
Mario Pavone	University of Catania, Italy
Riccardo Pecori	eCampus University, Italy
Stefano Piotto	University of Salerno, Italy
Clara Pizzuti	ICAR-CNR, Italy
Simone Righi	University of Venezia, Italy
Andrea Roli	University of Bologna, Italy
Federico Rossi	University of Siena, Italy
Johannes Schneider	Zurich University of Applied Sciences, Switzerland
Roberto Serra	University of Modena and Reggio Emilia, Italy
Alessandra Scotto di Freca	University of Cassino and Southern Lazio, Italy
Annalisa Socievole	ICAR-CNR, Italy
Giandomenico Spezzano	ICAR-CNR, Italy
Pasquale Stano	University of Salento, Italy
Thomas Stuetzle	Université Libre de Bruxelles, Belgium
Pietro Terna	University of Torino, Italy
Marco Villani	University of Modena and Reggio Emilia, Italy
Andrea Vinci	ICAR-CNR, Italy

Contents

Networks and Complex Systems

Collaborative Learning over Cellular Automata

Franco Cicirelli, Emilio Greco, Antonio Guerrieri, Giandomenico Spezzano, and Andrea Vinci(✉)

ICAR-CNR, Via P. Bucci 8/9C, 87036 Rende, Italy
{franco.cicirelli,emilio.greco,antonio.guerrieri, giandomenico.spezzano,andrea.vinci}@icar.cnr.it

Abstract. There are many real scenarios in which some correlated complex problems have to be addressed by different autonomous learners working in parallel. In such a scenario, the collaboration among the learners can be extremely useful since they can share acquired knowledge so as to reach a reduction in the learning time, an increase in the learning quality, or both of them. Anyway, in some cases, it is not always feasible to collaborate with other learners. This is because the problems to solve are not compatible or they can have dissimilar boundary conditions leading to very different problem solutions. In this paper, we propose an approach to collaborative learning which leverages cellular automata for efficiently solving a set of compatible and sufficiently similar problems. In this direction, the notion of compatibility and similarity between problems is also given and discussed. A case study based on the maze problem will show the effectiveness of the proposed approach.

Keywords: Collaborative Learning · Cellular Automata · Artificial Intelligence · Reinforcement Learning

1 Introduction

Promoting collaboration among distributed and autonomous learners is an important and hot topic in the domain of machine learning [19]. Collaboration is multi-objective, e.g., it aims at improving the quality of the overall learning process, reducing the time needed for learning a new task, or both of them. Collaboration becomes more important as the complexity of the learning task increases and as the availability of training data reduces or such data becomes difficult to obtain [10], e.g., when the data acquisition process takes a long time.

As an example, let's consider the case of a glazed building in which all the windows on the same facade are equipped with an automatic curtain system. An intelligent controller, one for each windows, is in charge of automatically adapt the height and inclination of a Venetian blind in order to avoid indoor glare and manage visual comfort while taking into consideration also the exchange of thermal energy with the external environment [7]. More in particular, the

C. De Stefano et al. (Eds.): WIVACE 2022, CCIS 1780, pp. 3–14, 2023.
https://doi.org/10.1007/978-3-031-31183-3_1

controller has to take into account that, for energy consumption concerns [6], it is preferable to keep the curtain as opened as possible during the winter and as closed as possible during the summer season [4]. The problem above is very complex, this is because it is very complicated to obtain a model describing the system, and the evolution of such model strongly depends by boundary conditions like window geometry, position and orientation, type and position of the furniture in the room, room color, season, and climatic conditions. All of this also implies that it is very difficult to have available data for training the controller, and that a viable solution for training is that of putting the controller in operation in the room and let the controller learn on-line with the data coming from the real environment.

Despite system complexity, though, some considerations can be done in order to deal with the training phase. A first consideration is that it is very likely that all the controllers operate in similar conditions, i.e., each windows is close to each other, and each windows has the same geometry, orientation, and external climatic conditions. This implies that, the physical nearness of windows can be, in this case, turned in a notion of *nearness* in the problems to solve. Taking into account only physical positions, though, is not enough to argument that a group of controllers have to deal with similar problems. As stated before, in fact, other factors like the room color and furniture heavily affect the boundary condition of the problem to solve. As a consequence, a suitable notion of *problem nearness* could be also introduced in order to estimate in which case a group of problems are *near to solve*. Once *similar problems* are identified, it is possible to consider that a kind of collaboration can be established among their relevant controllers during training [18].

The main contribution of this work is twofold: (i) a novel approach for promoting collaboration among learning entities is provided. Such an approach is a peer-to-peer one, i.e., each learner at the same time can absolve the function of students and teachers, in which each learner can both ask and provide suggestions to neighbors learners. These suggestions are used for improving the learning process for solving a problem, i.e., to determine the actions to perform in order to solving it. A learner rates the received suggestions and, on the base of such rates, it can quote the established collaborations; (ii) a problem to solve is defined as a labelled transition system for which a *similarity metric* is defined. Such a metric is exploitable to asses the similarity among problems so as to establish the learners which have to collaborate together. The proposed metric does not ensure that two or more problems are *semantically* similar, the metric introduced a *syntax-based* similarity. In the case the metric would fails in determining when problems are really similar, the quoting mechanism on the received suggestions will guide a learner to not consider anymore the suggestion received by a specific learner. An interesting aspect of our approach is that the jointly use of a peer-to-peer collaboration schema paired with the cellular automata paradigm, naturally opens to the exploitation of the approach in a parallel and decentralized context thus fostering scalability in dealing with large problems.

The proposed approach is experimented in the domain of maze problems. A cellular automata of maze solvers is considered, each of them exploiting a reinforcement learning algorithm to find problem solutions. The mazes to solve are randomly generated, and the adopted generation algorithm is able to generate mazes at a given distance with respect to the introduced metric. In such a way each solver in the cellular automata will have in its neighborhood other solvers which are solving similar problem, i.e., problems at a fixed distances. Experimental results show how the proposed approach is really effective in improving the learning phase and problem solving capabilities of the system.

The structure of the paper follows. Section 2 proposes some related works. Section 3 describes the proposed approach and related key concepts, while Sect. 4 shows the chosen case study. Finally, Sect. 5 concludes the paper with an indication of on-going and further work.

2 Related Work

In the last decades, worldwide research effort was given to develop new machine learning algorithms for the prediction of future events in a system and, furthermore, to add self-learning and self-adapting capabilities to such systems, making them intelligent. Such machine learning algorithms are able to processes both labeled and unlabeled data, and the resulting models are, in general, exploited for specific contexts and applications [4–6,20].

The current trend is to take advantage of the models already trained somewhere else (i.e., exploiting the knowledge already acquired) so as to speed up the learning process in some specific cases so solving new problems faster or with better solutions. For example, the so-called *transfer learning* has been defined as "the ability of a system to recognize and apply knowledge and skills learned in previous tasks to novel tasks" [17].

Together with the transfer learning, great attention is lately given to the *federated learning* [1]. Federated learning, initially developed by Google [12] to solve problems related to data privacy, communication costs, and regulations, is a distributed machine learning strategy in which models are trained in distributed nodes and are then merged in a centralized one. Such merged model is then scattered again to the distributed nodes that can take advantage of a new model performing better than the single trained models [9,11].

Other forms of *cooperative/collaborative learning* have been also applied so far [15,18,19]. More in detail, the authors in [18] review papers about *team learning* and *teammate learning.* In team learning only one learner is devoted to discovering behaviors for a team of agents [3,13] while in teammate learning each agent on the team independently learns how to improve its performance and the team's performance as a whole [8,14].

In [23] the authors used the concept of collaborative learning to realize a system for resource-constrained IoT edge devices. In particular, they used a partitioned model training approach applied to a hybrid deep learning model. Their results proved that the approach used gives better performance than the on-device training approach.

The authors of [15] introduce a very interesting form of collaborative learning inspired by how humans learn from other humans. They use cellular learning automata (CLA) [2] (namely, cellular automata where cells host learning automata [16]) to realize a parallel multi-agent learning entity. Anyway, all the agents involved in [15] collaborate to learn together the best way to solve a problem. Our work also wants to use a collaborative learning approach by taking advantage of cellular automata but, in our case, we consider different problems for each cell, and each cell hosts a problem that is regarded as similar with respect to other problems in its neighborhood. In this way, each cell can ask for suggestions from its neighbors to solve its problem.

3 Collaborative Learning

This work proposes an approach devoted to supporting collaborative learning among independent and autonomous learners hereafter also called agents. The basic assumption is that a set of *compatible* problems can be learnt and solved by a set of agents, each one learning on a different, yet compatible, problem. The overall goal is to improve the quality of the whole learning process. For these purposes, in the following we define what we consider as a *problem* and what properties a set of problems needs to own in order to be solved by the agents in a collaborative way. In this paper, we focus on collaborative learning for problems framed in the Reinforcement Learning [21] scenario. More in particular, this preliminary proposal is designed to exploit the well-known Q-Learning algorithm, described below, to support collaboration.

3.1 Problem Definition

In this paper, we consider Reinforcement Learning (RL) problems defined as an automata

$$P_{RL} = (S, A, Start, C, f(s,a), R(s,a)) \tag{1}$$

in which:

- S is the finite set of states;
- A is the finite set of actions;
- $Start \subset S$ is the set of initial states;
- C is a set of terminal conditions;
- $f(s,a) \rightarrow s'$ is the transition function, i.e., $s' \in S$ is the state obtained when the action $a \in A$ is executed in a state $s \in S$;
- $R(s,a)$ is a function which defines the rewards obtained when the action $a \in A$ is executed in the state $s \in S$.

A policy function for a problem P is a function $\pi(s) \rightarrow a$ which gives, for each state $s \in S$ the action $a \in A$ to perform. Solving the problem corresponds to determine the best policy function $\hat{\pi}(s)$ that maximizes the cumulative reward obtained during the evolution of the problem automata.

As our approach proposes to solve a set of problems in a collaborative way, we need to introduce some notions for assessing (i) if two problems are *compatible*, namely are able to be processed collaboratively by our approach, and (ii) an estimation of how much two problems are *similar*, that is, an estimation of how effective our collaborative approach could result in solving the two problems.

Problem Compatibility. Let P_1 and P_2 be two problems such that:

$$P_1 = (S_1, A_1, start_1, C_1, f_1(s,a), R_1(s,a))$$
$$P_2 = (S_2, A_2, start_2, C_2, f_2(s,a), R_2(s,a))$$

we state that P_1 and P_2 are *compatible* when the following condition hold:

$$S_1 = S_2 \land A_1 = A_2 \land start_1 = start_2 \land C_1 = C_2 \tag{2}$$

In other words, two problems are compatible when they share the same states, the same actions, the same starting state, and the same set of final conditions. Compatibility is a structural property that allows to use the same policy among problems.

Moreover, we define the notion of *similarity* between problems, as a measure of how an optimal policy for a problem P_1 could be a valuable policy also for a problem P_2.

Problem Similarity. Given two compatible problems P_1 and P_2, with $S = S_1 = S_2$ and $A = A_1 = A_2$, we define the *similarity index* $\sigma(P_1, P_2)$ as:

$$\begin{aligned}\sigma(P_1, P_2) = |\{(s,a) : (f_1(s,a) \rightarrow s' \land \neg f_2(s,a) \rightarrow s') \\ \lor (f_2(s,a) \rightarrow s'' \land \neg f_1(s,a) \rightarrow s'')\}| + |\{(s,a) : R_1(s,a) \neq R_2(s,a)\}|\end{aligned} \tag{3}$$

with $s, s', s'' \in S$, $a \in A$, and $|\cdot|$ being the cardinality of a set. Two compatible problems are identical if $\sigma = 0$ and increase their dissimilarity as σ grows. Intuitively, the defined similarity index *counts* the number of differences existing between the two considered transition functions and the two considered reward functions of the problems.

3.2 Q-Learning

An agent which solves an RL problem is an agent that *learns through experience* the best policy $\hat{\pi}$. The agent can run on the problem several times, starting from one of the initial states and moving among states by applying the transition function until a terminal condition holds. Each run is defined as an *episode*. Each time the agent moves from a state to another one through an action, it receives a reward according to the $R(s,a)$ function.

In the Q-Learning algorithm [22], the agent stores a matrix of Q-values, $Q : S \times A \rightarrow \mathbb{R}$, initialized to zero. When in a state s_t, the agent can choose, with a probability ϵ, a random action, or, with a probability $1 - \epsilon$,the action that corresponds to the $\max_{a \in A} Q(s_t, a)$. This randomness permits the agent to explore different solutions, avoiding getting stuck with local optima.

Every time the agent is in a state s_t, chooses an action a_t, executes it, retrieves the corresponding reward $r_t = R(s_t, a_t)$, and enters the new state $s_{t+1} = f(s_t, a_t)$, it updates the $Q(s, a)$ matrix according to the below reported Bellman equation, as a weighted average of the old value and the new information:

$$Q(s_t, a_A) \leftarrow Q(s_t, a_t) + \alpha \cdot (r_t + \gamma \cdot \max_a Q(s_{t+1}, a) - Q(s_t, a_t)) \quad (4)$$

where $\alpha \in [0, 1]$ is the learning rate and $\gamma \in [0, 1]$ is the discount factor.

After a number of episodes, the $Q(s, a)$ matrix is expected to converge, and the best policy found is given by $\hat{\pi}(s) = \max_a Q(s, a)$.

3.3 Proposed Approach

The presented approach relies on cellular automata and Q-Learning to support collaborative learning for solving a set of *compatible* problems solved through reinforcement learning.

Given a set of problems, each problem and a corresponding learning agent are assigned to a different cell of a cellular automata.

Each agent is responsible of learning the best policy of the assigned problem, by running a modified version of the Q-Learning which takes into account the CA structure. In particular, during the learning process, each agent can exploit the knowledge gathered by its neighbors. The CA will complete its evolution when all the cells reach the state of convergence.

The agents execute according to the following behavior:

1. Each agents is initialised with $Q(s, a) = 0$, and a rating value associated to all its neighbors $rt_i = 0$, where i identifies the n neighbors of the agent, given a neighborhood function (e.g., Von Neumann, Moore).
2. Each agent runs several episodes of the associated RL problem.
3. An agent chooses a random action with probability ϵ, asks to its neighbors with probability β, or otherwise chooses the action a that maximizes its $Q(s, a)$ value.
4. If an agent asks to its neighborhood, it chooses the neighbor associated to the highest rating rt_i, then checks the neighbor Q matrix and chooses the action a that maximizes such $Q(s, a)$ value. If more than one neighbor has the higher rate, it chooses the first. If the selected action a is not admissible, the associated rt_i is decreased by one and an admissible action is randomly chosen.
5. The agent executes the selected action a, and updates its $Q(s, a)$ according to Eq. 4.
6. If the selected action a is chosen by asking a neighbor, the agent updates the rating associated to the selected neighbor. If the suggested action produces an

increment of $Q(s, a)$, the rate rt_i of the corresponding neighbor is increased. If the suggested action produces a decrement of the $Q(s, a)$, the rt of the corresponding neighbor is decreased by 1.

7. The agent runs on several episodes until it does not reach the convergence, that is, until it does not update its $Q(s, a)$ matrix for 30 episodes. In the latter case the agent terminates the learning process.

When all the cells in the CA terminate, the whole learning process ends. The best policies found for each reinforcement learning problem will be given by the $Q(s, a)$ matrix computed by the corresponding agent.

4 Case Study

In order to asses the effectiveness of the proposed approach, we considered the resolution of several sets of compatible and similar maze problems. Each set is associated to a 5×5 cellular automata. In the following, we first define the maze problem as a RL problem. Then, we introduce the experimental setting, the baseline exploited for comparison purpose, and the evaluation metrics. Finally, we show and discuss the obtained experimental results.

4.1 Definition of the Maze Problem

A maze is defined on a grid of $M \times N$ cells as a non oriented graph $\mathcal{M} = \langle V, E, v_o \rangle$ where:

- each $v_{i,j} \in V$ is a cell on the grid, with $0 \leq i < N$ and $0 \leq j < M$,
- there exists a $v_o \in V$ that is the exit of the maze,
- there exists an edge $(v_{i,j}, v_{l,m}) \in E$ if an agent can move from $v_{i,j}$ to $v_{l,m}$. An edge $(v_{i,j}, v_{l,m})$ exists if $v_{i,j}$ is adjacent to $v_{l,m}$ and there is no "wall" between $v_{i,j}$ and $v_{l,m}$. Two cells $v_{i,j}$ and $v_{l,m}$ are adjacent if $j = m \wedge \|i-l\| = 1$ or if $i = l \wedge \|j - m\| = 1$.
  Since $\mathcal{M}$ is non-oriented, $(v_{i,j}, v_{l,m}) \in E \rightarrow (v_{l,m}, v_{i,j}) \in E$

Given a maze $\mathcal{M} = \langle V, E, v_o \rangle$, the reinforcement learning problem $P_{\mathcal{M}}$ for solving the maze $\mathcal{M}$ is defined as a tuple $\langle S, A, start, C, f(s,a), R(s,a) \rangle$ where:

- S is the set of states such that $\exists s_{i,j} \in S, \forall v_{i,j} \in V$. We also define the terminal state $s_o = s_{i,j}$ if $v_o = v_{i,j}$.
- $A = \{a_n, a_s, a_w, a_e\}$ is the set of available actions, which corresponds, respectively, to moving in the north, south, west, and east directions. We also define the concept of *opposite action* such that $a_n^{-1} = a_s$, $a_s^{-1} = a_n$, $a_w^{-1} = a_e$, and $a_e^{-1} = a_w$.
- $start = S \backslash \{s_o\}$, where $s_o \in S$, that is, each state $s_{i,j}$ can be a starting state but s_o, which is the terminal state;
- $C = \{s_o\}$, the game ends when the current state is the terminal state $s_o \in S$;

- $f(s, a)$ is the transition function, and is built as follows:

$$\begin{aligned} f(s_{i,j}, a_n) &\rightarrow s_{i-1,j},\ if\ \exists(v_{i,j}, v_{i-1,j}) \in E \\ f(s_{i,j}, a_s) &\rightarrow s_{i+1,j},\ if\ \exists(v_{i,j}, v_{i+1,j}) \in E \\ f(s_{i,j}, a_w) &\rightarrow s_{i,j-1},\ if\ \exists(v_{i,j}, v_{i,j-1}) \in E \\ f(s_{i,j}, a_e) &\rightarrow s_{i,j+1},\ if\ \exists(v_{i,j}, v_{i,j+1}) \in E \end{aligned} \tag{5}$$

- $R(s, a)$ is defined as follow:

$$R(s, a) = \begin{cases} 100, & if\ f(s, a) \rightarrow s_o \\ -1, & if\ f(s, a) \rightarrow s' \neq s_o. \end{cases} \tag{6}$$

The reward $R(s, a)$ is such that an high prize is given when the goal is reached, and a penalty is instead given for each other action. In such a way, the learning agents are pushed to find the shortest path toward the exit.

Solving the maze $\mathcal{M}$ consists in finding the best policy function $\hat{\pi}_{\mathcal{M}}(s) \rightarrow a$ that finds the path between any $s \in start$ to s_o and that maximizes the cumulative reward obtained during the evolution of the problem automata.

It is worth noting that, given a maze problem $P_{\mathcal{M}_1}$, adding a wall between $v_{i,j}$ and $v_{l,m}$ generates a new maze problem $P_{\mathcal{M}_2}$ such that $P_{\mathcal{M}_1}$ and $P_{\mathcal{M}_2}$ are *compatible* (see Sect. 3.1) and $\sigma(P_{\mathcal{M}_1}, P_{\mathcal{M}_2}) = 4$. This is because adding the wall means changing only the values of $f(v_{i,j}, a_x) \rightarrow v_{l,m}$, $f(v_{l,m}, a_x^{-1}) \rightarrow v_{i,j}$, $R(v_{i,j}, a_x)$, and $R(v_{l,m}, a_x^{-1})$, thus the compatibility is verified since only the transition function $f(s, a)$ and the reward function $R(s, a)$ change, and the value of the similarity is clearly given by applying Eq. 3. The same holds when a wall is removed from a maze.

4.2 Experimental Setting - Case Study Scenario

We validated our approach by solving mazes of size 5×5. Some example mazes are depicted in Fig. 1. We apply our method to different maze structures, obtained by randomly adding K walls to a basic maze that has no walls.

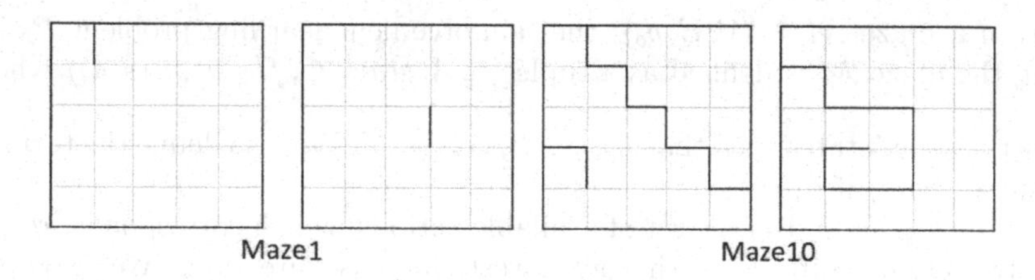

Fig. 1. Two different mazes with 1 wall (on the left) and two mazes with 10 walls (on the right).

With *maze0* we indicate a structure without internal walls. With *maze1* we indicate a structure with a single internal wall, randomly arranged in the internal perimeter. Simply, with *mazeK* we indicate a maze with K internal walls added in a random way. The structure of the maze is automatically generated by an algorithm that checks for the existence of an admissible solution. It is worth noting that if we generate a number of *mazeK* mazes, we guarantee that the similarity between each couple of the set is upper bounded to be less or equal than $2 \cdot K \cdot 4$.

The experiments are conducted by choosing sets of $W = 25$ maze problems having the same number of walls, arranged in a CA of size 5×5. The considered sets of mazes are generated each with a different $K \in \{0, 1, ..., 10\}$. The following preliminary results are gathered by comparing three cases:

- *Baseline*: obtained by applying an instance of classic Q-learning to each problem to solve without using collaboration, and $\alpha = 1$, $\gamma = 0.8$, and $\epsilon = 0.2$;
- *Collaborative Learning - Moore (CL-Moore)*: obtained by applying the proposed collaborative approach and considering the Moore neighborhood (8 neighbors) in the cellular automata and $\alpha = 1$, $\gamma = 0.8$, $\epsilon = 0.2$ and $\beta = 0.4$;
- *Collaborative Learning - von Neumann (CL-VN)*: obtained by applying the proposed collaborative approach and considering the von Neumann neighborhood (4 neighbors) in the cellular automata, and $\alpha = 1$, $\gamma = 0.8$, $\epsilon = 0.2$ and $\beta = 0.4$.

4.3 Comparison Metrics

For comparison purpose, the following metrics are defined.

- The *Average Training Episodes (ATE)* for solving a set of problems

$$ATE = \frac{1}{W} \sum_{i=1}^{W} T_i \tag{7}$$

where W is the number of issued problems and T_i is the number of episodes required for the agent training on the problem i to reach the convergence. The ATE indicates how many iterations are required to solve the whole set of problems. Obviously, lower is this value, more efficient is the training.
- The global *score*, which is the average for each problem of the *qscore* computed on each resulting $Q(s, a)$ matrix. The *qscore* computed on a $Q(s, a)$ matrix is the sum of the values of the $Q(s, a)$ normalized on its maximum value *maxQ*. When we apply two different training processes on the same set of problems, if the resulting solutions are similar, we expect to have similar scores.

$$qscore(Q) = \sum_{s \in S} \sum_{a \in A} \frac{Q(s, a)}{maxQ} \tag{8}$$

$$score = \frac{1}{W} \sum_{i=1}^{W} qscore(Q_i) \tag{9}$$

where W is the number of issued problems and Q_i is the $Q(s, a)$ matrix obtained for the problem i when the related agent has reached convergence.

4.4 Experimental Results

Given the baseline, CL-Moore, and CL-VN cases introduced in Sect. 4.2 we execute 30 runs of the training process for each set of mazes with $K \in 0, 1, ..., 10$. In the following, we discuss the resulting ATE and score metrics obtained by averaging the 30 executions.

Figure 2 shows the ATE needed to solve the whole set of problems. For each point, also the 95% confidence interval is shown as a bar. For the baseline case (without collaboration), we see that the ATE remains around 9500 episodes for all the cases. When our collaborative learning approach is applied, the ATE is lower than the baseline and increases with the number of walls in the mazes. In addition, it is worth of noting that also the confidence interval is more narrow, thus implying that the collaboration permits a more stable behavior. The best results are obtained in the CL-Moore case, thus witnessing the fact that an higher number of collaborating neighbors leads to a more efficient learning process.

Figure 3 shows that the achieved score remains almost the same in all the considered case. By crossing the results shown in Fig. 2 and Fig. 3 we can affirm that the proposed approach permits to speed-up the training process without impairing the learning quality.

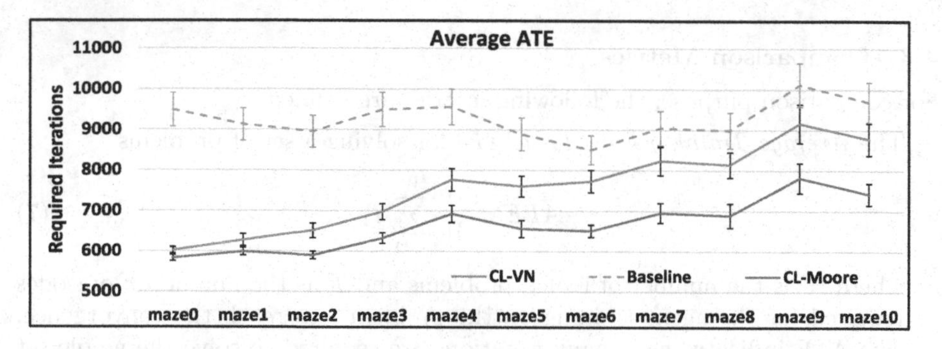

Fig. 2. Comparison of the average ATE calculated on all the problems by increasing dissimilarity, considering the baseline, CL-VN, and CL-Moore cases.

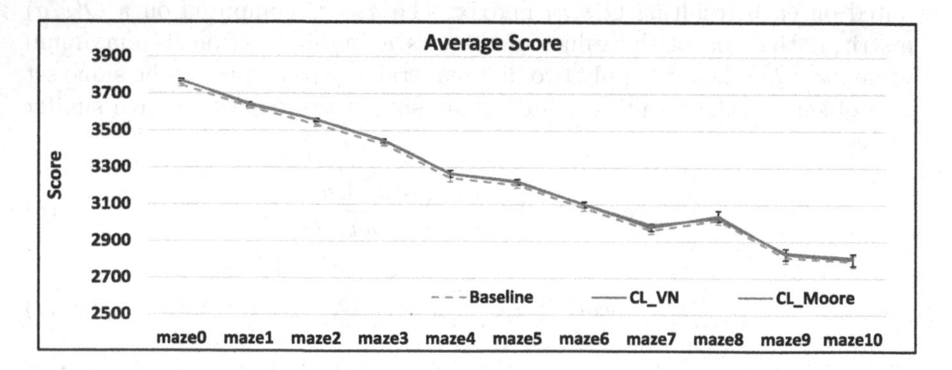

Fig. 3. Comparison of the average score metric calculated on all the problems by increasing dissimilarity, considering the baseline, CL-VN, and CL-Moore cases.

5 Conclusion

The paper presented a Collaborative Learning approach leveraging Q-Learning and Cellular Automata for solving sets of *compatible* and *similar* problems. The notions of compatibility and similarity between problems are also introduced.

The reported experiments, made by comparing our collaborative learning approach with a set of independent q-learning executions, have highlighted how our proposal obtains better results in terms of total number of episodes required to solve all the problems in a set, while maintaining the same quality in the solutions found, even when the dissimilarity grows for the considered problems.

Ongoing/future work is devoted to: (i) extending the set of experiments to a wider range of states of CA; (ii) improving the algorithm by providing suitable techniques for arranging CA structure in order to maximize the similarity among neighbor problems; (iii) assessing the approach taking into account other reinforcement learning algorithms which consider both a non finite state space or a non finite action space (e.g., Deep Q-Learning and Asynchronous Advantage Actor-Critic (A3C) Algorithm).

Acknowledgment. This work has been partially supported by the COGITO (A COGnItive dynamic sysTem to allOw buildings to learn and adapt) project, funded by the Italian Government (PON ARS01 00836) and by the CNR project "Industrial transition and resilience of post-Covid19 Societies - Sub-project: Energy Efficient Cognitive Buildings (TIRS-EECB)".

References

1. Aledhari, M., Razzak, R., Parizi, R.M., Saeed, F.: Federated learning: a survey on enabling technologies, protocols, and applications. IEEE Access **8**, 140699–140725 (2020)
2. Beigy, H., Meybodi, M.: A mathematical framework for cellular learning automata. Adv. Complex Syst. **7**(3–4), 295–319 (2004). https://doi.org/10.1142/S0219525904000202
3. Burke, E., Gustafson, S., Kendall, G.: A puzzle to challenge genetic programming. In: Foster, J.A., Lutton, E., Miller, J., Ryan, C., Tettamanzi, A. (eds.) EuroGP 2002. LNCS, vol. 2278, pp. 238–247. Springer, Heidelberg (2002). https://doi.org/10.1007/3-540-45984-7_23
4. Cicirelli, F., Guerrieri, A., Spezzano, G., Vinci, A.: IoT Edge Solutions for Cognitive Buildings. Internet of Things, Springer, Heidelberg (2022). https://doi.org/10.1007/978-3-031-15160-6
5. Cicirelli, F., Gentile, A.F., Greco, E., Guerrieri, A., Spezzano, G., Vinci, A.: An energy management system at the edge based on reinforcement learning. In: 2020 IEEE/ACM 24th International Symposium on Distributed Simulation and Real Time Applications (DS-RT), pp. 1–8. IEEE (2020)
6. Cicirelli, F., Guerrieri, A., Mastroianni, C., Scarcello, L., Spezzano, G., Vinci, A.: Balancing energy consumption and thermal comfort with deep reinforcement learning. In: IEEE 2nd International Conference on Human-Machine Systems (2021)

7. Greco, E., Spezzano, G.: Human-centered Reinforcement Learning for Lighting and Blind Control in Cognitive Buildings. In: Cicirelli, F., Guerrieri, A., Spezzano, G., Vinci, A. (eds.) IoT Edge Solutions for Cognitive Buildings. Internet of Things, pp. 285–303. Springer, Cham (2022). https://doi.org/10.1007/978-3-031-15160-6_13. Chap 13
8. Iba, H.: Evolutionary learning of communicating agents. Inf. Sci. **108**(1–4), 181–205 (1998)
9. Khan, I., Guerrieri, A., Spezzano, G., Vinci, A.: Occupancy prediction in buildings: an approach leveraging LSTM and Federated Learning. In: The IEEE 2022 DASC/PiCom/CBDCom/CyberSciTech. IEEE (2022)
10. Li, L., Fan, Y., Tse, M., Lin, K.Y.: A review of applications in federated learning. Comput. Industr. Eng. **149**, 106854 (2020)
11. Li, T., Sahu, A.K., Talwalkar, A., Smith, V.: Federated learning: challenges, methods, and future directions. IEEE Signal Process. Mag. **37**(3), 50–60 (2020)
12. McMahan, B., Moore, E., Ramage, D., Hampson, S., y Arcas, B.A.: Communication-efficient learning of deep networks from decentralized data. In: Artificial Intelligence and Statistics, pp. 1273–1282. PMLR (2017)
13. Miconi, T.: A collective genetic algorithm. In: Proceedings of the 3rd Annual Conference on Genetic and Evolutionary Computation, pp. 876–883 (2001)
14. Miconi, T.: When evolving populations is better than coevolving individuals: the blind mice problem. In: IJCAI, pp. 647–652. Citeseer (2003)
15. Mozafari, M., Shiri, M.E., Beigy, H.: A cooperative learning method based on cellular learning automata and its application in optimization problems. J. Comput. Sci. **11**, 279–288 (2015). https://doi.org/10.1016/j.jocs.2015.08.002
16. Narendra, K.S., Thathachar, M.: Learning automata-a survey. IEEE Trans. Syst. Man Cybern. **SMC-4**(4), 323–334 (1974). https://doi.org/10.1109/TSMC.1974.5408453
17. Pan, S.J., Yang, Q.: A survey on transfer learning. IEEE Trans. Knowl. Data Eng. **22**(10), 1345–1359 (2009)
18. Panait, L., Luke, S.: Collaborative multi-agent learning: a survey. Department of Computer Science, George Mason University, Technical report (2003)
19. Panait, L., Luke, S.: Cooperative multi-agent learning: the state of the art. Auton. Agent. Multi-Agent Syst. **11**(3), 387–434 (2005)
20. Sarker, I.H.: Machine learning: algorithms, real-world applications and research directions. SN Comput. Sci. **2**(3), 1–21 (2021)
21. Sutton, R.S., Barto, A.G.: Reinforcement Learning: An Introduction. MIT Press, Cambridge (2011)
22. Watkins, C.J., Dayan, P.: Q-learning. Mach. Learn. **8**(3), 279–292 (1992)
23. Welagedara, L., Harischandra, J., Jayawardene, N.: Edge intelligence based collaborative learning system for IoT edge. In: 2021 IEEE 12th Annual Information Technology, Electronics and Mobile Communication Conference (IEMCON), pp. 0667–0672 (2021). https://doi.org/10.1109/IEMCON53756.2021.9623215

An Agent-Based Model for Crowd Simulation

Carolina Crespi, Georgia Fargetta, Mario Pavone(✉), and Rocco A. Scollo

Department of Mathematics and Computer Science, University of Catania, v.le A. Doria, 6, 95125 Catania, Italy
{carolina.crespi,georgia.fargetta,rocco.scollo}@phd.unict.it, mpavone@dmi.unict.it

Abstract. In this paper, we propose an agent-based model for crowd simulation. It is made up of two types of agents that act differently: collaboratives, which share information about the paths and/or repair the ones that have been damaged, and defectors, who share no information, destroy some paths and/or nodes, and take advantage of the collaborative agents' guidance. The aim of the model is to investigate how the agents who engage in these two activities affect one another and, ultimately, on the collective behavior of the crowd. For both kinds of agents, three evaluation metrics have been considered: (*i*) the number of agents that have reached the exit; (*ii*) paths' costs; and (*iii*) exit times. According to the data analysis, the presence of defectors is essential in improving the agents' performance. Indeed, when both collaborators and defectors are present in the crowd, more agents exit, and their average path costs and average exit times are lower than when only collaborative agents are present.

Keywords: agent-base model · ant colony optimization · swarm intelligence · crowd simulation · collective behaviour

1 Introduction

Simulating crowd behavior has become one of the most challenging topics of the last few years. This is because there are lots of exciting and practical applications in different areas. In emergency management, for instance, it is crucial to know in advance or estimate how people would behave to establish efficient escape plans [11]. Understanding and outlining the different psychological aspects of human behaviors under dangerous circumstances is also a hot topic in social science studies [16] to improve evacuation plans [15]. Crowd simulations are also used for architectural design purposes, like for instance, where to place the exits in a building, in a cinema, in a theatre, in a stadium, in a school, and so on, to have an orderly maximum flow, and how to realize corridors to avoid bottlenecks [13]. Thanks to the increase in computers' performance, crowd simulations

C. De Stefano et al. (Eds.): WIVACE 2022, CCIS 1780, pp. 15–26, 2023.
https://doi.org/10.1007/978-3-031-31183-3_2

have become a challenge also in the entertainment industry [19]; battle scenes and crowd movements in movies and video games are just an example of this kind of application.

In most of these cases, however, there isn't enough information or real data to understand and forecast what will happen and how people will behave. This is due to several factors, the most important of which is that carrying out real experiments is not easy task as it requires to involve a large number of participants. Moreover, data taken from evacuation drills are not properly accurate since people know that they are not in danger and, on the other hand, is not ethical to put people in danger just to observe how they act and react [9]. Information about this kind of situation may be obtained in studies in which at first a model as detailed as possible is created, and then it is compared to a real-life event that happened in the past, adapting or defining *ad hoc* the parameters needed [1]. In this context, simulations are not only valid support but also a necessary tool to recreate, in a controlled environment, different kinds of situations otherwise difficult or impossible to analyze. Crowds' models can be very different from one to another and this depends on what one wants to focus on and on the framework used. In general, crowd models can be classified into two main categories, macroscopic and microscopic models [14]. In macroscopic models, people in a crowd are represented collectively as a homogeneous flow and they are mostly used to represent large-scale scenarios, even if they are not so accurate because they are not able to explain human emergent behaviors. Microscopic models, on the contrary, focus on individuals' characteristics and are then more accurate albeit more computationally expensive. There exists a third category in which falls all those models that have at the same time macroscopic and microscopic features, that is the hybrid or mesoscopic models. Those models try to exploit the best features of both categories or even features from different areas, combining them into a new unified framework.

Following this research line, we have realized an agent-based model for crowd simulations in which we have merged two techniques: (*i*) the *agent-based approach*, that is one of the most powerful tool to model individual decision-making and investigate the emergent behaviours [8]; and (*ii*) the *Ant Colony Optimization* (ACO) algorithm's principles to simulate the agents' movements and environment setup, as swarm intelligence algorithms can efficiently solve optimization problems [2,3,5,7,10,12], and show collective behaviors in model crowds' dynamics [4,6,17]. In the proposed model, the agents are put in an uncertain and dynamic virtual environment with the aim of reaching the exit, starting from a chosen point. Two different behavioral strategies for each agent are adopted: *collaborative*, which consists in sharing information on the paths and/or on the nodes and repairing the damaged ones; and *defector* that do not share any information, work individually, may accidentally destroy some paths and/or nodes, and takes advantage of the help of the collaborative agents as well. The model aims to understand how agents adopting these two different behaviors affect each other and, consequently, the crowd's overall behavior. To do this, three different evaluation metrics have been simultaneously compared for each kind of agent:

(*i*) the number of agents who reached the exit; (*ii*) the costs of the paths, and (*iii*) the exit times. According to the findings, the presence of a small percentage of defectors improves overall crowd performance by increasing the number of agents who reach the exit in the shortest time and at the lowest cost of the pathways. This happens, likely, because the defectors' activities encourage the other agents, particularly the collaborative ones, to better explore the environment in search of more promising pathways, utilizing the traces and information released into the environment.

2 The Crowd Model

In this work, the Ant Colony Optimization (ACO) algorithm framework has been considered to model agent dynamics and environment features. The ACO algorithm is a well-known metaheuristic generally used to solve different kinds of combinatorial optimization problems. In this context, we have exploited its capabilities to show collective dynamics. The environment is modeled as a weighted undirected graph $G = (V, E, w)$, where V is the set of vertices, $E \subseteq V \times V$ is the set of edges, and $w\colon V \times V \rightarrow \mathbb{R}^+$ is a *weighted* function that assigns to each edge of the graph a positive cost. The *weighted* function indicates how difficult is for the agents to cross an edge. Let be $A_i = \{j \in V : (i,j) \in E\}$ the set of vertices adjacent to vertex i and $\pi^k(t) = (\pi_1, \pi_2, \ldots, \pi_t)$ a non-empty sequence of vertices, with repetitions, visited by an agent k at the time-step t, where $(\pi_i, \pi_{i+1}) \in E$ for $i = \{1, \ldots, t-1\}$. Starting from a given point, a population of N agents, divided into Γ groups, explores the environment and tries to reach the destination as fast as possible, using a path with the lowest cost. Each group begins its exploration at regular intervals, as it was a delayed exiting process [14]. Indeed, it is assumed that in some contexts people do not all evacuate at the same time, but they organize themselves to evacuate in an orderly fashion, e.g., in schools, public offices, and especially in a recent pandemic plan to avoid Covid-19 spread. This situation was modeled by specifying that each group would begin its tour after a specific time.

At this stage, we applied the (ACO) algorithm's proportional transition rule to model the agents' path-choosing mechanism. At a given time t, an agent k on a vertex i selects one of its neighboring vertices j as a destination point with the probability $p_{ij}^k(t)$ determined by:

$$p_{ij}^k(t) = \begin{cases} \frac{\tau_{ij}(t)^\alpha \cdot \eta_{ij}(t)^\beta}{\sum_{l \in J_i^k} \tau_{il}(t)^\alpha \cdot \eta_{il}(t)^\beta} & \text{if } j \in J_i^k, \\ 0 & \text{otherwise,} \end{cases} \tag{1}$$

where $J_i^k = A_i \backslash \{\pi_{t-1}^k\}$ are all the possible displacements of the agent k; $A_i = \{j \in V : (i,j) \in E\}$ is the set of vertices adjacent to vertex i; $\pi^k(t) = (\pi_1, \pi_2, \ldots, \pi_t)$ is the set of vertices visited by an agent k; $\tau_{ij}(t)$ is the trace intensity on the edge (i,j); $\eta_{ij}(t) = 1/w_{ij}$ is the desirability of the edge (i,j); α and β are parameters that regulate the importance of trace intensity with respect to the desirability of an edge.

The *trace intensity* τ_{ij} on edge (i, j) indicates how many times an edge has been crossed. It is a piece of passive information left unintentionally by both agents and can help new agents make a decision based on the actions of the previous ones. It does not have a physical meaning, but it can be thought of as knowledge about the state of the environment and what other agents did before. Then, we used the reinforcement rule to model the interaction between the agents and the environment. When an agent crosses an edge, the trace $\tau_{ij}(t)$ on that edge is increased by a constant K, that is:

$$\tau_{ij}(t+1) = \tau_{ij}(t) + K, \tag{2}$$

with K a user-defined parameter. Finally, we used the global updating rule to model the effect of time on the environment. In particular, the more time passes, the more the trace, and so the knowledge about the state of the environment, changes. Every T ticks, i.e. the time unit used that corresponds to a single movement of all agents, the trace changes according to:

$$\tau_{ij}(t+1) = (1-\rho) \cdot \tau_{ij}(t), \tag{3}$$

where ρ is the evaporation decay parameter.

The *desirability* $\eta_{ij}(t)$ in Eq. 1 indicates, instead, how much an edge (i, j) is promising. This information is intentionally released by the collaborative agents just after crossing an edge, so it is strictly related to the discovery process. It is equal to the inverse of the weight of the edge (i, j), that is $\eta_{ij}(t) = 1/w(i, j)$. The lower the cost is, the greater the desirability will be, and vice versa. Note that the desirability is asymmetric because this information is present on the vertices, that is $\eta_{ij}(t) \leq \eta_{ji}(t)$ at a given time t. Then, in summary, Eq. 1 determines how the agents move into the environment and establish that an agent k on a node i chooses as destination a node j weighting the trace intensity $\tau_{ij}(t)$, which is reinforced at every passage according to Eq. 2 and changes in time according to Eq. 3, and the desirability $\eta_{ij}(t)$ that is equal to the inverse of the weight of the edge (i, j).

The agents interact with the environment also in different ways, according to their behavior. In particular, collaborators primarily behave by aiding all other agents in reaching the target place as quickly as feasible. They traverse an edge or vertex cautiously taking care to not damage it and to fix it if this was harmed. Furthermore, they leave information $\eta_{ij}(t)$ on how difficult it is to cross a certain edge, which the other agents might use in their own choices. Defectors, on the other hand, mostly behave in a hurried manner, carrying out acts that have the potential to disrupt the surrounding environment. After passing a node or an edge, they may accidentally destroy it, reducing the possibility of other agents investigating the environment. This action may have an impact not just on the cooperative agents but also on themselves, especially if the damaged path is critical to reaching the target place. Following this, the actions performed by the two types of agents can be summarized as follows:

- *collaborators* C: they leave an information $\eta_{ij}(t)$, after crossing an edge to help other agents in the search process. They may also repair a destroyed edge and/or a destroyed vertex before crossing that path with a probability P_e^C and P_v^C, respectively;

- *defectors* D: they do not leave any information after crossing an edge. They may accidentally destroy an edge and/or a vertex after their passage with a probability P_e^D and P_v^D, respectively. If a node or an edge is destroyed, then it is no more crossable by other agents; in the specific case of a node, this still becomes reachable but not crossable.

It follows that these two behaviors affect not only the agents but also the structure of the environment from time to time, thereby rendering it dynamic and increasing uncertainty. Consequently, focusing just on one of the three evaluation metrics would have been inaccurate, because a good path might not be the best in terms of cost, agent success rate, or exit time. A good path is one that lowers its cost and exit time while also increasing the number of agents who exit through it. The path cost $\pi(t) = (\pi_1, \pi_2, \dots, \pi_t)$ is calculated as $\sum_{i=1}^{t-1} w(\pi_i, \pi_{i+1})$, with π_1 and π_t starting and destination points, respectively. All agents in the environment complete a single movement (from one vertex to another) at each tick. It follows that the exit time is determined as the number of movements made by an agent throughout its investigation and corresponds to the length of the path $\pi(t)$.

3 Experimental Setup

The *NetLogo* multi-agent programmable modeling environment [18] was used to create the proposed model and perform the simulations. As stated in Sect. 2, the environments have been described as graphs having a topology similar to grid networks, where each node can be associated with its 8-neighbors. Node's connectivity with its neighbors is determined by two parameters: $0 \leq p_1 \leq 1$, which represents the probability of creating horizontal and vertical edges, and $0 \leq p_2 \leq 1$, which indicates the probability of creating oblique edges. Each edge's weight is a real number with a uniform distribution in the range $[1, 100]$.

For the experiments, three different configurations were considered, each of them generated setting $p_1 = 0.6$, and $p_2 = 0.2$:

- *scenario* $A2$ with $|V| = 100$ and $|E| = 213$;
- *scenario* $B2$ with $|V| = 225$ and $|E| = 495$;
- *scenario* $C2$ with $|V| = 400$ and $|E| = 889$.

In all test cases, $N = 1000$ agents were considered, separated into $\Gamma = 10$ groups; in this way each group consists of 100 agents. Of these, f are collaborative, and $(1-f)$ defectors, depending on the value of the user-defined parameter $f \in [0, 1]$, named *collaborative factor*. It follows that both or only one type of agent may be present in each group. If $f = 0.0$, only defector groups are examined; if $f = 0.5$, each group is created by half collaborative and half defectors; and if $f = 1.0$, only collaborative groups are considered.

Except for the first group, which obviously starts at the time $T_e = 0$, each group begins its exploration at a specific time $T_e = |V|$ from the group that precedes it. In general, the i-th group will start its exploration at $T_e \times (i-1)$. All the agents must reach the exit within a time limit T_{max} given by $T_{max} = 2 \times \Gamma \times T_e$, where Γ is the number of the groups, and 2 is a fixed parameter. The time range in which each agent must reach the exit is measured from the start of its exploration to the overall maximum time T_{max} (i.e. $T_{max} - (T_e \times (i-1))$).

This equation means that the first groups have more time to examine the environment than the others because the model permits groups to begin their exploration even if agents from earlier groups are still present in the environment. This implies that agents in the same group might exit at different times (always within their time range), but those in the first group have always more time to reach the exit. The trace released along the path deteriorates with time so that every $T_d = 50$ ticks the global updating rule, given by Eq. 3, is applied, with evaporation rate $\rho = 0.10$. The trace is initially set to $\tau_{ij}(t = 0) = 1.0$ for all edges $(i, j) \in E$; and the parameters α and β that regulate the importance of the trace and desirability in Eq. 1 are both set to 1.0. The destruction-repair probabilities of a vertex and/or an edge are the same for both kinds of agents and are respectively set to $P_e^C = P_e^D = 0.02$ and $P_v^C = P_v^D = 0.02$. Finally, 10 independent simulations have been performed for each value of f, ranging from 0.0 to 1.0 with a 0.1 step.

4 Results

As mentioned in Sect. 1, we simultaneously compared three evaluation metrics to understand how collaborative and/or defector strategies affect the crowd's overall behavior: (1) number of agents reaching the exit (Success Rate – SR); (2) exit times; and (3) cost of the paths for reaching the exit. In particular, we evaluated how the SR affects the exit times and the path cost by normalizing these last with respect to the former. In the following, we report the results obtained for all the configurations used.

4.1 Absolute and Relative Counting

In Fig. 1 are plotted the absolute counts of the number of agents for scenarios $A2$, $B2$, and $C2$ for different values of the collaborative factor f. Both kinds of agents are considered, and different colors are used to distinguish them: coral for collaborative agents and turquoise for defectors. When the overall behavior is considered, it seems that the crowd performance improves with respect to f, reaching a maximum value, and subsequently drop in all three scenarios. The highest number of exited agents is obtained, respectively, at $f = 0.7$ in scenario $A2$ (almost 650 - Fig. 1a), at $f = 0.6$ in scenario $B2$ (almost 180 – Fig. 1b), and at $f = 0.7$ in scenario $C2$ (almost 360 – Fig. 1c). Looking at the maximum values it appears that the crowd performs best in scenario $A2$, followed in scenario $C2$, and last in scenario $B2$. This may be determined by the scenario's complexity and the presence of specific node and edge configurations. When $f = 1.0$, that is when only collaborative agents are present, the overall performance is the worst. The same findings are obtained when $f = 0.0$, that is when there are just defectors, but this is not unexpected because they damage and block parts of their pathways with a given probability, making them physically unable to proceed toward the exit.

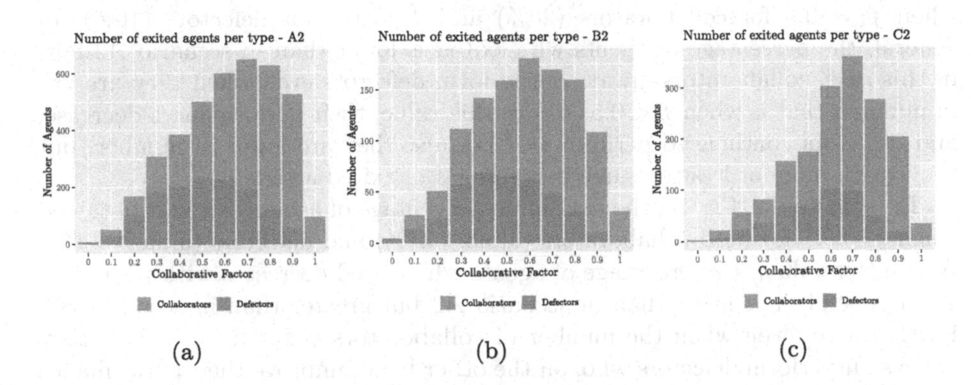

Fig. 1. Absolute counts for (a) scenario **A2**, (b) scenario **B2**, and (c) scenario **C2**.

In Fig. 2 are plotted the relative counts. Unlike the previous plots, these ones show the percentage of agents of each type, collaborators, and defectors, that reach the exit. The percentage is calculated over the total number of agents N, that is $N = 1000$. Also in this case, it seems that the performances of the agents increase with respect to f but with some differences depending on their type and scenario considered. Indeed, in scenario $A2$, (Fig. 2a), the maximum percentage of agents exited is obtained when $f = 0.7$ both for collaborators and defectors. Indeed, for this value, more than 60% of both kinds of agents reach the exit. Above this value of f, the performances of both agents get worse, but it is a note of interest that even when the defectors are significantly lower in number ($f = 0.9$), they still maintain good performances and outperform

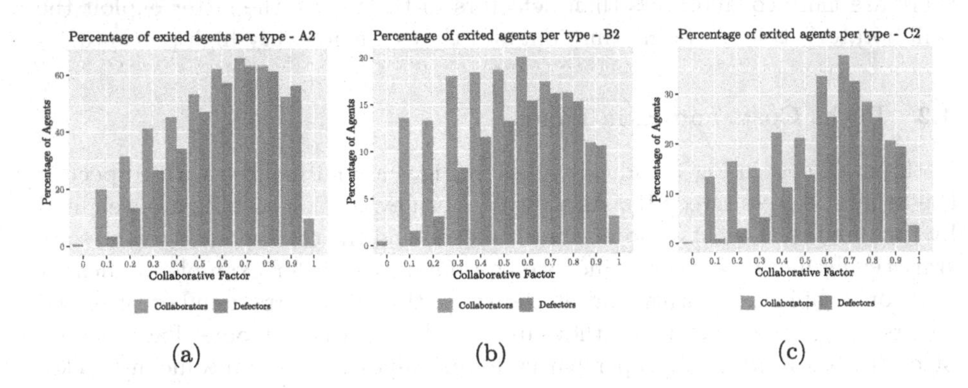

Fig. 2. Relative count for (a) scenario **A2**, (b) scenario **B2** and (c) scenario **C2**.

the collaborators. In general, collaborators performances improve as f increases, reaching a peak, and then decrease until the minimum for $f = 1.0$. For this last value of f, only 10% of collaborators reach the exit. The same is true for defectors but their performances do not drop as the collaborators ones.

In scenario $B2$, Fig. 2b, the maximum percentage of agents exited is obtained when $f = 0.6$ for collaborators (20%) and $f = 0.7$ for defectors (16%). In general, the percentage of agents who exited is lower than in scenario $A2$ but, in this case, collaborators always outperform defectors even when they are low in number, that is when $f < 0.5$. Above this value, their performances decrease, and defectors continue to perform well because they are lower in number, but they reach more or less the same percentages of exited agents.

In scenario $C2$, Fig. 2c, the maximum percentage of agents exited is obtained when $f = 0.7$ for both collaborators (almost 37%) and defectors (almost 32%). Above this value, the percentage of agents who exited decreases with respect to f. In general, it is lower than in scenario $A2$ but greater than in scenario $B2$. Furthermore, even when the number of collaborators is small ($f < 0.5$), they always outperform defectors who, on the other hand, improve their performance in the specular situation of numerical inferiority ($f > 0.5$). Above this value, their performances decrease, and defectors continue to perform better than collaborators because they are lower in number, but they reach more or less the same percentages of exited agents. In all scenarios, the percentage of collaborators that reach the exit when $f = 1.0$, that is when the crowd is completely collaborative, is extremely inferior to the percentage of the same kind of agents for lower values of f.

Based on the results in Fig. 1, we may conclude that not all the $N = 1000$ agents reach the exit. This depends on the setup of each scenario as well as the fact that the agents have a limited amount of time to reach the exit. In general, the more complicated the configurations, more difficulties the agents find in reaching the exit. The agents performance improves due to the presence of defectors: when the crowd is mixed ($0.0 < f < 1.0$), that is, when it contains both collaborators and defectors, more agents exit than when it consists only of collaborators ($f = 1.0$). In fact, Fig. 2 shows that when $f > 0.5$, and thus when there are more collaborators than defectors in the crowd, the latter exploit the collaborators' traces and information, and they exit in a very high percentage.

4.2 Paths Costs and Exit Times

Both paths' cost in Fig. 3 and exit times in Fig. 4 are normalized with respect to the agents' success rate (SR). As a result, both evaluation metrics are weighted by the number of agents who reached the exit, and the corresponding values are statistically more significant since we can determine if, for example, a cheaper path or a rapid exit correspond at the same time to a large number of exited agents. In general, the lower they are, the better the outcome; lower values of path costs represent cheaper paths; lower values of exit times mean quicker agents.

About the paths costs, in scenario $A2$, Fig. 3a, the costs of collaborators' and defectors' paths decrease with respect to f, but in opposite ways. Defectors considerably reduce the costs of their pathways in relation to f, and the minimum value is reached when $f = 0.7$. Collaborators gradually reduce the cost of their pathways in relation to f, achieve the minimum for $f = 0.7$, and then rapidly worsen in performance for $f = 1.0$, the value for which the path cost increases dramatically. In scenario $B2$, Fig. 3b, the costs of defectors' pathways increase with respect to f while the costs of collaborators worsen. Defectors, in particular, significantly improve their performances and achieve the minimum when $f = 0.6$. Collaborators find a good path for $f = 0.3$ and then go worse until they reach their worst performance for $f = 1.0$. The pattern in scenario $C2$, Fig. 3c, is analogous to the previous ones. Defectors and collaborators find their best minimum path for $f = 0.7$: the former continue to improve their results in terms of f; collaborators, on the other hand, gradually improve and then suddenly drop for $f = 1.0$.

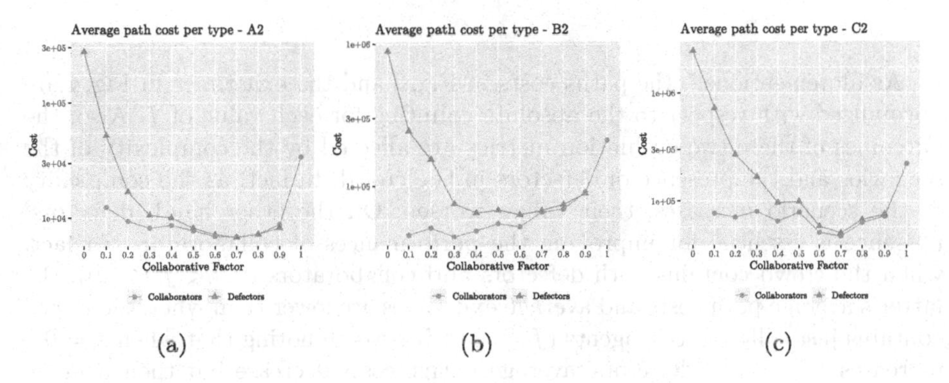

Fig. 3. The path cost for (a) scenario **A2**, (b) scenario **B2** and (c) scenario **C2**.

About the exit times, in scenario $A2$, Fig. 4a, the defectors exit faster when $f = 0.0$ than when $f = 0.1$. After that, at the increasing of $f > 0.1$, their performances improve until to achieve the lowest exit time with $f = 0.8$. For $f = 0.9$ the exit time is higher than the previous two cases. Even collaborators gradually improve their performances, which means they exit quickly even though they are few. They achieve the minimum for $f = 0.8$, exactly as the defectors, while, strangely, the worst exit time compared to all f values is reached for $f = 1.0$, i.e. when the crowd is all composed only of collaborators. Scenario B2, Fig. 4b, has similar features to the previous one. Defectors perform poorly for $f = 0.1$ and then improve above this value, lowering their exit time until they reach the minimum for $f = 0.7$. Collaborators improve their performances until $f = 0.6$, after which point they worsen. They have their worst results when $f = 1.0$, also on this scenario. Even scenario C2, Fig. 4c, is similar to the others. At $f = 0.7$, both collaborators and defectors reach the minimum exit time. Collaborators exit slowly at $f = 1.0$, but unlike the previous two scenarios, their performance at this value of f is better than at $f = 0.1$.

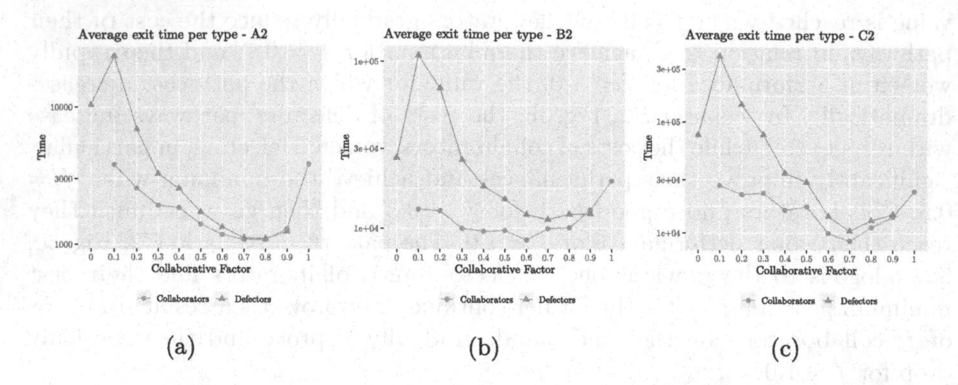

Fig. 4. The exit time for (a) scenario **A2**, (b) scenario **B2** and (c) scenario **C2**.

As aforementioned, the paths costs in Fig. 3 and the exit times in Fig. 4 are normalized with respect to the absolute counting for each value of f. Also, the outcomes of these two evaluation metrics are affected by the complexity of the scenarios and the presence of defectors in the crowd. In fact, as the complexity of the scenario increases, their values worsen. On the other hand, defectors' presence is essential for improving the performances of collaborators. In fact, when the crowd contains both defectors and collaborators ($0.0 < f < 1.0$), the latter's average path costs and average exit times are lower than when the crowd contains just collaborative agents ($f = 1.0$). It's worth noting that when $f = 0.0$ increases to $f = 0.1$, defectors' average paths' costs decrease but their average exit times increase. Given that the number of exited agents increases for the same f values, we may deduce that for $f = 0.0$, defectors who reached the exit first are simply the lucky ones who did not become trapped in the environment.

5 Conclusions

In this paper, we present an agent-based model for crowd simulation. It consists of two different types of agents that behave differently: (*i*) collaboratives, which share information about the paths and/or repair the ones that have been destroyed, and (*ii*) defectors, which share no information, destroy some paths and/or nodes, and take advantage of the collaborative agents' guidance. The model aims to show how the agents who adopt these two different behaviors influence one other and, as a result, the crowd's overall behavior. To do this, we compared three evaluation metrics for both types of agents simultaneously: (1) number of agents that have reached the exit; the (2) paths costs, and (3) exit times. We deduce from the data analysis that the features of the scenario and the presence of defectors influence the agents' performances, and hence the evaluation metrics used. In the first case, as complexity increases, the evaluation metrics values worsen, because the agents have more paths to explore.

In the second case, when the crowd comprises both collaborators and defectors ($f \neq 0$), more agents exit, and their average path costs and average exit times are lower than when the crowd contains just collaborative agents ($f = 1.0$). We believe that the defectors' actions, such as blocking nodes and/or edges, aid the agents in reducing the complexity of the scenario while also steering the search process toward more promising pathways by exploiting the collaborators' traces and information. The proposed model and the obtained findings are currently being investigated and require more qualitative and quantitative validations. In this regard, our efforts are already geared toward making the model as realistic as feasible. Ongoing projects include simulations with the collaboration parameter f changing over time; simulations in scenarios with multiple starting points and endings; and sensitivity analysis of the parameters used. In future works, we would study different values and combinations as $\alpha = 0.5$ and $\beta = 1$, $\alpha = 1$ and $\beta = 0.5$ and $\alpha = 0.5$ and $\beta = 0.5$, to investigate all cases for the middle and highest value of each parameter and each combination.

References

1. Bosse, T., Hoogendoorn, M., Klein, M.C.A., Treur, J., van der Wal, C.N., van Wissen, A.: Modelling collective decision making in groups and crowds: integrating social contagion and interacting emotions, beliefs and intentions. Auton. Agent. Multi-Agent Syst. **27**(1), 52–84 (2013). https://doi.org/10.1007/s10458-012-9201-1
2. Consoli, P., Collerà, A., Pavone, M.: Swarm intelligence heuristics for graph coloring problem. In: 2013 IEEE Congress on Evolutionary Computation, pp. 1909–1916 (2013). https://doi.org/10.1109/CEC.2013.6557792
3. Consoli, P., Pavone, M.: O-BEE-COL: optimal BEEs for COLoring graphs. In: Legrand, P., Corsini, M.-M., Hao, J.-K., Monmarché, N., Lutton, E., Schoenauer, M. (eds.) EA 2013. LNCS, vol. 8752, pp. 243–255. Springer, Cham (2014). https://doi.org/10.1007/978-3-319-11683-9_19
4. Crespi, C., Fargetta, G., Pavone, M., Scollo, R.A.: an agent-based model to investigate different behaviours in a crowd simulation. In: Mernik, M., Eftimov, T., Črepinšek, M. (eds.) BIOMA 2022. LNCS, vol. 13627, pp. 1–14. Springer, Cham (2022). https://doi.org/10.1007/978-3-031-21094-5_1
5. Crespi, C., Fargetta, G., Pavone, M., Scollo, R.A.: How a different ant behavior affects on the performance of the whole colony. In: Di Gaspero, L., Festa, P., Nakib, A., Pavone, M. (eds.) MIC 2022. LNCS, vol. 13838, pp. 187–199. Springer, Cham (2023). https://doi.org/10.1007/978-3-031-26504-4_14
6. Crespi, C., Fargetta, G., Pavone, M., Scollo, R.A., Scrimali, L.: A game theory approach for crowd evacuation modelling. In: Filipič, B., Minisci, E., Vasile, M. (eds.) BIOMA 2020. LNCS, vol. 12438, pp. 228–239. Springer, Cham (2020). https://doi.org/10.1007/978-3-030-63710-1_18
7. Crespi, C., Scollo, R.A., Pavone, M.: Effects of different dynamics in an ant colony optimization algorithm. In: 2020 7th International Conference on Soft Computing Machine Intelligence (ISCMI 2020), pp. 8–11. IEEE (2020). https://doi.org/10.1109/ISCMI51676.2020.9311553

8. DeAngelis, D.L., Diaz, S.G.: Decision-making in agent-based modeling: a current review and future prospectus. Front. Ecol. Evol. **6**, 237 (2019). https://doi.org/10.3389/fevo.2018.00237
9. Drury, J., et al.: Cooperation versus competition in a mass emergency evacuation: a new laboratory simulation and a new theoretical model. Behav. Res. Methods **41**, 957–970 (2009). https://doi.org/10.3758/BRM.41.3.957
10. Khamis, N., Selamat, H., Ismail, F.S., Lutfy, O.F., Haniff, M.F., Nordin, I.N.A.M.: Optimized exit door locations for a safer emergency evacuation using crowd evacuation model and artificial bee colony optimization. Chaos Solitons Fractals **131**, 109505 (2020). https://doi.org/10.1016/j.chaos.2019.109505
11. Peng, Y., Li, S.W., Hu, Z.Z.: A self-learning dynamic path planning method for evacuation in large public buildings based on neural networks. Neurocomputing **365**, 71–85 (2019). https://doi.org/10.1016/j.neucom.2019.06.099
12. Pintea, C.-M., Matei, O., Ramadan, R.A., Pavone, M., Niazi, M., Azar, A.T.: A fuzzy approach of sensitivity for multiple colonies on ant colony optimization. In: Balas, V.E., Jain, L.C., Balas, M.M. (eds.) SOFA 2016. AISC, vol. 634, pp. 87–95. Springer, Cham (2018). https://doi.org/10.1007/978-3-319-62524-9_8
13. Shi, X., Ye, Z., Shiwakoti, N., Tang, D., Lin, J.: Examining effect of architectural adjustment on pedestrian crowd flow at bottleneck. Phys. A **522**, 350–364 (2019). https://doi.org/10.1016/j.physa.2019.01.086
14. Siyam, N., Alqaryouti, O., Abdallah, S.: Research issues in agent-based simulation for pedestrians evacuation. IEEE Access **8**, 134435–134455 (2020). https://doi.org/10.1109/ACCESS.2019.2956880
15. van der Wal, C.N., Formolo, D., Robinson, M.A., Minkov, M., Bosse, T.: Simulating crowd evacuation with socio-cultural, cognitive, and emotional elements. In: Mercik, J. (ed.) Transactions on Computational Collective Intelligence XXVII. LNCS, vol. 10480, pp. 139–177. Springer, Cham (2017). https://doi.org/10.1007/978-3-319-70647-4_11
16. Varghese, E.B., Thampi, S.M.: Towards the cognitive and psychological perspectives of crowd behaviour: a vision-based analysis. Connect. Sci. **33**(2), 380–405 (2021). https://doi.org/10.1080/09540091.2020.1772723
17. Wang, S., Liu, H., Gao, K., Zhang, J.: A multi-species artificial bee colony algorithm and its application for crowd simulation. IEEE Access **7**, 2549–2558 (2019). https://doi.org/10.1109/ACCESS.2018.2886629
18. Wilensky, U.: NetLogo. Center for Connected Learning and Computer-Based Modeling, Northwestern University, Evanston, IL (1999). http://ccl.northwestern.edu/netlogo/
19. Yücel, F., Sürer, E.: Implementation of a generic framework on crowd simulation: a new environment to model crowd behavior and design video games. Mugla J. Sci. Technol. **6**, 69–78 (2020). https://doi.org/10.22531/muglajsci.706841

Chemical Neural Networks and Semantic Information Investigated Through Synthetic Cells

Lorenzo Del Moro[1], Beatrice Ruzzante[1], Maurizio Magarini[1], Pier Luigi Gentili[2], Giordano Rampioni[3], Andrea Roli[4,5], Luisa Damiano[6], and Pasquale Stano[7](✉)

[1] Department of Electronics, Information and Bioengineering, Politecnico di Milano, Milan, Italy

[2] Department of Chemistry, Biology and Biotechnology, Università degli Studi di Perugia, Perugia, Italy

[3] Sciences Department, University of RomaTre, Rome, Italy

[4] Department of Computer Science and Engineering, Campus of Cesena, Università di Bologna, Cesena, Italy

[5] European Centre for Living Technology, Venice, Italy

[6] Università IULM, Milan, Italy

[7] Department of Biological and Environmental Sciences and Technologies (DiSTeBA), University of Salento, Lecce, Italy

pasquale.stano@unisalento.it

Abstract. In a previous contribution we briefly sketched novel topics that lie at the interface between synthetic biology (SB) and artificial intelligence (AI). In particular, we discussed (a) the possibility of engrafting chemical AI-like tools in bottom up synthetic cell systems, and (b) the investigation of fundamental concepts of information theory – such as the "semantic" information – by means of synthetic cells. Here we intend to report on the progress done by our groups in these fields and shortly devise future steps for theoretical and experimental approaches.

Keywords: Synthetic Biology · Synthetic Cells · Chemical Neural Networks · Semantic Information

1 The Synthetic Cell Platform

The interest toward synthetic cells (SCs), especially when referred to bottom-up approaches, has significantly increased in the past decade. SCs are simplified cell-like structures that can be built from scratch by employing specific materials such as primitive or modern biomolecules, or fully artificial molecules as well, and specific techniques, typically based on a combination of self- and guided-assembly. In contrast with the so-called top-down SCs, often designed to perform specific

L. Del Moro and B. Ruzzante—These Authors contribute equally to this work.

C. De Stefano et al. (Eds.): WIVACE 2022, CCIS 1780, pp. 27–39, 2023.
https://doi.org/10.1007/978-3-031-31183-3_3

tasks in applied sciences (e.g., bioproduction, biosensoring, bioremediation, etc.), a peculiarity of bottom-up SCs refers to their potential role as tools – or, better, as platforms – that can be devoted to investigate fundamental scientific questions directly or indirectly related to theories of living systems. For example, the recent directions that highlight how SCs can be interfaced with biological cells thanks to chemical signaling [12] have prompted further investigation about what communication really is at the most basic level, and when, or if, SCs can be considered "cognitive" [2]. Actually, this and other related questions have stimulated our interest toward SCs in order to further developing the "Sciences of Artificial" in a novel and quite original direction: the wetware domain [1]. On the other hand, the theoretical investigations about what SCs actually are and how they contribute to model life and cognition from an embodied perspective [2,3,17] can generate innovative proposals that exploit the unique capability they have in handling molecular information and communication.

Here we will shortly report on the progress made in two recently started research lines: (a) the possibility of engrafting chemical artificial intelligence (AI)-like tools in bottom-up SCs (in particular, chemical phospho-neural networks [7]); and (b) the undergoing investigation on quantification of semantic information in SC-environment systems based on the Kolchinsky-Wolpert approach [9]. This contribution, clearly, is not intended to be conclusive about any of these aspects. Rather, its aim is to highlight unexplored avenues of research that, we believe, can offer fruitful idea to explore, develop, discuss and therefore further advance the field. In particular, it calls for focusing on those aspects of information and communication theories that can be applied to chemical systems, in order to generate radically novel approaches to artificial life and artificial intelligence.

2 Phospho-Neural Networks [7] Inside SCs

As a first task, we have considered the implantation of an upstream AI-inspired "module" inside SCs to perform embodied perceptive tasks and ultimately to control gene expression. In particular, in the CIBB 2021 conference we have suggested the construction of a chemical neural network (CNN). We have identified three open questions, namely: (i) whether it is technically possible such an implantation, (ii) which SC "perceptive" properties it can generate, and (iii) what kind of modeling approach best fits with the embodied vision we promote.

(i) To answer the first question we firstly reviewed the state-of-the-art of CNNs, discovering that many reports focus on their modeling, not to their construction. A very limited number of CNN have been constructed, mainly exploiting the "strand displacement" DNA technique. On the other hand, several reports discuss how signaling and regulatory networks, *in vivo*, can be interpreted as CNNs. A specific class of signaling networks attracted our attention: the bacteria two-component signaling (TCS) systems, based on protein phosphorylation. We have presented the idea in a couple of recent publications [6,16]. *In vivo*, TCS systems enable bacteria to sense, respond, and adapt to their environments, letting the cell perceive chemical signals present in their surroundings. They are composed of two macromolecular elements: a membrane sensor (S), and a cytosolic response regulator (R). Upon binding of an extracellular signaling molecule A to S, the S:A complex autocatalyzes its own phosphorylation on the intracellular side. Next, the phosphorylated S transfers the phosphate group to R. In turn, phosphorylated R interacts with DNA, modifying the gene expression pattern. Importantly, several sensors and response regulators coexist in cells, and sometimes the signaling pathways are not insulated as explained above, but can cross-talk and generate convergent and/or divergent neural network-like signaling patterns [7]. According to our analysis, reconstituting cross-talk TCS systems inside SCs is an achievable – yet challenging – task. In [6] we have made a preliminary plan about which strategies should be adopted to identify the exact molecular species to be employed in future experimental approaches.

(ii) In order to answer the second questions (i.e., which SC "perceptive" properties it can generate), we will make use of modeling techniques, constraining the network architecture within the realm of practical feasibility, in terms of number of components and their connectivity. For example, a realistic target can be made 2–3 S/R pathways with full or partial cross-talk between the "channels".

(iii) With respect to the third question, our orientation points again to the direction of chemical embodiment. In order to keep into account the chemical nature of the neural network elements, we need to recall that CNN elements are proteins and thus can display conformational diversity [5], which, in turn, implies a diversity in the physico-chemical parameters describing their behavior, such as binding constants and kinetic constants (moreover, when a molecule has a limited number of conformers due to restricted structural degrees of freedom, it is possible to extend its responsiveness by embedding it in distinct micro-environments [4,5]). Therefore, we maintain that the preferred modeling of CNNs starts from the idealized two-state (yes/no) logic, but finally moves toward the continuum spectrum (gray-scale) of possibilities, which can be approached via different methods, such as soft Ising spins [8] and fuzzy logic. In the case of fuzzy logic, the physico-chemical input and output variables that best fits to the nature of chemical responsive elements are granulated into fuzzy sets. We expect that this approach will provide an accurate modeling and a gain of more insightful conclusions about the CNN dynamics inside SCs.

3 Syntactic and Semantic Information

The first formal attempt to define a quantitative measure of information is usually attributed to C. E. Shannon, whose work is recognized as the birth of information theory [15]. Shannon was mainly interested to the engineering problem of reproducing a message from a point to another, therefore the meaning of a transmitted message (semantic) was intentionally not considered. He wrote: *"The fundamental problem of communication is that of reproducing at one point either exactly or approximately a message selected at another point. Frequently the messages have meaning; that is they refer to or are correlated according to some system with certain physical or conceptual entities. These semantic aspects of communication are irrelevant to the engineering problem"*.

It is worth noting that, although Shannon's interest was related to the transmission of information in communication systems, his theory can be also used to study many physical systems, especially the concept of syntactic information that has a clear physical meaning. In fact, a generic physical system can be modeled by a random variable, since it may be present in different states, each one with a probability of presentation. In addition, it has also a possible correlation with other physical systems. This correlation is derived by the channel that models a physical causal law (or more laws) that describes the interaction between two or more systems.

Since the appearance of the work of Shannon, not all members of the information science community shared the idea that the semantic aspect was irrelevant. One of these was MacKay who thought information was related to the change in the receiver's mindset. He stated: *"Information is a distinction that makes a difference"*. This definition of information influenced then Bateson, who formulated his famous idea of information as *"The difference that makes a difference"* [10]. As a consequence, differently from the Shannon syntactic information, which refers only to statistical aspects, the semantic information refers to the amount of statistical correlation that conveys a meaning, or in other words, that affects in some way the system that receive the information.

However, due to the fact that many engineering problems are addressed by the mathematical tools defined by Shannon, such as entropy, mutual information, channel capacity and so on, only a little interest has been given to the development of a more comprehensive theory that concerns also with the semantic aspect. Consequently, until now, common communication systems, whose target is an error-free and high-speed data transmission, have been optimized over classical key performance indicators (KPIs), such as throughput and delay, and built over the Shannon paradigm of a reliable transmission of sequences of symbols.

4 The Kolchinsky-Wolpert Approach

According to the Kolchinsky and Wolpert (KW) definition [9], semantic information is defined as the fraction of total syntactic information a physical system has about its environment that is causally necessary for the system to maintain its own existence. This characteristic is quantified by introducing the *viability* function $V(P(x_\tau))$, where $P(x_\tau)$ is the system distribution at time τ, calculated at a certain pre-defined time scale τ (more details can be found in [9,13]).

4.1 A Route to Measure Semantic Information

In a recent work, the present authors proposed a study about the quantification of semantic information that a simple SC extracts from its environment [13]. This investigation was supported by motivations of diverse nature. The most impelling one was related to the interest elicited in the community of "Molecular Communication" (MC) engineers about the progressive advancements made in communicating SCs [11,12]. Typically, MC approaches refer to *syntactic* aspects of information, the ones that are classically associated with well-established concepts in ICT, as defined by Shannon. These refer to a full arsenal of concepts such as the modeling of a communication system, in terms of abstract layers like encoding, transmitting, propagating, receiving, and decoding elements, the definition of information content of a source (expressed in bits per symbol), and the reliable transfer of information, which is measured by the channel capacity (expressed in bits per channel use).

With specific reference to semantic information, the proposal by KW attracted our attention [11] because of its *operative* definition in a theoretical framework that can be readily applied to physical agents situated in an environment, and based on system dynamics (these are peculiar features that resonate well with embodied approaches to model autonomy, life-and-cognition, and with the role that synthetic biology can have as a wetware science of artificial [2,3]). The ideas behind the KW approach are based on the concept of autonomous agents, which turn out to be fundamental. In fact, there are other approaches to the semantic aspects of information, which anyway are not fully satisfying: the KW approach, in fact, gives the possibility to apply the concept to many physical systems, both living and not, and in such a way that the meaning is intrinsic into them.

The KW approach looks at an autonomous agent (i.e., the physical system) whose intrinsic goal is to actively maintain its own existence within a certain environment. They are a "*system* $\mathcal{X}$" with a state space X and an "*environment* $\mathcal{Y}$" with a state space Y. Derived from physics, a state is a possible configuration in which a system can be found, and consequently the state space represents all the possible configurations a system may assume. For the sake of simplicity, the state space is hereafter assumed to be discrete and finite, even though also a continuous state space can be considered. The coupled state space of system and environment, in case of separate degree of freedom, is $X \times Y$. Moreover, both the system and the environment may be isolated from the rest of the world or not. From a mathematical point of view, a state space X is modeled as a random variable with a discrete mass distribution $P_X(x)$ ($P(x)$ for short), while the joint state space defined by the system and its environment $X \times Y$ are modeled as a pair of random variables with a joint mass distribution $P_{X,Y}(x,y)$ ($P(x,y)$ for short). The communication channel through which the system receives information from its surroundings is described by the conditional probability $P_{X|Y}(x|y)$ ($P(x|y)$ for short). Eventually, the system and the environment are subject to joint dynamics. From now on that dynamics will be stochastic, discrete time, and first order Markovian [9].

Another noteworthy aspect is the causal contribution intrinsic into the definition of KW of the semantic information, which can be quantified by means of counter-factual interventions. Loosely speaking, an intervention is a counterfactual operation that modifies the syntactic information between the system and its environment. After the intervention, the viability function of a system subjects to a dynamic might change at a certain time instant, and this is a clue that in the modification of the syntactic information some semantic information is lost or earned. For example, a person living in a city has a strong correlation with that environment. Intervening by modifying the environment from the city to a rain forest could decrease the viability function, meaning that much of the correlation between the person and the city was meaningful.

4.2 The Viability Function

Although different definitions may be adopted for the viability function, KW suggest using the negative value of the Shannon entropy, which is a concept deeply tied to thermodynamics. Let X_τ denote the random variable describing the state of the system $\mathcal{X}$ at time τ. The viability function is defined as

$$V\left(P(x_\tau)\right) \stackrel{\text{def}}{=} -H\left(X_\tau\right), \tag{1}$$

where $P(x_\tau)$ is the probability of being in state x_τ and

$$H\left(X_\tau\right) \triangleq -\sum_{x_\tau \in \mathcal{X}_\tau} P(x_\tau) \log_2\left(P(x_\tau)\right) \tag{2}$$

is the entropy computed over the whole set of states $\mathcal{X}_\tau$.

The definition in (1) is taken because of the relationship between information theory and non-equilibrium statistical physics, as observed by Schrodinger in [14] (living beings must maintain itself in a low entropy state to stay alive). In fact, a system in a low entropy state is characterized by a high degree of order and every living being is seen as a system with high degree of coherence between particles constituting it. The application of this concept requires to assign a value of internal entropy to each state x_τ according to what it represents. For instance, in [9] the internal entropy of each state is set to 0 with exception of the state associated with the death of the system; in that case the internal entropy is H_{dead} bits. According to the above-mentioned considerations, the dead state has a large internal entropy [9], the system equilibrates instantly within the dead state, and once dead, the system cannot evolve again in time.

4.3 Evaluation of Semantic Information

KW introduced two types of semantic information. The first, called *stored* semantic information, is the amount of semantic information a system initially has about its environment. The second, called *observed* semantic information, represents the semantic information acquired during the ongoing dynamics and that, differently from the stored one, is not initially present in the system.

Stored Semantic Information. It is computed from the mutual information between system and environment at time $\tau = 0$ as

$$I(X_0, Y_0) = \sum_{x_0, y_0} P(x_0, y_0) \log_2 \left(\frac{P(x_0, y_0)}{P(x_0) P(y_0)} \right) \tag{3}$$

by a scrambling (modification) of the initial mutual information defined by a particular intervention, which is called *optimal* intervention. Before introducing the formalism behind the optimal intervention and stored semantic information, we briefly discuss: *i*) what an intervention is, *ii*) why it is needed to determine an optimal intervention, and *iii*) how it could modify the initial mutual information and consequently the viability function at the reference time τ.

An intervention refers to a possible coarse-graining function applied to the state space of the environment $Y' = \phi(Y)$. The description of the joint state space of the system and of the environment is initially given by the distributed $P(x_0, y_0)$. An intervention may modify this distribution, and consequently also the initial mutual information. In fact, the operation of mapping due to the intervention implies that the system cannot distinguish anymore some different states of the environment as said before. In this way, there is a straightforward change in the initial channel law $P(x_0|y_0)$ that becomes an intervened channel law under the intervention ϕ

$$\hat{P}^{\phi}(x_0|y_0) \triangleq P(x_0|\phi(y_0)) = \frac{\sum_{y_0':\phi(y_0')=\phi(y_0)} P(x_0, y_0')}{\sum_{y_0':\phi(y_0')=\phi(y_0)} P(y_0')}. \tag{4}$$

The sum over y_0' means that the intervened function may map different states in only one and that the system cannot recognize the granularity of the states space of the environment. The coarse-graining of the communication channel determines which distinctions the system can make about the environment under each partial intervention.

Moreover, the joint distribution after an intervention ϕ is called intervened joint initial distribution and it is defined as

$$\hat{P}^{\phi}(x_0, y_0) \triangleq P(x_0, \phi(y_0)) = P(x_0|\phi(y_0))P(y_0). \tag{5}$$

The most destructive intervention, named '*fully scrambled*', possibly leads to a situation where the system and the environment are independent and so $\hat{P}^{\phi}(x_0, y_0) \triangleq \hat{P}^{full}(x_0, y_0) = P(x_0) P(y_0)$. In this particular case the initial mutual information is clearly equal to zero. In fact, the system cannot anymore distinguish the states of the environment and $P(x_0|\phi(y_0)) = P(x_0)$.

In case of full intervention, if the coupled dynamic of the system and of the environment is left unaltered (remember that the dynamic is considered first-order Markovian) after time τ we get a new joint distribution $\hat{P}^{full}(x_\tau, y_\tau)$. By means of this intervened joint distribution and of the actual joint distribution, it is possible to compute firstly, the viability at time τ as $V(\hat{P}^{full}(x_\tau))$ and $V(P(x_\tau))$, after an obvious marginalization over Y_τ, and consequently the stored viability value as

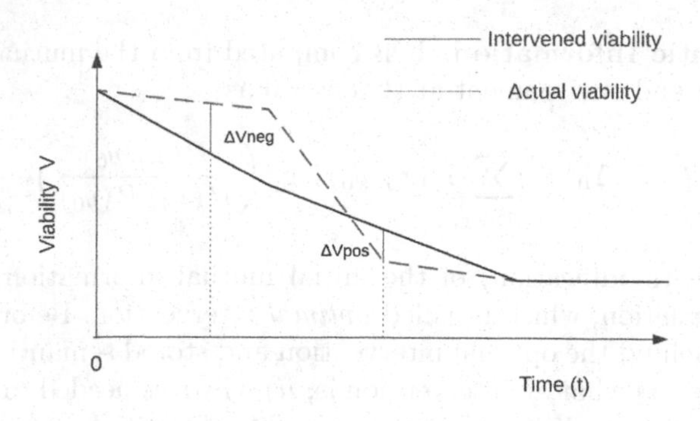

Fig. 1. Qualitative example of the actual and intervened viability.

$$\Delta V_{tot}^{stored} = V\left(P(x_\tau)\right) - V(\hat{P}^{full}(x_\tau)). \tag{6}$$

This difference represents how much degree of existence has been lost due to the specific full intervention. A qualitative example of the actual and intervened viability is shown in Fig. 1. A negative value of ΔV_{tot}^{stored} means that the intervention affects positively the viability of the system. When ΔV_{tot}^{stored} is instead positive the intervention affects negatively the viability of the system. Nevertheless, the function $V\left(P(x_\tau)\right)$ at the end of the dynamics of both the full intervention and the non-intervened distribution settles to minimum value. This is due to the fact that for the non-intervened case the system reaches the dead state associated with a maximum predefined value of internal entropy, as explained in Sect. 4.2, and the full intervention does not affect the final behavior of the system.

Anyway, through the use of the full intervention there is no possibility to understand which part of the initial mutual information is actually meaningful. In fact, there might exist other interventions less destructive that could bring to have a bigger initial mutual information (since system and environment are not anymore fully uncorrelated) and with a higher viability compared to the full intervention. In other words, the full intervention could destroy also part of the meaningless information. This is the reason why we need an optimal intervention, i.e., an intervention that destroys only the part of the syntactic information (mutual information) that is meaningless, thus leaving constant the viability at time τ as in case of no intervention. The value of initial mutual information found by means of this optimal intervention is therefore the stored semantic information. If, after an optimal intervention, we further decrease the initial mutual information, we inevitably affect the meaningful content of the initial mutual information decreasing the viability.

As briefly depicted before, each partial intervention might modify the initial mutual information and possibly also the viability. As first step KW in their work defines the viability curve as

$$D_{stored}(R) \triangleq \max_{\phi \in \Phi} V(\hat{P}^{\phi}\left(x_\tau\right)) \quad s.t \quad I_{\hat{P}^{\phi}}\left(X_0, Y_0\right) = R, \tag{7}$$

where $\hat{P}^{\phi}(x_\tau)$ is the intervention that brings to the amount of mutual information R. It represents the maximum achievable viability under different values of initial mutual information, due to different interventions. In addition, they define the optimal intervention as that intervention that achieves the same viability at time τ as the not intervened case while giving the smallest value of initial mutual information

$$\hat{P}^{opt}_{X_0,Y_0} \triangleq \underset{\hat{P}^{\phi}_{X_0,Y_0}:\phi\in\Phi}{\arg\min}\ I_{\hat{P}^{\phi}_{X_0,Y_0}}(X_0,Y_0) \quad s.t \quad V(\hat{P}^{\phi}(x_\tau) = V(P(x_\tau)). \tag{8}$$

The stored semantic information is then defined as the initial mutual information under the optimal intervention

$$I_{stored} \triangleq I_{\hat{P}^{opt}_{X_0,Y_0}}(X_0,Y_0). \tag{9}$$

Thanks to the optimal intervention we are able to identify what are the environment states that affect the behavior of the system.

Observed Semantic Information. It refers to the amount of syntactic information that is acquired during the dynamics of system and environment and that causally contributes to maintain the existence of the system. In this case, the interventions are applied to the dynamics of the system since the aim is to perturb the interaction between system and environment. In order to compute this dynamic flow of information KW suggest to use the transfer entropy over time. Transfer entropy from the environment Y_t to the system X_{t+1} reflects the uncertainty that is reduced about X at time $t+1$, by observing Y at time t and knowing X again at time t. Intuitively, if there is an information flow during the dynamics from a system Y and another system X, by observing Y_t we could remove part of the uncertainty about X_{t+1}, knowing X_t, since the system X can increase its correlation with Y. Transfer entropy is defined as

$$T_p(Y_t \to X_{t+1}) = I_p(X_{t+1};Y_t|X_t) = H(X_{t+1}|X_t) - H(X_{t+1}|X_t,Y_t), \tag{10}$$

where

$$H(X_{t+1}|X_t) = \sum_{x_{t+1}}\sum_{x_t} P(x_{t+1},x_t)\log\frac{1}{P(x_{t+1}|x_t)} \tag{11}$$

and

$$H(X_{t+1}|X_t,Y_t) = \sum_{x_{t+1}}\sum_{x_t}\sum_{y_t} P(x_{t+1},x_t,y_t)\log\frac{1}{P(x_{t+1}|x_t,y_t)} \tag{12}$$

are the conditional entropy of X_{t+1} given X_t and the conditional entropy of X_{t+1} given (X_t,Y_t), respectively. From (10) it can be observed that the transfer entropy corresponds to the conditional mutual information when the side information is represented by the random variable X_t.

Since the goal is to compute the transfer entropy over time from 1 to τ, the overall syntactic information flow is simply the sum over time of each dynamic flow from an instant to another, so that we want to compute

$$\sum_{t=0}^{\tau-1} T_p(Y_t \to X_{t+1}) = \sum_{t=0}^{\tau-1} I_p(X_{t+1}; Y_t|X_t). \tag{13}$$

Concerning instead interventions, they are still coarse-grain function. Since the overall dynamics is defined by $P_{X_{t+1},Y_{t+1}|X_t,Y_t} = P_{X_{t+1}|X_t,Y_t} P_{Y_{t+1}|X_{t+1},X_t,Y_t}$ we apply the intervention only to the first term since, in this manner, only the information flow between the environment and the system is scrambled and not the opposite flow of information.

The intervened conditional probability is then computed as

$$\hat{P}^{\phi}(x_{t+1}|x_t, y_t) \triangleq P(x_{t+1}|x_t, \phi(y_t)) = \frac{\sum_{y'_t:\phi(y'_t)=\phi(y_t)} P(x_{t+1}|x_t, y'_t)\hat{P}^{\phi}(x_t, y'_t)}{\sum_{y'_t:\phi(y'_t)=\phi(y_t)} \hat{P}^{\phi}(x_t, y'_t)}, \tag{14}$$

where the sum over y'_t means that the intervention coarse-grains the environment state and the system cannot recognize that granularity.

All the definitions used for the stored semantic information can be introduced also in case of the observed semantic information. We have a full intervention in case of constant coarse-grain function, since the system X_{t+1} is independent of Y_t given X_t. It is possible also to evaluate the so called viability value in case of the observed semantic information exactly as done in (6). In addition, also the viability curve is defined in analogous way. Since different interventions may bring to the same value of transfer entropy but different viability levels, KW define the viability curve as the maximum value of viability under those different interventions that have the same value of transfer entropy as

$$D_{obs}(R) \triangleq \max_{\phi \in \Phi} V(\hat{P}^{\phi}(x_t)) \quad s.t. \quad \sum_{t=0}^{\tau-1} T_{\hat{P}^{\phi}}(Y_t \to X_{t+1}) = R. \tag{15}$$

Analogously to the considerations done for the observed semantic information, an optimal intervention must be found. This optimal intervention is defined as the one that preserves the viability function of the system, while reducing as much as possible the transfer entropy. In this case the remaining part of transfer entropy is that part of information that is meaningful for the system existence. Thus, the optimal intervention is defined as

$$\hat{P}^{opt}_{X_{0..\tau},Y_{0..\tau}} \in \underset{\hat{P}^{\phi}_{X_{0..\tau},Y_{0..\tau}}:\phi\in\Phi}{\arg\min} \sum_{t=0}^{\tau-1} T_{\hat{P}^{\phi}}(Y_t \to X_{t+1}) \quad s.t \quad V(\hat{P}^{\phi}(x_\tau)) = V(P(x_\tau)), \tag{16}$$

where $\hat{P}^{\phi}_{X_{0..\tau},Y_{0..\tau}}$ describes the joint distributions over time, from 0 to τ. Once the optimal intervention $\hat{P}^{opt}_{X_{0..\tau},Y_{0..\tau}}$ is obtained, it is straightforward to find the amount of observed semantic information as

$$I_{observed} \triangleq \sum_{t=0}^{\tau-1} T_{\hat{P}^{opt}}(Y_t \rightarrow X_{t+1}). \tag{17}$$

5 The Synthetic Cell Case

To calculate the amount of stored semantic information I_{stored}, we propose the following strategy:

1. define an SC structure and its internal mechanisms. These should be designed either at the sole scope of self-maintenance (viability/existence in a non-equilibrium autopoietic state), or – more modestly but more realistically – for accomplishing predefined tasks (e.g., producing a compound of interest, display a target behavior, etc.);
2. define an environment E that includes effectors for the SC, according to a well-defined spatial-temporal pattern (called actual distribution P^*, which can be designed in order to be the optimal distribution – see [9], and correspondingly leading to the maximal value V^* of the viability function);
3. given P^*, simulate SC/E dynamics at time τ, and consequently calculate the optimal value V^* of the viability function, and the corresponding mutual information I^* (expressed in bits) exchanged between SC and E;
4. repeat the previous step for all possible "intervened" distributions P_i, obtained by scrambling/cross-graining the environmental variables and their pattern; calculate the corresponding values V_i of the viability function, and the mutual information I_i as well, for each P_i;
5. rank the entire set of values (I_i, V_i) to determine the minimal I_i value (now called I_{stored}, in bits) that does not reduce the value of the viability function below its optimal value V^* (corresponding to the actual distribution P^*, which is also the optimal one);
6. according to KW, I_{stored} defines the amount of stored semantic information in the dynamical SC/E whole-system.

In our initial proof-of-concept study [13], we devised a simple SC whose task is the production of a toxin, e.g., SC as an agent in a smart drug delivery scenario. A signal molecule, present in the environment, can permeate into the SC and activate the production of a toxin. The SC behavior is simply ruled by two processes: toxin production and decay of SC active components, while signal molecules permeation is considered very fast. Given a certain time scale, it will be possible to calculate the maximum value of mutual information I^* together with its semantic part (I_{stored}).

References

1. Cordeschi, R.: The Discovery of the Artificial. Behavior, Mind and Machines Before and Beyond Cybernetics. Studies in Cognitive Systems, Mind and Machines Before and Beyond Cybernetics. Springer, Dordrecht (2002). https://doi.org/10.1007/978-94-015-9870-5
2. Damiano, L., Stano, P.: Synthetic biology and artificial intelligence. Grounding a cross-disciplinary approach to the synthetic exploration of (embodied) cognition. Complex Syst. **27**, 199–228 (2018). https://doi.org/10.25088/ComplexSystems.27.3.199
3. Damiano, L., Stano, P.: A wetware embodied AI? Towards an autopoietic organizational approach grounded in synthetic biology. Front. Bioeng. Biotechnol. **9**, 873 (2021). https://doi.org/10.3389/fbioe.2021.724023
4. Gentili, P.L.: The fuzziness of the molecular world and its perspectives. Molecules **23**(8), 2074 (2018). https://doi.org/10.3390/molecules23082074
5. Gentili, P.L.: Establishing a new link between fuzzy logic, neuroscience, and quantum mechanics through Bayesian probability: perspectives in artificial intelligence and unconventional computing. Molecules **26**(19), 5987 (2021). https://doi.org/10.3390/molecules26195987
6. Gentili, P.L., Stano, P.: Chemical neural networks inside synthetic cells? A proposal for their realization and modeling. Front. Bioeng. Biotechnol. **10**, 927110 (2022). https://doi.org/10.3389/fbioe.2022.927110
7. Hellingwerf, K.J., Postma, P.W., Tommassen, J., Westerhoff, H.V.: Signal transduction in bacteria: phospho-neural network(s) in Escherichia coli? FEMS Microbiol. Rev. **16**(4), 309–321 (1995). https://doi.org/10.1111/j.1574-6976.1995.tb00178.x
8. Horiguchi, T.: Spin model with fuzzy Ising spin. Phys. Lett. A **176**(3), 179–183 (1993). https://doi.org/10.1016/0375-9601(93)91031-Y
9. Kolchinsky, A., Wolpert, D.H.: Semantic information, autonomous agency and non-equilibrium statistical physics. Interface Focus **8**, 20180041 (2018). https://doi.org/10.1098/rsfs.2018.0041
10. Logan, R.K.: What is information?: why is it relativistic and what is its relationship to materiality, meaning and organization. Information **3**(1), 68–91 (2012). https://doi.org/10.3390/info3010068
11. Magarini, M., Stano, P.: Synthetic cells engaged in molecular communication: an opportunity for modelling Shannon- and semantic-information in the chemical domain. Front. Commun. Netw. **2**, 48 (2021). https://doi.org/10.3389/frcmn.2021.724597
12. Rampioni, G., D'Angelo, F., Leoni, L., Stano, P.: Gene-expressing liposomes as synthetic cells for molecular communication studies. Front. Bioeng. Biotechnol. **7**, 1 (2019). https://doi.org/10.3389/fbioe.2019.00001
13. Ruzzante, B., Del Moro, L., Magarini, M., Stano, P.: Synthetic cells extract semantic information from their environment. IEEE Trans. Mol. Biol. Multi-Scale Commun. **9**(1), 23–27 (2023). https://doi.org/10.1109/TMBMC.2023.3244399
14. Schrodinger, E.: What is Life? Cambridge University Press, Cambridge (1944)
15. Shannon, C.E.: A mathematical theory of communication. Bell Syst. Tech. J. **27**(3), 379–423 (1948). https://doi.org/10.1002/j.1538-7305.1948.tb01338.x
16. Stano, P., Rampioni, G., Roli, A., Gentili, P.L., Damiano, L.: En route for implanting a minimal chemical perceptron into artificial cells. In: Holler, S., Löffler, R., Bartlett, S. (eds.) Proceedings of the Conference on Artificial Life; Online, 18–22 July 2022, pp. 465–467. MIT Press, Cambridge (2022)

17. Stano, P.: Exploring information and communication theories for synthetic cell research. Front. Bioeng. Biotechnol. **10**, 927156 (2022). https://www.frontiersin.org/articles/10.3389/fbioe.2022.927156

Spread of Perturbations in Supply Chain Networks: The Effect of the Bow-Tie Organization on the Resilience of the Global Automotive System

Elisa Flori[1], Yi Zhu[2], Sandra Paterlini[2], Francesco Pattarin[1], and Marco Villani[3,4](✉)

[1] Department of Economics "Marco Biagi", University of Modena and Reggio Emilia, Modena, Italy
[2] Department of Economics and Management, University of Trento, Trento, Italy
[3] Department of Physics, Informatics and Mathematics, University of Modena and Reggio Emilia, Modena, Italy
marco.villani@unimore.it
[4] European Centre for Living Technology, Venice, Italy

Abstract. Many real-world systems are subject to external perturbations, damages, or attacks with potentially ruinous consequences. The internal organization of a system allows it to effectively resist to such perturbations with more or less success. In this work, we study the resilience properties of the global automotive supply-chain by considering the bow-tie structure of the directed network stemming from customer-supplier relationships. Data have been retrieved by Bloomberg supply chain database between 2018 to 2020. Our analysis involves 3,323 companies connected by 11,182 trade links and spanning 135 economic sectors. Our results indicate that the size of propagation of a perturbation depends on the area of the bow-tie structure in which it initially originates. Also, it is possible to identify resistance structures within some bow-tie areas. Thus, we provide insights into the fragility and resilience of different network components and the diffusion paths of perturbations across the system. Interestingly, the level of abstraction used allows our results to generalize beyond the case in question to many systems that can be represented through directed graphs.

Keywords: Network analysis · perturbation spread · bow-tie model · supply chain · automotive sector

1 Introduction

Many real-world systems are potentially subject to failures or external perturbations that cause disruptions and harmful consequences; for example, adverse events affecting the Internet, power grids, organizations, and natural systems [2, 18, 26, 28, 31, 39, 40]. The failure of a single node or link can cause a blackout in a power grid or the collapse of an ecological system [9, 14]. Analyzing the size of the impacts shocks cause on complex

C. De Stefano et al. (Eds.): WIVACE 2022, CCIS 1780, pp. 40–57, 2023.
https://doi.org/10.1007/978-3-031-31183-3_4

networks, in terms of how many nodes and links are affected, gives interesting clues about the statics and dynamics of such systems [29, 37]. Also, understanding how the structure of a system determines its response to shocks is important, because it sheds light on its local and global resilience and how they change after a shock [11, 18, 33, 38].

Several systems can be effectively represented by directed graphs, where component entities are "nodes" and links between them are "arcs". Arcs have directions determined by the flows of artifacts that go from one node to another—e.g. information, energy and commodities. Supply chain networks are an example of directed graphs. Suppliers are connected to customers that receive intermediate or final goods and services from them. This is an interesting empirical field for studying the propagation of perturbations in economic systems and to identify some regularities that may have general significance with respect to the representation and investigation of networks [6, 23].

In this article, we first check if nodes in the global automotive supply chain network can be effectively represented by a bow-tie structural model [8], initially introduced in studies about the world-wide-web [3, 8, 34, 36] and extended, albeit with slightly different meanings, to other areas, like biological systems and the management of complex social organizations [12, 16, 19, 25, 35]. Our analyses suggest that the bow-tie model is a good representation of the supply chain network.

Furthermore, we are interested in the influence the bow-tie structure has in shaping the paths of perturbations across the network. In fact, the analysis of the structure of the system could allow the study of its resilience: static structural features and dynamic responses to perturbations are closely connected [15]. Therefore, we carried out several simulations of perturbations under different scenarios and identified some significant determinants of resilience in the supply-chain, bow-tie like, network. While it is well known that having (few) big hubs improves network resilience when perturbations start by randomly hitting some nodes, we find that the position of nodes —or relative density of them— in the areas of the bow-tie structure is also very important. We argue that this is because of the spread of perturbations through sequences of directed arcs connecting nodes ("paths") depends on their connections and, therefore, on the area they belong to. To our knowledge, we are the first to suggest such an analysis in the scientific literature.

By relying on our unique dataset of 3,323 companies (nodes) from the global automotive sector, connected by 11,182 trade relations (arcs) and collected from the Bloomberg platform, we provide new insights into the characteristics of the global automotive supply chain network (Sect. 2). In particular, the directionality of supplier-customer relationships leads to the spontaneous emergence of five major areas, as described in [8, 17] (Sect. 3). By applying such taxonomy, we can better understand the percolation of avalanches and shocks within the system and then evaluate network resilience and robustness (Sect. 4). This allows us to characterize some stylized facts about the global automotive network. Eventually, we examine the network robustness to avalanches and shocks at different levels and show that the automotive supply chain network is more resilient than random networks and networks with the same degree distribution, while also quantifying the impact delivered by nodes belonging to different bow-tie areas.

2 Data

The data collection procedure started by identifying and retrieving all automobile, motorcycle, construction machinery and heavy-duty vehicles manufacturers available in the Bloomberg Supply Chain database in April 2021; such companies are called "focal companies", and we recovered 165 of them. Then, the final dataset was built (1) by identifying and selecting the main suppliers and customers of focal companies and then (2) by identifying and selecting the main suppliers and customers of the latter[1,2] (the network therefore includes the focal nodes and the nodes which can be reached staring from them in 1 or at most 2 steps - see Fig. 1).

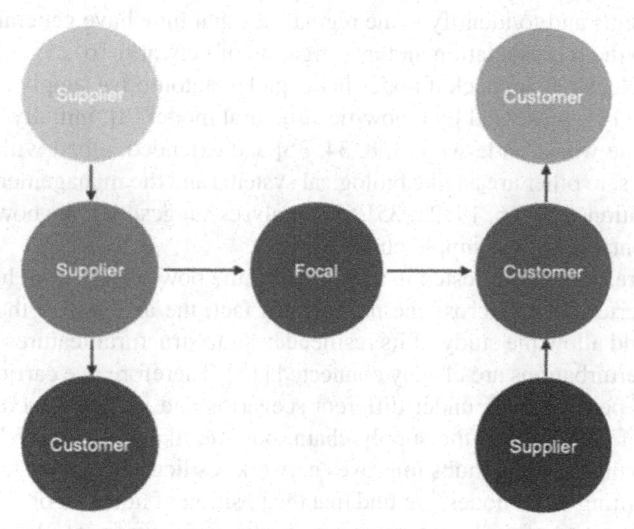

Fig. 1. Exemplification of focal company, supplier, and customer. Circles are nodes and arrows are directed arcs connecting them; the direction of any arc follows the flow of goods and services from a company to another.

Notice that some companies are both suppliers and customers—they have both incoming and outgoing arcs— some are only suppliers—do not have any incoming arc—and some are only customers—do not have any outgoing arc (see Table 1).

The final supply chain network dataset consists of 3,323 companies (nodes) connected by 11,182 trade relations (arcs). For all companies we know: (a) their size, measured by the average annual sales (USD millions) over the 2018–2020 fiscal years; (b) the region they are legally established in; (c) their economic activity sector according

[1] Bloomberg defines "main" as top-ranking suppliers and customers according to revenues or cost of goods sold. The number of main suppliers and customers is not the same for every company and typically is in the range of a few to twenty units.

[2] This approach follows Bellamy et al. [7]. They used Bloomberg's Supply Chain Relationship Database (SPLC) to recreate a network of connections between 3,106 companies to understand how administrative environmental innovations (AEIs) relate to environmental disclosure.

Table 1. Distribution of network nodes by role.

Type of node	Absolute frequency	Relative frequency %
Supplier and customer	1,192	35.9
of which: Focal companies	92	2.8
Only supplier	1,201	36.1
of which: Focal companies	25	0.8
Only customer	930	28.0
of which: Focal companies	48	1.5

to Morgan Stanley Capital International and Standard & Poor's GICS® taxonomy.[3] So, besides the number of nodes and arcs that characterize the automotive supply chain network, we can provide a measure of its size based on average 2018–2020 total sales. This amounts to about 27 billion USD, out of which focal companies generate 2,890 billion USD (11%). We estimate the market share of focal companies in the global automotive market to be about 79%.[4] The network is weakly connected by construction, as we discarded isolated nodes.

The network diameter is 20, which means that the longest of all shortest (or "geodesic") paths connecting any two nodes is 20 arcs.

Another way to look at the network is with respect to in- and out-degree distributions. By using a schema based on Maximum-Likelihood [10][5], we fit three probability distributions to in- and out-degrees: Poisson, Log-normal and Power law. Comparisons are commonly done in three ways: (1) eye-balling the log-log scale cumulative degree distributions to see if there is any law that fits well to the empirical distribution; (2) compute the Kolmogorov-Smirnoff distance (KS) between the fitted and empirical distributions and see what their differences are; (3) statistical testing of such differences, to see if they are significant beyond sampling errors [20]. We made all such comparisons for in- and out-degrees. The pictures of fitted and empirical degree distributions are in Fig. 2.

It is apparent that the Poisson law has a very poor fit in both cases, so we can claim that the global supply-chain network is not a random one. Vice versa, Power- or Log-normal law fit better [4, 32]. This observation, also present in other works, argues in

[3] The Global Industry Classification System (GICS) consists of 11 sectors, 24 industry groups, 69 industries and 158 sub-industries, where each level is more granular than its predecessors so that, for example, an industry is a more specific definition of economic activity that an industry group (Global Industry Classification Standard (GICS®) Methodology, retrieved 7 June 2021). We collected data on sub-industries.

[4] According to IBISWorld, Global Car & Automobile Sales (retrieved 12 August 2021), the average size of the global automotive market over the same period was 3,690 billion USD.

[5] We define "tail nodes" according to their in- and out-degrees; all nodes that have a degree higher than the first quartile of the overall distribution are tail nodes. For in degree, this means that a tail node has at least five upstream edges, while for out degree the threshold of downstream edges is four.

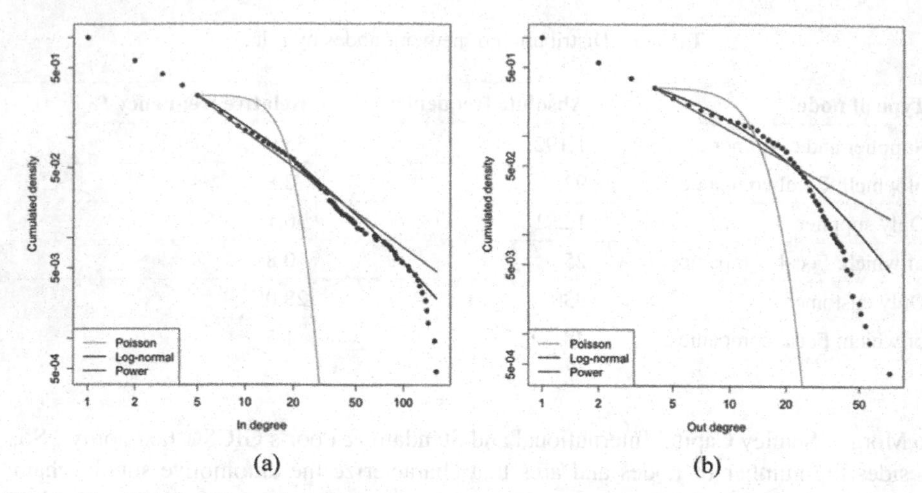

Fig. 2. Empirical and fitted degree distributions on log-log scale. (a) In-degrees. (b) Out-degrees. Fitting is performed by Maximum Likelihood.

favor of formation mechanisms where the connectivity of each node plays a role, in a sort of "preferential attachment" [1]. Indeed, preferential attachment mechanisms generate situations where some nodes play a crucial role in intermediating connections across the network because they act as hubs.

As we turn to examine the Kolmogorov-Smirnoff (KS) distance statistic, it is apparent that Poisson's fit is inferior to Log-normal and Power laws by far (Table 2).[6] The log-normal law seems to be generally preferred, but for in-degrees the KS for Power and Log-normal are quite close; therefore, we computed likelihood-ratio statistics (LR) to compare them.

As Clauset et al. [10] and Gillespie [20] explain, the LR is a one-directional test; therefore, its outcome depends on the way we cast the alternative hypothesis against the null. The null hypothesis is:

H0: Log-normal and Power law distributions are indistinguishable.

Table 2. KS distances between theoretical probability laws and empirical degree distributions.

Theoretical distribution	In-degree	Out-degree
Poisson	0.5182	0.3945
Log-normal	0.0342	0.0909
Power	0.0384	0.1219

The two alternatives, marked as "1" and "2" are:

[6] The lowest the measure, the better the fit.

H1: Log-normal is better than Power law.
H2: Power Law is better than Log-normal law.

Therefore, if we reject H0 against H1 and do not reject it against H2, then we can safely conclude that the Log-normal law is a better representation of the degree distribution than the Power law. The marginal probability values of these tests for in-degrees are, respectively, 0.019 and 0.981; therefore, Log-normal is favored to Power Law.[7]

3 Bow-Tie Organization of the Automotive Network

In the automotive system, the directionality of supplier-customer relationships leads to the spontaneous emergence of a network bow-tie structure. The bow-tie structure comprises five major areas, as described in Broder and co-authors [8] and Fujita and co-authors [17]. We thus examine what the network components are according to this taxonomy. Here, the "strongly connected component" (Scc) is the largest collection of nodes that can be reached from any other node in the network through directed arcs.[8] The "in component" (In) is the set of nodes that cannot be reached from the Scc but point to it. Nodes that belong to the "out component" (Out) can be reached from the Scc but do not point to it. The in- and out-component nodes may be connected through "tube" nodes (Tubes), bypassing the Scc. Also, some marginal nodes called "tendrils" (Tendrils) are connected to in- and out-component nodes, respectively, with outgoing and incoming arcs respectively (see Fig. 3).[9]

The features of the automotive supply chain network are in Table 3. Tubes are a residual component (21 nodes or 0.6% of the total), which means that most connections are intermediated by nodes belonging to the Scc. This component sums up to 755 nodes (22.7% of the total) with 58 focal nodes. Also, in- and out-components are both important, together with tendrils (2,547 nodes or 76.6% of the total).

To determine whether the identified areas are large or small, we compare the bow-tie organization of systems where there is not any specific construction rule. Here, we use networks consisting of the same number of nodes and connections as the automotive's, but where: (i) connections are uniformly distributed, as in Erdòs-Reny or "random" network –Rnd hereafter– or (ii) the connectivity distribution of each node is preserved, but each link is randomly redirected, which we call "redirected networks"—Rid in the following.

[7] For the out-degree distribution, Log-normal is favored to Poisson- and Power-law at < 0.0001 significance level of the test. Also, Log-normal is favored to Poisson at < 0.0001 significance level of the test for in-degree distribution.

[8] Some strongly connected components may exist within other areas. In this case, however, they are only within them and therefore share part of their characteristics (see Sect. 4). Thus, we call Scc the main area, and references to any minor Scc's will be correctly disambiguated.

[9] Since our network is weakly connected by construction, there are no isolated components. Also, note that the in- and out-component areas may have nodes interconnecting inside them, but it is not possible to reach any node from any other within them. The same applies to tubes and tendrils areas.

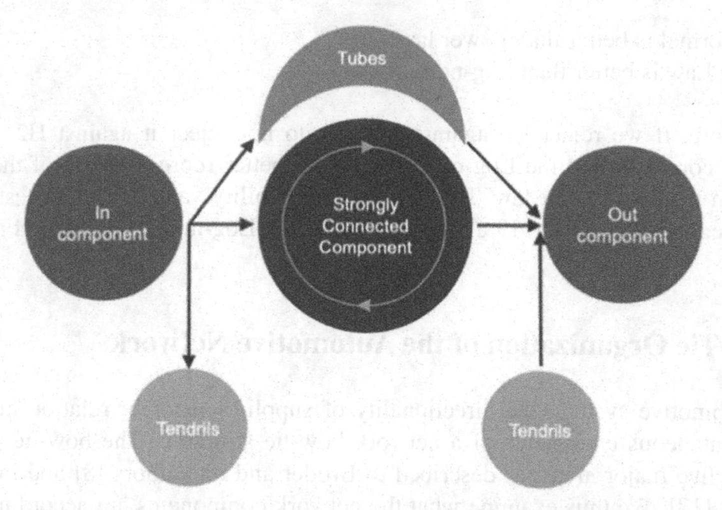

Fig. 3. Taxonomy of network components according to the bow tie model. The five network components are represented by circles. Arrows exemplify the directed connections of nodes belonging to each component. Scc is the largest area where from each node it is possible to reach any other node; In, the area that has only one directional paths towards the Scc; Out, the area reachable from the Scc where paths lead to nodes that have not any further exit; Tubes, the zone that directly connects In and Out zones; Tendrils, one or more sets of nodes that start from the In zone and directly reach the border of the system, or that reach the Out zone starting from the border.

As shown in Table 3, in Rnd networks, the Scc is very large, comprising more than 90% of the network, while every other zone occupies a fraction of less than 4% of the system, resulting in a network configuration quite distant from the actual automotive network. On the contrary, the Rid graph has Scc, In and Out areas of similar size to those of the actual network. These facts support the hypothesis that the distribution of the connectivity of nodes determines the size of the bow-tie areas. We notice that the automotive system has small number of strongly connected components which do not belong to the main Scc area; this is not the case in either Rnd or Rid artificial networks.

It is interesting that the extension of the In and Out areas of the automotive network are both larger than 25% of the whole system. Indeed, a very similar structure was found in early World-Wide-Web related searches [8]. Our interpretation is that the extension of such zones depends on the nature of the actual system, but also on the procedure we followed to map it (see Sect. 2). We think the latter aspect is important and worth further investigation, because similar sampling schemes are common—e.g. Dong and co-authors [13]. We do not pursue this line here, but we deem that this topic is not enough studied in the extant published research, as far as we know.

Table 3. Comparison of actual and simulated Rnd and Rid networks morphology. In the first two columns we show the count and percent frequency of nodes by type of network component they belong to; figures in brackets are the number of focal nodes. In the last two columns, we show the percent frequency of nodes by type of network component they belong to. In the case of Rnd networks and scale-free Rid networks, averages on simulated ensembles of 10 networks are shown; the ± margin is three standard deviations from the mean.

Components	Actual		Rnd	Rid
	Count	Frequency %	Frequency %	Frequency %
Scc	755 (58)	22.7	92.4 ± 0.2	29.0 ± 0.9
In	1,071 (18)	32.2	3.9 ± 0.5	35.0 ± 0.9
Out	883 (72)	26.6	3.7 ± 0.4	25.0 ± 0.9
Tubes	21 (1)	0.6	0 ± 0.0	0.36 ± 0.3
Tendrils	593 (16)	17.8	0.03 ± 0.02	10.5 ± 0.9

4 Diffusion of Perturbations in the Automotive System

4.1 Introduction

In this section we discuss the propagation of perturbations across the supply chain network. All simulated perturbations start from a single node of the system; this way, we examine the risk connected to each node and map the local and global resilience characteristics of the network.

We envision several scenarios that differ because of the resiliency of nodes to perturbations. First, in the case of a "rigid" production technologies, every supply relationship of a node is essential, so that the loss of any incoming link leads to a node stop operating and to its failure. This position corresponds to situations in which the missing supplier supplies essential and otherwise not replaceable parts; alternatively, one can think of short-term scenarios, in which agents do not have enough time to look for alternative suppliers. Second, in the case of "flexible" technology, we assume that a node stays operative even if it loses some of its suppliers and does not fail. In this way it is possible to simulate situations in which a node can deal (at least partially and/or up to a certain point) with supply shortages - for example, by making use of small local suppliers, or by producing the necessary utilities by itself.

The difference between the two technological scenarios involves the maximum size of the affected part of the system, but the transmission mechanism is very similar. In the following we can therefore refer to the rigidity scenario.

Under these assumptions, we examine the resilience of the network over the short time, where firms cannot effectively respond to perturbations by changing their technology, suppliers, or customers.[10]

---

[10] This is in line with most extant research, such as Inoue and Todo [22]. Investigating proactive responses to shocks is an interesting topic but beyond the scope of this article. We leave it for future investigation.

Given the nature of the perturbations we propose, the bow-tie organization of the system influences its resilience properties; indeed, the position and the final size of the network regions involved by perturbations strongly depends on the bow-tie area where the initial disturbance happens. We assume two types of perturbations:

1. "Avalanches", where the propagation of perturbations spreads along the supplier-customer direction; any perturbation hitting a node involves its outgoing links—i.e. it goes downstream.
2. "Shocks", the same as above plus upstream propagation of perturbations, because if a node loses all its main customers, then it fails; in this case, the customer-supplier relationships also play a role.

The asymmetry in the propagation of perturbations, downstream only (1) or downstream and upstream (2), is significant, because it allows to investigate the resilience of the network in diverse conditions. Beyond the specific case of the automotive supply-chain we examine, our analyses may provide insights about the effects of perturbations in similar systems, such as socio-technological networks, chemical reaction systems, or gene regulation networks.

4.2 Avalanches

In our simulations, the cascade of avalanche perturbations starts from a single node; we hit every node in the network, one at a time, and see what the outcomes are. The main evidence we collected is:

1. As expected, if the first perturbated node belongs to the Scc area, then every other Scc node is hit. The failure of Scc nodes leads to perturbations in the Out component, so that all its nodes are obliterated. This does not depend on which specific Scc node is first hit; any perturbation in the Scc leads to the failure of all Out nodes.
2. If the first perturbated node belongs to the In zone, then all downstream nodes are involved until the cascade of perturbations reaches the Scc area. Then, the avalanche continues as above, so that the size of the disruption is the same plus some nodes in the In zone.[11]
3. If the initial node belongs to the Out zone, then the cascade of perturbations will only hit downstream nodes until it reaches the edge of the system: here the avalanche ends. Consequently, the cascades of perturbations starting from the Out zone are very small in size.
4. If the initial node belongs to a Tube, the avalanche involves only some of the nodes of that zone and their downstream in the Out area. Thus, the cascades of perturbations starting from this zone are very small in size.

[11] Sometimes it happens that an avalanche starting in the In zone hits some nodes belonging to Tubes, because one or more of the failed nodes in the In zone are suppliers of one or more nodes belonging to the Tubes zone. However, tubes have few nodes (see Table 3), so they are not worth discussing.

5. Finally, if the initial node belongs to a Tendril connected to the In zone, then the avalanche involves only the nodes of such area downstream. If the initial node belongs to the Tendril area connected to the Out zone, then the avalanche involves the downstream nodes of that structure and the nodes of the Out zone downstream of the Tendril structure.

Because of the size of the bow-tie regions in the automotive supply-chain, the size distribution of avalanches is bimodal, with many large avalanches (cases 1 and 2) and many small ones (cases 3, 4 and 5) as shown in Fig. 4. The typical size of the avalanches is very close —sometimes the same— to the sum of the size of the Scc and Out zones.

Therefore, it looks like that the extent of the damage of a single node perturbation can cause to the system strongly depends on the area in which the node is located. As already observed in Fig. 3, nodes in Scc area are central, so when an avalanche hit them, the impact is very large. This is also the case when In nodes are hit, as the directed graph propagate the avalanche to the Scc components. Hence, sensitive nodes belong both to the Scc and the In area of the network.

Another interesting statistics concerns the number of times each node is involved in one of 3,323 possible avalanches; we call this "susceptibility", in line with Serra and co-authors [30]. There are nodes often involved, while other almost never are. As expected, Scc and Out nodes exhibit large susceptivity, while In, Tubes and Tendrils are basically unaffected by avalanches (see Fig. 5).

The same analysis can be applied also to random networks (Rnd) and redirected networks (Rid). The size distribution of the avalanches is again bimodal, and the observation that the typical size of the avalanches is close to the sum of the size of the Scc and Out zones remain valid. However, the large size of the Scc zone of Rnd networks leads to a particular bimodality, in which more than 96% of avalanches involve a very narrow range close to 96% of the system. The automotive network is therefore much more robust than a corresponding Rnd system. On the other hand, the size distribution of avalanches of a Rid system is substantially similar to that of the automotive network, despite the median size of avalanches is slightly higher than its, because the larger size of the Scc and Out zones.

4.3 Shocks

Shocks are like avalanches, but they are characterized by an additional process involving possible retro-propagations of perturbations in the customer-supplier direction, which results in more nodes potentially involved. The main evidence we get for shocks is:

1. If the first hit node belongs to the Scc area, all nodes of the Scc and Out zone are involved, as for the case of avalanches. Furthermore, most if not all the nodes belonging to the In zone often lose their customers and are then perturbated. This does not happen in avalanches. Also, some Tendril nodes originating from the in component fail. Finally, nodes in the Out zone are also perturbated, causing the involvement of all their customers in the Tendrils. However, the cascade of perturbations can be interrupted by resistance structures present in Tubes.

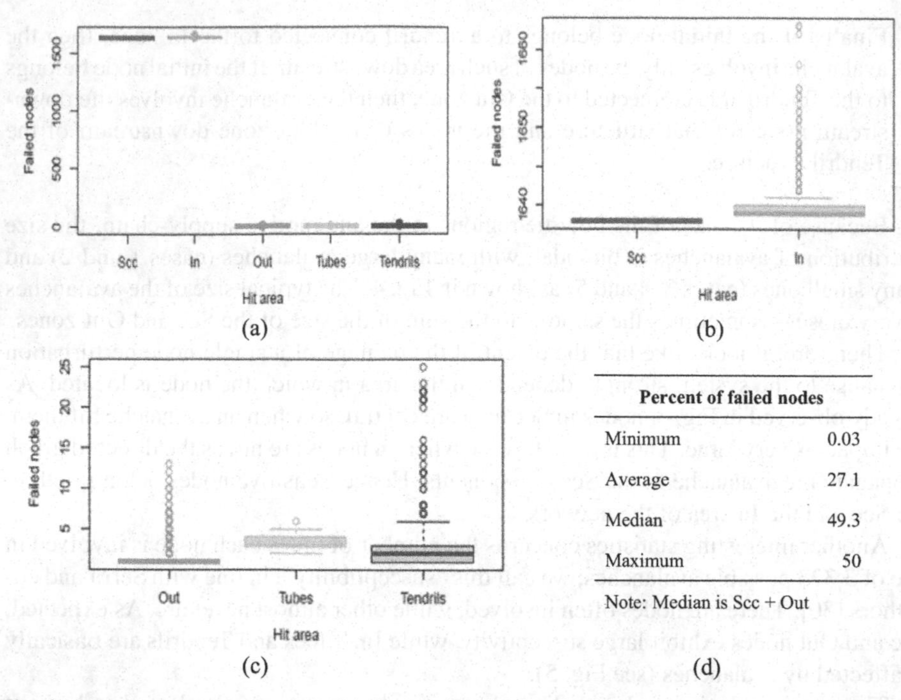

Percent of failed nodes	
Minimum	0.03
Average	27.1
Median	49.3
Maximum	50
Note: Median is Scc + Out	

Fig. 4. Distribution of avalanches. (a) Count of nodes involved by avalanches starting in the Scc, In, Out, Tubes and Tendrils areas. (b) Details of avalanches starting in the Scc and In zones, their variances are very small. (c) Details of avalanches starting in the Out, Tubes and Tendrils zones. The little number of outliers suggests that there are different propagation paths within some small zones. (d) Statistics about the size of the avalanches, as fractions of hit nodes in the whole network.

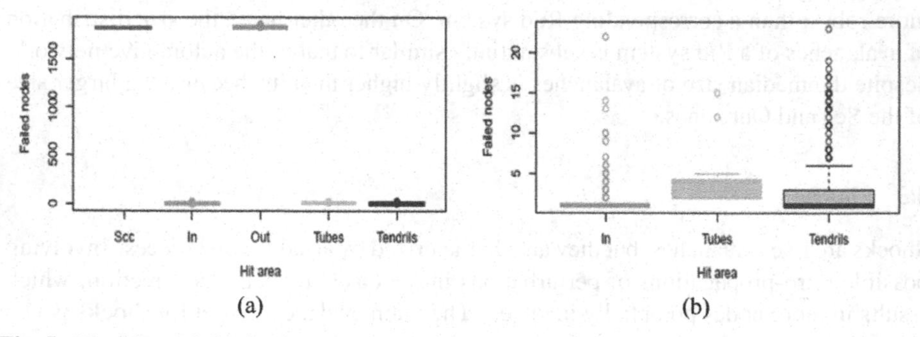

Fig. 5. The bimodal distribution of susceptivities in case of avalanches. (a) The "box-plot" graph in which we indicate the fraction of times each node is involved in one of 3323 possible avalanches, calculated based on the node belonging to Scc, In, Out, Tubes and Tendrils areas. (b) A magnification focused only on nodes belonging to In, Tubes and Tendrils zones. The presence of outliers (however very small - see part (a) of this figure) indicates the possibility of different avalanches propagation paths.

2. When the first hit node belongs to the In zone, all downstream nodes are perturbated until the cascade reaches the Scc: from here, the shock will continue as already described, involving the whole Out zone and parts of Tubes and Tendrils. So, shocks starting from the In zone typically have a bigger size than shocks starting in the Scc area. Again, the cascade of perturbations can be stopped by resistance structures present in Tubes.
3. If the first hit node belongs to the Out zone, then the cascade of perturbations affects all downstream nodes until it reaches the edge of the system: here the cascade ends. Nodes belonging to the Scc zone are not involved because they have at least one additional customer in the Scc zone not hit by any perturbation. Nodes in the Tendril area of the Out zone can be involved, but sometimes the propagation can be stopped by resistance structures. Nodes in Tubes can be involved, and in the absence of resistance structures the propagation of perturbations can reach the border of the In zone, where it stops because the nodes belonging there have at least one customer in the same zone or in the Scc area. Therefore, shocks originating in the Out area necessarily limited in size.
4. When the first hit node belongs to a Tube, the shock involves its nodes downstream and also the downstream nodes of the Out area. In the absence of resistance structures, the propagation can reach the border of the In zone, but here it stops because the nodes belonging to the In zone have at least one customer in the In zone or in the Scc area. So, perturbations starting in Tubes are very small in size.
5. Finally, if the first hit node belongs to a Tendril connected to the In zone, then the shock involves only the nodes of that area downstream—with exceptions in case of resistance structures. If the first hit node belongs to a Tendril connected to the Out zone, then the perturbation involves the Tendril and Out nodes downstream: in some cases, some Out nodes lose all customers and back-propagate the perturbation. In any case, perturbation ends once it reaches the Scc area. The shocks originating in a Tendril area are limited in size.

Obviously, shocks have larger effects than avalanches (see Fig. 6). However, the asymmetry of the propagation of perturbations creates the conditions for the emergence of resistance structures that can interrupt the retro-propagation of a shock along the supply chain network.

A first large resistance structure is the Scc area. While initial-seeded shocks in the In zone can invade the Scc area, which is downstream, none of the initial-seeded shocks in the Out zone can involve Scc nodes via the back-propagation of perturbations in the customer-supplier sense.[12] This resistance is due to the fact that each node belonging to the Scc has at least one customer belonging to the same area; therefore, the back-propagation in the upstream direction is ineffective. It is interesting to note that this property could allow to block the propagation of shocks in the customer-supplier direction, even if it were not necessary to lose all customers to damage a node.

[12] Although, back propagations through Tubes can reach the In zone because they bypass the Scc. Frontier (which is upstream with respect to the Scc zone), but once they have reached the frontier, they cannot invade the In zone, because the nodes of this zone have at least one customer in the In zone itself or in the Scc, areas not reached by the shock.

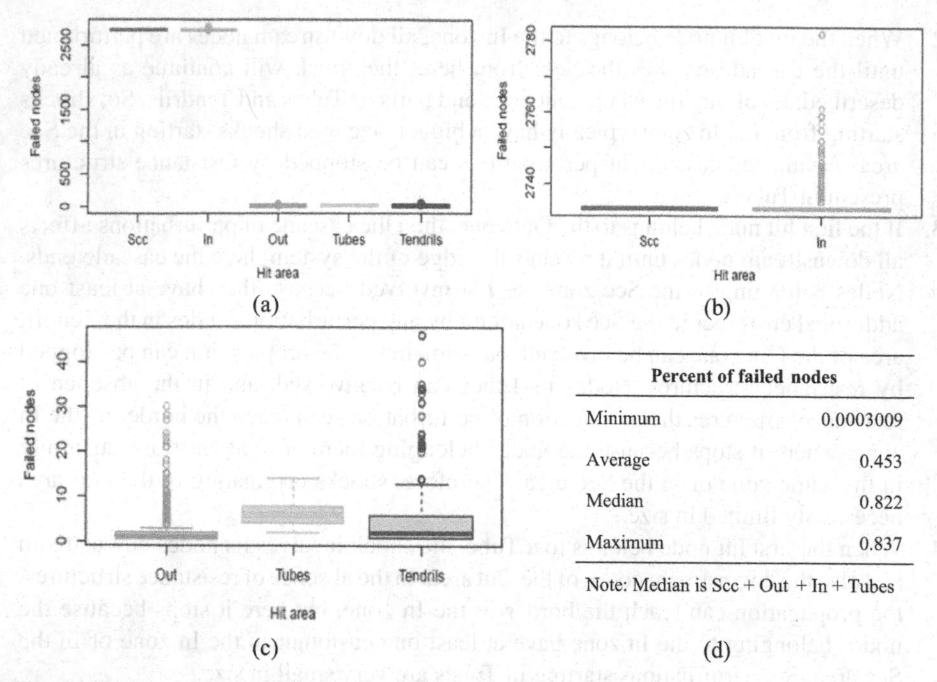

Percent of failed nodes	
Minimum	0.0003009
Average	0.453
Median	0.822
Maximum	0.837

Note: Median is Scc + Out + In + Tubes

Fig. 6. The bimodal distribution of shocks. (a) The "box-plot" graph in which we indicate the fraction of the system involved by the shocks with initial source in the Scc, In, Out, Tubes and Tendrils areas. (b) A magnification focused only on shocks started in the Scc and In zones (the variances are very small). (c) A magnification focused only on shocks started in the Out, Tubes and Tendrils zones. The presence of outliers (however very small - see part (a) of this figure) indicates the possibility of different propagation paths within the various (small) zones. (d) Statistics about the size of the shocks. Obviously, given the bimodality of the distribution, the average is an abstraction (no avalanche size corresponds to the average value); the median size is close to the sum of the dimensions of the Scc, Out, In and Tubes areas (most part of the avalanches affect these areas).

A second-level resistance is *local* Scc's in the Out, Tubes and Tendrils areas which, because of the conformation of the structure they are immersed in, in some cases they are not bypassed by back-propagation (see Fig. 7).[13] Once groups of nodes are able to resist back-propagation, they can in turn become bases for defending their customers upstream, and thus preserve large areas. As anticipated, given the size and sparseness characteristics of the systems we simulated, these structures are unlikely in the case of random linking; indeed, we only found them in the actual automotive network.

Summing up, it never happens that an initial perturbation hitting a node in the Out zone gives rise to large shocks. In case of shocks, the risk (susceptibility) connected to nodes belonging to the Scc and Out zones are basically the same as for avalanches, while the susceptibility of a big part of the nodes belonging to other areas considerably

[13] Again, in case it is not necessary to lose all customers to damage a node. Also, secondary Scc's may be the core of larger resistance structures.

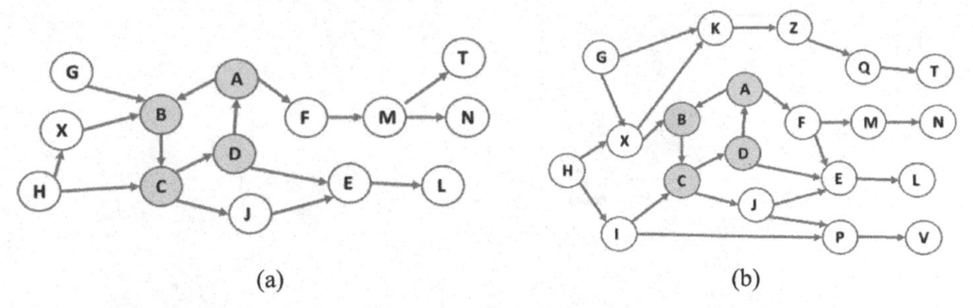

Fig. 7. Resistance structures (in yellow) immersed in an area (In, Out, Tube or Tendril) where the flow of material globally proceeds in only one direction. From left to right: (a). Back propagations coming from T, N or L are blocked, because sooner or later they only hit nodes that have at least one client not involved in the shock; (b) back propagations are blocked. It should be noted that nodes X and I do not belong to the local SCC, but the SCC guarantees that they have at least one not failed customer. (Color figure online)

increases. This effect is visible in the boxplots of Fig. 8, where we notice a strong bimodality of susceptibility for In, Tubes and Tendrils areas; this is because a subset of the nodes is protected by the resistance structures—definitely more effective for nodes belonging to the Tubes and Tendrils areas.

The shock propagation analysis can be repeated for random networks (Rnd) and redirected networks (Rid). The size distribution of the shocks is again bimodal, and the observation that the typical size of the avalanches is close to the sum of the size of the Scc, Out, In and Tubes areas remains valid. However, the large size of the main Scc zone of the Rnd networks and the small size of Tubes and Tendrils lead to a situation in which more than 96% of shocks involve 3,321 nodes out of 3,323. The automotive network is therefore more robust than a corresponding Rnd configuration.

On the other hand, the size distribution of shocks of a Rid system is substantially like that of the automotive network: the size of 55% of the shocks is close to 2,973 (3,001 being the sum of the Scc, Out, In and Tubes zones) up to a maximum size of 2,986,[14] while the remaining 45% is composed by small shocks, with a maximum size of 19. No major shocks are generated from the Out area.

5 Discussion and Conclusion

In this study, we aim to understand the diffusion path of perturbation by examining the properties of the global automotive supply-chain sampled from the Bloomberg database between 2018 to 2020. The directionality of supplier-customer relationships leads to the spontaneous emergence of five major areas according to the bow-tie organization [8, 17]. The application of such taxonomy —a large Scc, In, Out, Tubes, and Tendrils— allows us to better understand the diffusion of perturbations, be them avalanches or shocks, across the system and to evaluate the network resilience and robustness.

[14] The typical size of shocks for Rid is larger than that for the automotive system, because the slightly larger size of the Scc, In and Out zones.

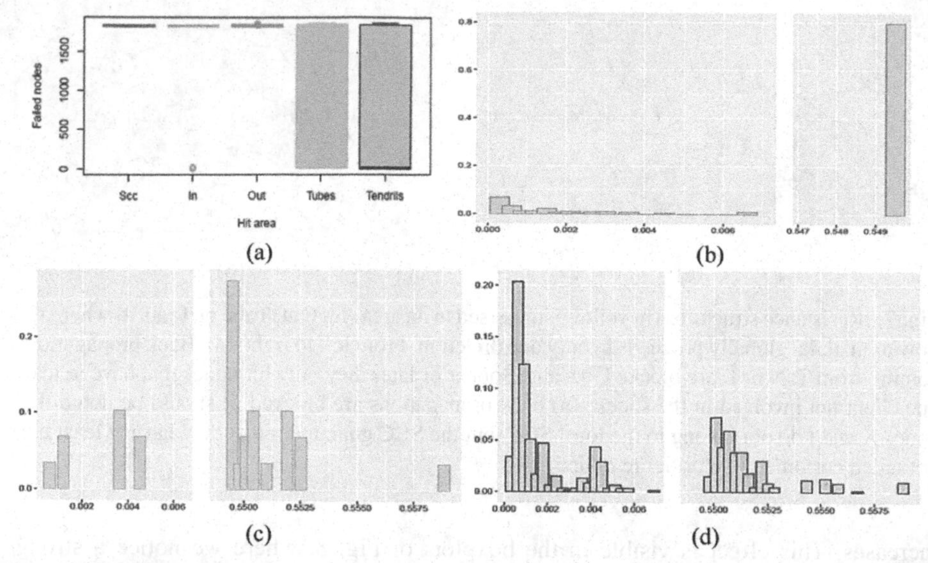

Fig. 8. The distribution of susceptivities in case of shocks: (a) box plot of the percent fraction of cases where each node is involved in one of 3,323 possible shocks, by the zone it belongs to in the bow-tie model. The appearance of the boxplots relating to the In, Tubes and Tendrils areas, indicates a bimodal distribution. (b) The bimodal nature of the susceptibility present in the In area, where a small fraction of nodes have low susceptibilities, while the majority of nodes are hit by perturbations. (c) The bimodal nature of the susceptibility present in the Tubes area. (d) The bimodal nature of the susceptibility present in the Tendrils area. Here, the resistance structures are particularly effective.

The effects of perturbations are strongly dependent on the different areas of the bow-tie where they originate. We can identify Scc and In nodes as those that generate the greatest avalanches and shocks. Still, when looking at susceptivity, avalanches strongly affect Scc and Out nodes, which are the largest involved areas; shocks might also impact In, Tubes and Tendrils. Scc nodes are central in the network, thus we expect them to be capable of producing a larger impact on the entire system when hit, while Out nodes are largely affected by avalanches and shocks. It is interesting to note that In nodes, although not as central as those in the Scc area, are the origin of large avalanches and shocks. On the contrary, an initial a perturbation generated in the Out zone never gives rise to large avalanches or shocks.

Our results indicate that a shock, as expected, involves more nodes than an avalanche, due to an additional process involving possible back-propagations of perturbations in the customer-supplier direction. However, it is interesting to note that in this case it is possible to identify interesting resistance structures, based on zones that appear locally and are organized as Scc's—including but not limited to the main bow-tie Scc area. In general, when is not necessary to lose all clients to make a node fail, these Scc's may be the core of larger resistance structures.

Finally, we examined the network robustness to avalanches and shocks and show that the automotive supply chain network is more resilient than both random networks and redirected networks with the same degree distribution.

The capability of a system to recover quickly and effectively to a planned performance level after an unexpected disruption depends on its structure, as well as on the specific resistance of its nodes [5, 21]. In supply-chain systems, efforts to build resilience can be made both at the firm or supply-chain level. The latter level depends on the connections of firms in the network; this requires an understanding of its structure and of the way nodes interact with each other [21, 27]. In this sense, to improve supply chain resilience, it is imperative to understand the supply chain network topology [24].

In this work we used the division into "bow-tie" areas to better understand the consequence of perturbations applied in different parts of a system that can be represented by a directed graph.

We notice that the central area Scc, although significant, does not actually generate the largest perturbations (which start from the nodes belonging to the In area): it is instead the area most often affected by avalanches and shocks, being involved both by perturbations that depart from it, both by the perturbations coming from the In area. The same fate is shared by the Out zone, towards which all these perturbations converge. Interestingly, avalanches and shocks from the Out, Tubes an Tendrils zone never take on significant dimensions. Finally. The nodes belonging to Tubes and Tendrils are almost never involved in avalanches (while a fraction of them can be involved very frequently by shocks).

Based on the bow tie model, we can therefore identify the areas and nodes starting from which perturbations can have a great impact on the system and therefore indicate the zones where is useful to intensify the risk management and mitigation strategies. These insights may also apply to other real-world systems organized according to bow-tie structures.

References

1. Albert, R., Barabási, A.L.: Statistical mechanics of complex networks. Rev. Mod. Phys. **74**(1), 47 (2002)
2. Albert, R., Jeong, H., Barabási, A.L.: Error and attack tolerance of complex networks. Nature **406**, 378–382 (2000)
3. Avrachenkov, K., Litvak, N., Pham, K.S.: Distribution of PageRank mass among principle components of the web. In: Bonato, A., Chung, F.R.K. (eds.) WAW 2007. LNCS, vol. 4863, pp. 16–28. Springer, Heidelberg (2007). https://doi.org/10.1007/978-3-540-77004-6_2
4. Barabási, A.L.: Network Science, pp. 73–82. Cambridge University Press (2017)
5. Behzadi, G., O'Sullivan, M.J., Olsen, T.L.: On metrics for supply chain resilience. Eur. J. Oper. Res. **287**(1), 145–158 (2020)
6. Belhadi, A., Kamble, S., Jabbour, C.J.C., Gunasekaran, A., Ndubisi, N.O., Venkatesh, M.: Manufacturing and service supply chain resilience to the COVID-19 outbreak lessons learned from the automobile and airline industries. Technol. Forecast. Soc. Chang. **163**, 120447 (2021)
7. Bellamy, M.A., Ghosh, S., Hora, H.: The influence of supply network structure on firm innovation. J. Oper. Manag. **32**(6), 357–373 (2014)
8. Broder, A., et al.: Graph structure in the web. Comput. Netw. **33**(1–6), 309–320 (2000)

9. Buldyrev, S.V., Parshani, R., Paul, G., Stanley, H.E., Havlin, S.: Catastrophic cascade of failures in interdependent networks. Nature **464**, 1025–1028 (2010). https://doi.org/10.1038/nature08932
10. Clauset, A., Shalizi, C.R., Newman, M.E.J.: Power-law distributions in empirical data. SIAM Rev. **51**(4), 661–670 (2009)
11. Cohen, R., Erez, K., ben-Avraham, D., Havlin, S.: Resilience of the internet to random breakdowns. Phys. Rev. Lett. **85**(21), 4626–4628 (2000)
12. Culwick, M.D., Merry, A.F., Clarke, D.M., Taraporewalla, K.J., Gibbs, N.M.: Bow-tie diagrams for risk management in anaesthesia. Anaesthesia Intens. Care **44**(6), 712–718 (2016)
13. Dong, Q., Yu, E.Y., Li, W.J.: A network sampling strategy inspired by epidemic spreading. Secur. Commun. Netw. **2022**, 7003265 (2022). https://doi.org/10.1155/2022/7003265
14. Dunne, J.A., Williams, R.J.: Cascading extinctions and community collapse in model food webs. Philos. Trans. Roy. Soc. B: Biol. Sci. **364**(1524), 1711–1723 (2009)
15. Fiksel, J.: Designing resilient, sustainable systems. Environ. Sci. Technol. **37**(23), 5330–5339 (2003)
16. Friedlander, T., Mayo, A.E., Tlusty, T., Alon, U.: Evolution of bow-tie architectures in biology. PLoS Comput. Biol. **11**(3), e1004055 (2015)
17. Fujita, Y., Kichikawa, Y., Fujiwara, Y., Souma, W., Iyetomi, H.: Local bow-tie structure of the web. Appl. Netw. Sci. **4**(1), 1–15 (2019). https://doi.org/10.1007/s41109-019-0127-2
18. Gao, J., Barzel, B., Barabasi, A.L.: Universal resilience patterns in complex networks. Nature **530**, 307–312 (2016)
19. Ghosh Roy, G., He, S., Geard, N., Verspoor, K.: Bow-tie architecture of gene regulatory networks in species of varying complexity. J. R. Soc. Interface **18**(179), 20210069 (2021)
20. Gillespie, C.S.: The poweRlaw package: comparing distributions (2020). https://cran.r-project.org/web/packages/poweRlaw/index.html
21. Hearnshaw, E.J.S., Wilson, M.M.J.: A complex network approach to supply chain theory. Int. J. Oper. Prod. Manag. **33**(4), 442–469 (2013)
22. Inoue, H., Todo, Y.: Firm-level propagation of shocks through supply-chain networks. Nat. Sustain. **2**, 841–847 (2019)
23. Ivanov, D., Dolgui, A.: Viability of intertwined supply networks: extending the supply chain resilience angles towards survivability. A position paper motivated by COVID-19 outbreak. Int. J. Prod. Res. **58**(10), 2904–2915 (2020)
24. Kim, Y., Choi, T.Y., Yan, T., Dooley, K.: Structural investigation of supply networks: a social network analysis approach. J. Oper. Manag. **29**, 194–211 (2011)
25. Muniz, M.V.P., Lima, G.B.A., Caiado, R.G.G., Quelhas, O.L.G.: Bow tie to improve risk management of natural gas pipelines. Process Saf. Prog. **37**(2), 169–175 (2018)
26. Newman, M.: Networks. Oxford University Press, Oxford (2018)
27. Ozdemir, D., Sharma, M., Dhir, A., Daim, T.: Supply chain resilience during the COVID-19 pandemic. Technol. Soc. **68**, 101847 (2022)
28. Pastor-Satorras, R., Vespignani, A.: Epidemic spreading in scale-free networks. Phys. Rev. Lett. **86**, 3200–3203 (2001)
29. Serra, R., Villani, M., Graudenzi, A., Kauffman, S.A.: Why a simple model of genetic regulatory networks describes the distribution of avalanches in gene expression data. J. Theor. Biol. **246**(3), 449–460 (2007)
30. Serra, R., Villani, M., Semeria, A.: Genetic network models and statistical properties of gene expression data in knock-out experiments. J. Theor. Biol. **227**(1), 149–157 (2004)
31. Shai, P., Porter, M.A., Pascual, M., Kefi, S.: The multilayer nature of ecological networks. Nat. Ecol. Evol. **1**(4), 0101 (2017)
32. Sheridan, P., Onodera, T.: A preferential attachment paradox: how does preferential attachment combine with growth to produce networks with log-normal in-degree distributions? Scientific Rep. **8**(1), 2811 (2018). https://doi.org/10.1038/s41598-018-21133-2

33. Strogatz, S.H.: Exploring complex networks. Nature **410**(6825), 268–276 (2001)
34. Tawde, V.B., Oates, T., Glover, E.: Generating web graphs with embedded communities. In: Leonardi, S. (ed.) WAW 2004. LNCS, vol. 3243, pp. 80–91. Springer, Heidelberg (2004). https://doi.org/10.1007/978-3-540-30216-2_7
35. Tieri, P., et al.: Network, degeneracy and bow tie. Integrating paradigms and architectures to grasp the complexity of the immune system. Theor. Biol. Med. Modell. **7**(1), 1–16 (2010)
36. Timár, G., Goltsev, A.V., Dorogovtsev, S.N., Mendes, J.F.: Mapping the structure of directed networks: beyond the bow-tie diagram. Phys. Rev. Lett. **118**(7), 078301 (2017)
37. Villani, M., La Rocca, L., Kauffman, S.A., Serra, R.: Dynamical criticality in gene regulatory networks. Complexity **2018**, 1–14 (2018). https://doi.org/10.1155/2018/5980636
38. Watts, D.J.: A simple model of global cascades on random networks. Proc. Natl. Acad. Sci. **99**(9), 5766–5771 (2002)
39. Yang, Y., Nishikawa, T., Motter, A.E.: Small vulnerable sets determine large network cascades in power grids. Science **358**, N6365 (2017). https://doi.org/10.1126/science.aan318
40. Zhao, J.H., Zhou, H.J., Liu, Y.Y.: Inducing effect on the percolation transition in complex networks. Nat. Commun. **4**, 2412 (2013)

An Efficient Implementation of Flux Variability Analysis for Metabolic Networks

Bruno G. Galuzzi[1,2(✉)] and Chiara Damiani[1,2]

[1] Department of Biotechnology and Biosciences, University of Milano-Bicocca, Milan, Italy
{bruno.galuzzi,chiara.damiani}@unimib.it
[2] SYSBIO Centre of Systems Biology/ ISBE.IT, Milan, Italy

Abstract. Flux Variability Analysis (FVA) is an important method to analyze the range of fluxes of a metabolic network. FVA consists in performing a large number of independent optimization problems, to obtain the maximum and minimum flux through each reaction in the network. Although several strategies to make the computation more efficient have been proposed, the computation time of an FVA can still be limiting. We present a two-step procedure to accelerate the FVA computational time that exploits the large presence within metabolic networks of sets of reactions that necessarily have an identical optimal flux value or only differ by a multiplication constant. The first step identifies such sets of reactions. The second step computes the maximum and minimum flux value for just one element of each of set, reducing the total number of optimization problems compared to the classical FVA. We show that, when applied to any metabolic network model included in the BiGG database, our FVA algorithm reduces the total number of optimization problems of about 35%, and the computation time of FVA of about 30%.

Keywords: Metabolic networks · Flux Balance Analysis · Flux variability Analysis · Constrained-based modeling

1 Introduction

The study of cell metabolism is of paramount importance in various fields, including health, wellness, and bio-transformations [4]. Indeed, metabolism is related with most cellular processes and may act as an integrative readout of the pato-physiological state of a cell [18]. The first requirement to understand metabolism is the knowledge of metabolic fluxes, i.e. the velocities and the directions of all the biochemical reactions involved in all metabolic processes, such as glycolysis and oxidative phosphorylation.

Currently, direct determination of metabolic fluxes at the genome-wide level is not feasible. However, they can be predicted numerically, via integration of multiple -omics data (e.g., transcriptomics, proteomics, and metabolomics) into

C. De Stefano et al. (Eds.): WIVACE 2022, CCIS 1780, pp. 58–69, 2023.
https://doi.org/10.1007/978-3-031-31183-3_5

constraint-based stoichiometric metabolic models [3,4,13]. The starting point of constrained-based modeling is the information embedded in the metabolic network, which represents the set of all the possible biochemical reactions that can occur in a cell in a specific organism or tissue. Such information can be represented with a stoichiometric matrix S of dimension $m \times n$, where m is the number of metabolites and n is the number of reactions. In a constrained-based metabolic model, a steady-state condition is imposed, that is, the total production of any metabolite must equal to the total amount of its consumption. Hence, a possible metabolic flux configuration is represented by a vector $\vec{v}$, for which $S\vec{v} = 0$, i.e. the null space of the stoichiometric matrix. Additional constraints, such as thermodynamics or capacity constraints, can also be incorporated.

Despite the large number of constraints, a metabolic network typically includes more reactions than metabolites, resulting in a large space of possible feasible flux distributions. In this case, different strategies can be used to predict the target metabolic flux distribution. A possibility consists in using efficient flux sampling strategies [7,12] to explore the entire region of feasible flux solutions and obtaining information on the range of feasible flux solutions and on their probabilities. Another possibility is Flux Balance Analysis (FBA), which assumes that the cell behaviour is optimal with respect to an objective function, such as the biomass production rate. The objective function defines a reaction that must be maximized or minimized under the set constraints. FBA identifies a single solution among the set of possibly many alternative optimal solutions.

Alternatively, it is possible to study the range of each metabolic flux across the set of alternative optima by means of Flux Variability Analysis (FVA) [16]. In a nutshell, FVA consists in finding the minimum and maximum flux through each reaction in the network, given some constraints on the state of the network, e.g., imposing a minimum percentage of the maximal biomass production rate. Typical applications of FVA in systems biology include investigating network flexibility and network redundancy [24]. Recently, FVA has gained importance as a preliminary step for omics data integration. For example, in [5,8], to sensibly limit the flux of an internal reaction based on gene expression data, we first needed to compute the maximum and minimum flux through such reaction, based on nutrient availability constraints. In fact, limiting the flux boundaries defined *a priori* by the modeler might produce no effect on the flux if they are larger than the actual maximal flux determined by the environment.

Given a generic metabolic network with n reactions, FVA requires $2n$ optimization problems to be solved, maximizing and minimizing the flux of any reaction in the model. Given the high number of possible reactions and constraints characterizing a metabolic network, efficient computation of FVA is fundamental. The aim of this work consists of exploring a new way to partially reduce the computational time classically required to solve FVA.

1.1 State of the Art

The computation of FVA clearly can be distributed and is therefore ideally suited for high performance computing. For example, the current FVA implementations in Cobrapy [6], COBRA toolbox [22], and COBRAjl [11] split the optimization problem into multiple processes. Other implementation have been designed to reduce the computational time requested to find an initial feasible solution. For example, fastFVA [23], one of the most used efficient implementations of FVA, coded in C++, iterates through all the reactions and solves the two optimization problems using a specialized LP solver, such as GLPK or CPLEX, but without spending time effort in finding a feasible solution or pre-processing the linear system, since the feasible region is the same for all the optimization problems. Finally, implementations of FVA exist that include thermodynamic constraints [17] to remove unbounded fluxes through reactions contained in internal cycles.

1.2 Our Contribution

To the best of our knowledge, no current implementation of FVA exploits the fact that in a generic metabolic network, there are a lot of constraints of the form $av_i + bv_j = 0$. Hence, if $v_i^{\max}$ and $v_i^{\min}$ have already been computed, one can omit the two optimization problems for $v_j^{\max}$ and $v_j^{\min}$. This simple but effective consideration is expected to considerably reduce the total number of optimization problems for FVA of any metabolic network. To clarify this observation with an example, in Fig. 1, we show a small sub-network of the human metabolism related to the glycolysis and Pentose phosphate pathways, represented using the web-app Escher [14]. Each blue arrow represents a reaction and each node represents a metabolite involved in a reaction. Under the steady-state assumption, the total production of any metabolite must be equal to the total amount of its consumption. It can be noticed that many metabolites, such as 3-phospho-D-glycerate (3pg_c), D-glycerate 2-phosphate (2pg_c), and phosphoenolpyruvate (pep_c) are involved in two different reactions only. Hence, the steady-state assumption imposes that the flux through phosphoglycerate kinase (PGK) and phosphoglycerate mutase (PGM) must be equal, and that the flux of

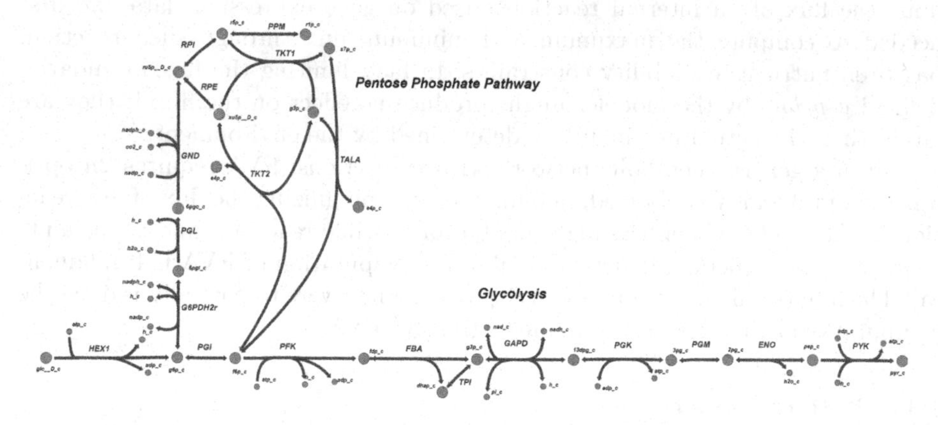

Fig. 1. Example of human metabolic sub-network related to glycolysis and Pentose phosphate pathways.

the Phosphoglycerate kinase (PGK) and Enolase (ENO) must be equal. Hence, once the maximum and minimum flux through the PGK reaction has been found, one can avoid to solve the optimization problems for the PGK and ENO reactions. In this work, we investigated the presence of these type of constraints involving only two fluxes in all the metabolic models publicly available in the BiGG model database [19]. Then, we used such information to implement a Python-based Cobrapy [6] extension that improves the computation of FVA including an efficient pre-FVA step to find all the possible reactions for which the optimization problems can be omitted. Such pre-FVA step can be used not only for FVA but also for all computational tools requiring the computation of several optimization problems for the same metabolic networks, such as finding blocked reactions, searching for essential reactions of a target objective function, and reaction deletion analysis.

2 Material and Methods

2.1 COBRA Model

Assuming that cell behavior is optimal with respect to an objective, optimization methods, such as Flux Balance Analysis (FBA) [21], can be used to calculate an optimal flux distribution with respect to a specific objective. In a nutshell, a metabolic network is associated with the following linear programming (LP) problem:

$$\begin{aligned} &\max \sum_{i=1}^{r} w_i v_i \\ &S \cdot \vec{v} = \vec{0} \\ &\vec{v_L} \leq \vec{v} \leq \vec{v_U} \end{aligned} \tag{1}$$

where w_i is the objective coefficient of flux i, and $\vec{v_L}$ and $\vec{v_U}$ represent the possible bounds used to mimic as closely as possible the biological process in the analysis.

2.2 Flux Variability Analysis

Flux Variability Analysis (FVA) [9,16] is a constraint-based modeling technique aimed at determining the maximal (and minimal) possible flux through any reaction of the model, to evaluate the cell's range of metabolic capabilities. FVA solves the following two linear programming optimization problems (one for minimization and one for maximization) for each flux v_j of interest, with $j = 1, \ldots, n$:

$$\begin{aligned} &\max / \min \; v_j \\ &S \cdot \vec{v} = \vec{0} \\ &\vec{v_L} \leq \vec{v} \leq \vec{v_U} \\ &\vec{v_i} \geq \gamma Z_0 \end{aligned} \tag{2}$$

where Z_0 is an optimal solution for Eq. 1, and γ is a parameter, which controls whether the analysis is done w.r.t. suboptimal network states ($0 \leq \gamma < 1$) or to the optimal state ($\gamma = 1$). FVA can be used also to search for blocked reactions in a network. A blocked reaction r_i in a metabolic network is a reaction that carries no flux in any feasible solution (i.e. $v_i = 0$). Practically, the blocked reactions are the subset of reactions for which $\max v_j = \min v_j = 0$.

2.3 Pre-step to Accelerate FVA

Our improvement of FVA computation is based on a pre-FVA operation to select the subset of reactions for which it is really necessary to solve the maximization and the minimization problems. We showed this step in Fig. 2. First, we considered the set of all the possible reactions $\{v_1, \ldots, v_n\}$, and we split the reactions in connected-subsets. Each connected-subset is formed by reactions whose flux value can be derived by multiplying any other flux value of the subset by a known constant (e.g., $2v_1 - 3v_2 = 0$), which may also take value 1 (e.g., $v_1 - v_2 = 0$). For the computation of the connected-subsets, we built a graph whose nodes represent the reactions $r_1, \ldots, r_n$ of the network. We connected two nodes r_i and r_j if and only if a constrain of the form $av_i + bv_j = 0$ exists. The connected-subsets corresponds with the connected components of such graph. Then, we created a set I formed by one arbitrary reaction for each of these subsets, and we performed FVA for the reactions in I only. Afterwards, we derived the rest of the FVA values using the direct relation with the FVA values computed before. This implies to solve a linear system in which the variables are the fluxes of the reactions not belonging to I.

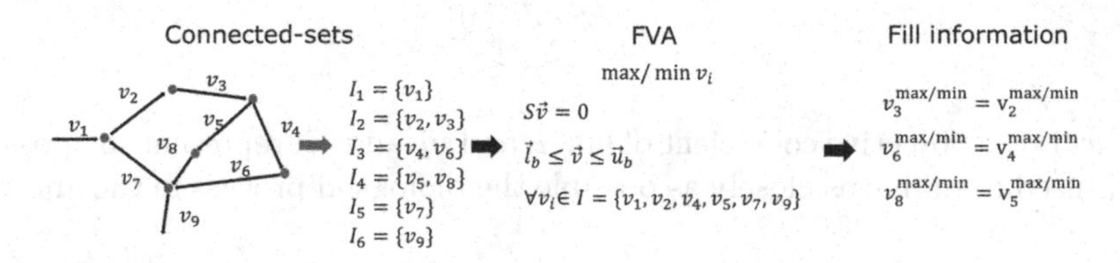

Fig. 2. Schematization of the main steps involved in the computation of efficient FVA.

2.4 Datasets

To demonstrate the applicability of our strategy to real-world metabolic networks, and the improvement as compared to classic FVA, we considered all the metabolic networks in the BiGG database [19]. BiGG Models integrates many published genome-scale metabolic networks into a single database with a set of standardized identifiers. BiGG models include information on the metabolic genes associated with reactions in the form of Gene Protein Reaction rules. Metabolites are linked to many external databases (e.g. KEGG, PubChem).

The total number of metabolic networks is 108 collected from 85 different organisms. The dimension of the models span from 95 reactions and 72 metabolites (e_coli_core [20]) to 10, 600 reactions and 5, 835 metabolites (Recon3D [2]).

2.5 Software Availability and Computational Architecture

To analyze the COBRA models and perform classical FVA, we used the functions provided by the Cobrapy toolkit [6]. To compute the connection-sets, we wrote a specific Python code based on the function *connected_components* provided by the Scipy library [26]. The code can be promptly integrated into the Cobrapy toolkit. All computations were performed on an Intel(R)@3GHz 32GB, using Gurobi as solver and one single CPU.

The source code and documentation are available at https://github.com/CompBtBs/efficientFVA.

3 Experimental Results

3.1 Metabolic Networks Present Many Connected Sets

We investigated the number of connected-sets of each metabolic network in the BiGG database. In particular, we computed the number of reactions, the number of connected-sets, the number of connected-sets formed by 2 reaction at least, and the dimension of the largest connected-set. In Table 1, we reported the results for 10 of the 108 networks. We also reported in Fig. 3a a histogram of the distribution of the ratio between the number of connected-sets and the number of reactions for all the 108 models.

At first instance, all the networks have less connected-sets than the number of reactions. The ratio between the total number of connected-subsets and the reactions ranges between 0.48 (iJB785 [1]) and 0.79 (iLB1027_lipid [15]), with mean 0.63 and standard deviation 0.04. This indicates that, in any network, there are a lot of constraints of the form $av_i + bv_j = 0$. Moreover, some models include very large connected-sets. For example, in the iCHOv1 model [10], the largest connected-set is made by 56 reactions. The presence of these connected-sets is due to the intrinsic nature of metabolic networks. Indeed, many metabolites are involved in just two reactions: one reaction consumes and the other produces such metabolite. Moreover, linear chains of reactions exist in which a series of reactions are connected linearly to produce/consume specific metabolites involved in just two of these reactions. From this analysis, we found that the total number of optimizations required for the FVA can be reduced by at least 35% in most of the networks, when considering a single reaction per connected set.

Table 1. Number of reactions, connected-sets (CS), connected-sets formed by 2 reaction at least (2-CS), and dimension of the largest connected-set (max-CS), for ten networks from the BiGG database.

Model	Reactions	CS	CS /Reactions	2-CS	max-CS
e_coli_core	95	62	0.65	26	4
iAB_RBC_283	469	291	0.62	98	16
iIS312_Amastigote	519	321	0.62	117	14
iAT_PLT_636	1008	645	0.64	176	21
iRC1080	2191	1620	0.74	298	26
iYL1228	2262	1353	0.60	512	15
iMM1415	3726	2346	0.63	668	38
RECON1	3741	2329	0.62	648	38
iCHOv1	6663	4793	0.72	927	56
Recon3D	10600	7414	0.70	1697	28

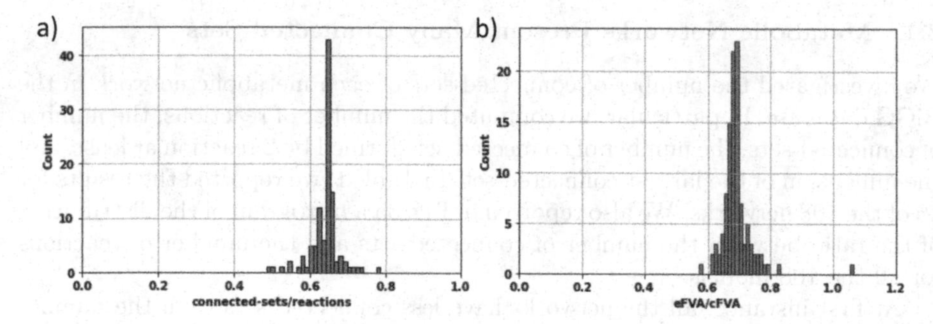

Fig. 3. a) Histogram of the distribution of the ratio between the number of connected-sets and the number of reactions for all the 108 models. b) Histogram of the distribution of the ratio between the mean computational time of our implementation of FVA (eFVA) and of classical FVA (cFVA), for all the 108 models.

3.2 Connected-Sets Reduce the Computational Time for FVA

We investigated the possible computational savings that can be achieved by computing the connected-sets of a metabolic network. In Table 2, we reported: the mean ± the standard deviation over five different runs of the computational time of classical FVA (cFVA), of our overall implementation (eFVA), and of the step for the identification of the connected-sets (pre-FVA) alone. Note that the reported computational time of eFVA includes the pre-FVA computational time. The table also reports the ratio between the mean computational time of our implementation of FVA(eFVA) and of the classical FVA(cFVA). We also reported in Fig. 3b an histogram of the ratio between eFVA and cFVA for all the 108 models.

Table 2. Computational time for cFVA, eFVA, pre-FVA step, and the ratio between eFVA and cFVA on ten metabolic networks from the BiGG database.

Model	cFVA(s)	eFVA(s)	pre-FVA step(s)	eFVA/FVA
e_coli_core	0.06 ± 0.0	0.07 ± 0.02	0.02 ± 0.0	1.06
iAB_RBC_283	0.63 ± 0.01	0.46 ± 0.0	0.06 ± 0.0	0.72
iIS312_Amastigote	0.83 ± 0.01	0.6 ± 0.01	0.07 ± 0.01	0.72
iAT_PLT_636	3.82 ± 0.02	2.58 ± 0.03	0.14 ± 0	0.67
iRC1080	15.72 ± 0.08	12.37 ± 0.1	0.31 ± 0.01	0.79
iYL1228	16.1 ± 0.04	10.99 ± 0.12	0.72 ± 0.05	0.68
iMM1415	53 ± 0.32	37.36 ± 0.24	2.35 ± 0.01	0.70
RECON1	41.16 ± 0.06	28.64 ± 0.05	2.27 ± 0.01	0.70
iCHOv1	150.08 ± 0.18	116.47 ± 0.21	6.1 ± 0.16	0.78
Recon3D	246.03 ± 0.24	204.11 ± 0.38	30.62 ± 0.06	0.83

In all the cases, we observed a standard deviation for cFVA, eFVA, and the pre-FVA step of at least one order of magnitude less then the corresponding mean values. The computational time required for eFVA results less than cFVA in all the cases, except for the smallest network (e_coli_core), for which the time required for cFVA and eFVA results similar. This is due to the additional time required to the computation of the connected-subsets that, for this case, represents about 30% of the entire eFVA time. For all the other cases, the time required for the pre-FVA step results one or more order of magnitude less than the entire eFVA time. More importantly, the ratio between the eFVA and cFVA ranges between 0.57 (iJB785) and 0.83 (Recon3D) with mean 0.69 and standard deviation 0.05.

We expected the number of metabolites, involved in just two reactions, to increase with the number of reactions, and hence the gain of using eFVA to be more evident for larger metabolic networks. Indeed, the Pearson correlation between the ratio of the number of connected-sets over the number of reactions and the ratio of the mean computational time of eFVA over cFVA, for all the 108 models is significant, namely 0.49 ($p_{value} < 0.001$), 0.67 ($p_{value} < 0.001$) if the outlier corresponding with smallest network (e_coli_core) is removed. The larger gain for larger networks can also be noticed in Fig. 4, where we reported the time required for cFVA, eFVA, and pre-FVA step, as a function of the number of reactions.

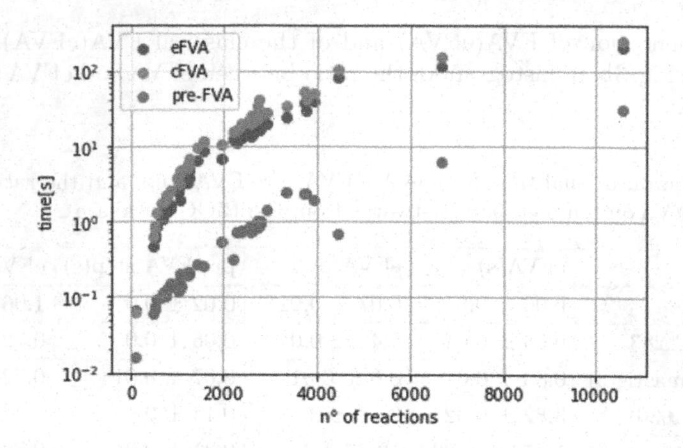

Fig. 4. Scatter-plot showing the time required for cFVA, eFVA, and pre-FVA step, as a function of the number of reactions.

3.3 Connected-Sets Improve the Search of Blocked Reactions in Large Metabolic Networks

The information of the connected-sets can be used not only for FVA, but also to accelerate the search for blocked reactions. To investigate this fact, we computed the subset of blocked reactions for the ten metabolic networks analyzed in Table 1. In Table 3, we reported the computational time required to search blocked reactions using the implementation provided by the Cobrapy library (cBR) and our version based on the pre-computation of connected-sets (eBR). Again, we reported the mean and standard deviation of five different runs. Surprisingly, all the networks, except iAT_PLT_636 [25] and Recon3D, show a non negligible number of blocked reactions varying between 4.31% to 59.92%. The

Table 3. Number of blocked reactions, and computational time for cBR, eBR, and pre-FVA step and the ratio between eFVAand cFVAon ten metabolic networks from the BiGG database.

Model	% blocked reactions	cBR(s)	eBR(s)	eBR/cBR
e_coli_core	8.42	0.02 ± 0.00	0.02 ± 0.01	1.44
iAB_RBC_283	3.41	0.19 ± 0.01	0.19 ± 0.06	1.02
iIS312_Amastigote	59.92	0.17 ± 0.01	0.18 ± 0.08	1.10
iAT_PLT_636	0.00	1.99 ± 0.08	1.46 ± 0.14	0.73
iRC1080	27.75	8.04 ± 0.24	6.62 ± 0.31	0.82
iYL1228	45.93	7.06 ± 0.02	5.51 ± 0.79	0.78
iMM1415	34.73	43.02 ± 2.18	34.9 ± 2.49	0.81
RECON1	34.06	36.65 ± 0.69	27.42 ± 2.48	0.75
iCHOv1	35.76	135.4 ± 17.66	121.36 ± 6.27	0.90
Recon3D	0.00	333.55 ± 37.57	275.48 ± 31.64	0.83

computational time required for eBR results less than cBR in seven out of ten networks. For moderate sizes of the metabolic network ($r < 1000$), the use of connected-sets does not improve or even worsen (e_coli_core) the time required to find the blocked reactions. Note that, for the computation of blocked reactions, the Cobrapy implementation has a pre-processing step in which a general optimization is solved setting randomly a target function. Then, all the reactions for which the fluxes of the optimal solution differ from 0 era excluded from the $2n$ optimization problems. This pre-processing step already speeds up the computation of blocked reactions, and so the time saving by our implementation results less important for small metabolic networks. On the contrary, for metabolic networks having more than 1000 reactions, the time required for the computation of the connected sets results one or more order of magnitude less than the entire eBR time and the ratio between the eBR and cBR results between 0.73 (iAR_PLT_636) and 0.83 (Recon3D).

4 Conclusions

The numerical computation of plausible metabolic fluxes using constrained-based modeling is acquiring increasingly relevance to understand the mechanisms related to the physio-pathological state of a cell or an organism. To this aim, efficient tools for the analysis of large genome-wide metabolic networks are mandatory. In this work, we have considered the computation of Flux Variability Analysis, and proposed a simple but effective method to accelerate such a computation by means of the connected-sets formed by reactions whose flux value can be directly derived from the flux value of any reaction in the same set. Of course, our modification of classical FVA is useful only when the number of connected-sets is less than the total number of reactions. However we verified that this fact holds for all the networks in the BiGG database. In our experiments, we were able to reduce the total number of optimization problems of about 35%, and the computation time of FVA of about 30%. Obviously, the quantity of saved computational time depends intrinsically on the quantity of constraints of the form $av_i + bv_j = 0$ in the network. However, we have shown that this quantity is non negligible in all current metabolic network reconstructions. Therefore, even considering the time required for the computation of the connected-sets in the overall computation time, our approach significantly improves the efficiency in practise.

In this work, to compute the flux variability of the reactions for which it necessarily must be computed after the identification of connected-sets, we relied on the FVA implementation in COBRApy, and used a single CPU. However, we remark that our implementation does not need a specific implementation of FVA for this step. Therefore, to further improve the computation process, one could use our approach in combination with either parallelization or more efficient versions of FVA, such as the one proposed in Thiele and Gudmundsson [23].

As a further work, we propose to exploit other possible specific constraints coming e.g. from the integration of -omics data, to reduce even more the total number of optimization problems necessary for FVA.

References

1. Broddrick, J.T., et al.: Unique attributes of cyanobacterial metabolism revealed by improved genome-scale metabolic modeling and essential gene analysis. Proc. Natl. Acad. Sci. **113**(51), E8344–E8353 (2016)
2. Brunk, E., et al.: Recon3d enables a three-dimensional view of gene variation in human metabolism. Nat. Biotechnol. **36**(3), 272–281 (2018)
3. Cavill, R., Jennen, D., Kleinjans, J., Briedé, J.J.: Transcriptomic and metabolomic data integration. Briefings Bioinf. **17**(5), 891–901 (2016)
4. Damiani, C., Gaglio, D., Sacco, E., Alberghina, L., Vanoni, M.: Systems metabolomics: from metabolomic snapshots to design principles. Curr. Opin. Biotechnol. **63**, 190–199 (2020)
5. Di Filippo, M., et al.: Integrate: model-based multi-omics data integration to characterize multi-level metabolic regulation. PLoS Comput. Biol. **18**(2), e1009337 (2022)
6. Ebrahim, A., Lerman, J.A., Palsson, B.O., Hyduke, D.R.: COBRapy: constraints-based reconstruction and analysis for python. BMC Syst. Biol. **7**(1), 1–6 (2013)
7. Fallahi, S., Skaug, H.J., Alendal, G.: A comparison of monte Carlo sampling methods for metabolic network models. PLoS ONE **15**(7), e0235393 (2020)
8. Galuzzi, B.G., Vanoni, M., Damiani, C.: Combining denoising of RNA-seq data and flux balance analysis for cluster analysis of single cells. BMC Bioinform. **23**(6), 1–21 (2022)
9. Gudmundsson, S., Thiele, I.: Computationally efficient flux variability analysis. BMC Bioinform. **11**(1), 1–3 (2010)
10. Hefzi, H., et al.: A consensus genome-scale reconstruction of Chinese hamster ovary cell metabolism. Cell Syst. **3**(5), 434–443 (2016)
11. Heirendt, L., Thiele, I., Fleming, R.M.: DistributedFBA. jl: high-level, high-performance flux balance analysis in Julia. Bioinformatics **33**(9), 1421–1423 (2017)
12. Herrmann, H.A., Dyson, B.C., Vass, L., Johnson, G.N., Schwartz, J.M.: Flux sampling is a powerful tool to study metabolism under changing environmental conditions. NPJ Syst. Biol. Appl. **5**(1), 1–8 (2019)
13. Hyduke, D.R., Lewis, N.E., Palsson, B.Ø.: Analysis of omics data with genome-scale models of metabolism. Mol. BioSyst. **9**(2), 167–174 (2013)
14. King, Z.A., Dräger, A., Ebrahim, A., Sonnenschein, N., Lewis, N.E., Palsson, B.O.: Escher: a web application for building, sharing, and embedding data-rich visualizations of biological pathways. PLoS Comput. Biol. **11**(8), e1004321 (2015)
15. Levering, J., et al.: Genome-scale model reveals metabolic basis of biomass partitioning in a model diatom. PLoS ONE **11**(5), e0155038 (2016)
16. Mahadevan, R., Schilling, C.H.: The effects of alternate optimal solutions in constraint-based genome-scale metabolic models. Metab. Eng. **5**(4), 264–276 (2003)
17. Müller, A.C., Bockmayr, A.: Fast thermodynamically constrained flux variability analysis. Bioinformatics **29**(7), 903–909 (2013)
18. Nielsen, J.: Systems biology of metabolism: a driver for developing personalized and precision medicine. Cell Metab. **25**(3), 572–579 (2017)
19. Norsigian, C.J., et al.: BiGG models 2020: multi-strain genome-scale models and expansion across the phylogenetic tree. Nucleic Acids Res. **48**(D1), D402–D406 (2020)
20. Orth, J.D., Fleming, R.M., Palsson, B.Ø.: Reconstruction and use of microbial metabolic networks: the core Escherichia coli metabolic model as an educational guide. EcoSal Plus 4(1) (2010)

21. Orth, J.D., Thiele, I., Palsson, B.Ø.: What is flux balance analysis? Nat. Biotechnol. **28**(3), 245–248 (2010)
22. Schellenberger, J., et al.: Quantitative prediction of cellular metabolism with constraint-based models: the cobra toolbox v2. 0. Nat. Protoc. **6**(9), 1290–1307 (2011)
23. Thiele, I., Gudmundsson, S.: Computationally efficient flux variability analysis. BMC Bioninform. **11**(489), 1–3 (2010)
24. Thiele, I., Fleming, R.M., Bordbar, A., Schellenberger, J., Palsson, B.Ø.: Functional characterization of alternate optimal solutions of Escherichia coli's transcriptional and translational machinery. Biophys. J . **98**(10), 2072–2081 (2010)
25. Thomas, A., Rahmanian, S., Bordbar, A., Palsson, B.Ø., Jamshidi, N.: Network reconstruction of platelet metabolism identifies metabolic signature for aspirin resistance. Sci. Rep. **4**(1), 1–10 (2014)
26. Virtanen, P., et al.: SciPy 1.0: fundamental algorithms for scientific computing in Python. Nat. Methods **17**(3), 261–272 (2020)

Exploring the Solution Space of Cancer Evolution Inference Frameworks for Single-Cell Sequencing Data

Davide Maspero[1,2], Fabrizio Angaroni[2], Lucrezia Patruno[2], Daniele Ramazzotti[3], David Posada[4,5,6], and Alex Graudenzi[2,7](✉)

[1] CNAG-CRG, Centre for Genomic Regulation (CRG), Barcelona Institute of Science and Technology (BIST), Barcelona, Spain
[2] Department of Informatics, Systems and Communication, University of Milan-Bicocca, Milan, Italy
alex.graudenzi@unimib.it
[3] Department of Medicine and Surgery, University of Milan-Bicocca, Monza, Italy
[4] Centre of Biomedic Investigation (CINBIO), University of Vigo, Vigo, Spain
[5] Department of Biochemistry, Genetics, and Immunology, Univ. de Vigo, Vigo, Spain
[6] Galicia Sur Health Research Institute (IIS Galicia Sur), SERGAS-UVIGO, Vigo, Spain
[7] Bicocca Bioinformatics, Biostatistics and Bioimaging Centre – B4, Milan, Italy

Abstract. In recent years, many algorithmic strategies have been developed to exploit single-cell mutational profiles generated via sequencing experiments of cancer samples and return reliable models of cancer evolution. Here, we introduce the COB-tree algorithm, which summarizes the solutions explored by state-of-the-art methods for clonal tree inference, to return a unique consensus optimum branching tree. The method proves to be highly effective in detecting pairwise temporal relations between genomic events, as demonstrated by extensive tests on simulated datasets. We also provide a new method to visualize and quantitatively inspect the solution space of the inference methods, via Principal Coordinate Analysis. Finally, the application of our method to a single-cell dataset of patient-derived melanoma xenografts shows significant differences between the COB-tree solution and the maximum likelihood ones.

Keywords: Cancer evolution · Single-cell sequencing · Markov Chain Monte Carlo

In cancer data science, many efforts are devoted to the design of methods for the reconstruction of cancer evolution models from sequencing data [2,3,16,21,22,27]. Indeed, such models are becoming essential to identify possible regularities and repeated evolutionary patterns across tumors, as well as to investigate the impact of therapeutic strategies [6]. In particular, single-cell DNA and

C. De Stefano et al. (Eds.): WIVACE 2022, CCIS 1780, pp. 70–81, 2023.
https://doi.org/10.1007/978-3-031-31183-3_6

RNA sequencing experiments, performed, e.g., on biopsies or on patient-derived models, are a priceless source of high-resolution data on individual tumors.

Despite the typically high levels of noise, mostly due to technical and experimental limitations [20], with such data it is possible to call genomic variants (e.g., single-nucleotide variants, indels, structural alterations, copy-number alterations) via consolidated pipelines [18,24,28]. Accordingly, the most widely used methods to reconstruct the evolutionary history of tumors from single-cell mutational profiles rely on robust statistical frameworks to return accurate models by reducing the impact of noise (see [16,22,23]).

In this lively field, key attention is devoted to the characterization of the solution space of the inference frameworks. This can be of help both from the theoretical and the application perspectives [1,8,27]. In brief, the approximate estimate of the computational complexity of the statistical frameworks proposed in some of the most recent works in the field, namely SCITE [16], LACE [22], VERSO [25], is $\mathcal{O}(nm^3 \log(m))$ to reach MCMC convergence, where n is the number of cells/samples (observations), and m is the number of mutations (variables), as initially discussed in [17]. Thus, for mutation tree reconstructions, it is evident that the complexity mainly depends on the number of mutations included in the final model. Unfortunately, as this number increases (i.e., more mutations are present) it may becomes unfeasible to reach convergence. In addition, in many cases, the algorithm may return equivalent solutions, which share the same likelihood value, but with different topologies.

For these reasons, in this work we aim at: (*i*) characterizing the space of solutions explored during the MCMC inference of state-of-the-art algorithms for the reconstruction of clonal trees from single-cell mutational profiles; (*ii*) summarising the collection of solutions, so to return a unique *Consensus Optimum Branching* tree (COB-tree), instead of the Maximum Likelihood one (ML tree). In other words, the goal is to design an algorithm that takes as input a collection of trees sampled during the MCMC of an arbitrary method for clonal tree inference, and exploits the regularities of the solution space to return a unique COB-tree solution.

Note that similar approaches have already been employed in classical phylogenetic studies. For example, BEAST 2 [5] uses the Maximum Clade Credibility method, whereas in [19] the authors propose to employ the Majority Rule. Such approaches could not be directly employed in our analyses due to the intrinsic differences between phylogenetic trees and clonal trees. In particular, the former are binary trees, and thus the number of edges is fixed and depends on the number of samples. This is not valid in clonal trees, where any node could have an arbitrary number of outgoing edges.

We also point out that the opportunity of computing consensus trees in cancer phylogeny is debated. For example, in [1] the authors argue that summary methods returning only a single tree may not accurately represent the topological features of the solution space. By assuming that the solution space is rugged and includes different local minima related to clonal trees displaying distant topologies, the authors suggest to cluster the tree solutions and, successively, apply a summary method for each cluster. Such approach is not computationally feasible for our goal, because the clustering step is limited to a small number of (small) trees, which is orders of magnitudes lower than the trees sampled during an MCMC. We also note that, in their conclusion, the authors state that a proper characterization of the solution space under an error model of single-cell mutation profiles has not been presented yet.

1 Materials and Methods

COB-tree: A New Algorithm for Clonal Tree Inference. In this work, we introduce a new algorithm for the inference of clonal trees from binary mutational profiles. The general idea is to return a unique consensus tree, obtained by exploiting the solutions explored during the MCMC of an arbitrary algorithm for clonal tree reconstruction, such as LACE [22], VERSO [25] or SCITE [16].

In detail, given an edge-weighted digraph in which any weight is the number of times that a given parental relation is returned during the MCMC (i.e., the frequency of such edge as sampled by the MCMC, which underlying its posterior probability), we identify a unique consensus tree by applying algorithms for optimum branching. The outcome COB-tree model includes all nodes connected with a set of edges that maximises the weight sum. To do this, in our case we employed the efficient implementation of Tarjan [29] of the optimum branching tree method originally proposed in [9,12]. This method is analog to the minimum spanning tree problem, but when considering a directed graph. Note that our algorithm is: (*i*) deterministic; (*ii*) computationally efficient, to handle the vast number of trees sampled during an MCMC; (*iii*) independent of the order of the input tree list.

The algorithmic steps are detailed in the following:

- The COB-tree algorithm takes as input a binary data matrix D, with n rows representing samples (i.e., single cells or biological samples) and m columns representing genomic mutations. Each entry of D is equal to 1 if the mutation is present in a given sample, 0 if it is absent, NA if the information is missing (e.g., due to low coverage). Such data format is widely used in cancer phylogeny studies. For example, the same data format was used in [22].
- In the first step, a generic algorithm for the reconstruction of clonal/mutational trees (e.g., LACE, VERSO or SCITE) is applied to input data D, recording all the solutions sampled during the MCMC. In the output tree T nodes represents mutations (clones) and edges represent parental relations, as in [22].
- For each tree T^p sampled during the MCMC, we generate the corresponding adjacency matrix M^p with dimension $m \times m$, where $p \in [1, \dots r]$ and r is the number of MCMC iterations.
- We compute a weighted adjacency matrix W with dimensions $m \times m$, where each entry is defined as: $W_{i,j} = \frac{\sum_{p=1}^{r} M_{i,j}^p}{r}$. So, W_{ij} stores a weight that corresponds to the frequency by which the mutation i is found as parent of mutation j in all sampled trees. W represents a edge-weighted digraph.

- Finally, we apply the Tarjan algorithm to W in order to find the COB-tree model.

Synthetic Data Generation. In order to both characterize the solution space and test the performance of the COB-tree algorithm, we generated a number of simulated datasets with the following procedure. We first randomly generated a number of ground-truth topologies T_{gt}. In particular, given a specific number of nodes m (i.e., mutations or clones), 20 topologies are created. Starting from the root, we attached a random number (between 2 to 5) of children nodes. We then selected one of the children nodes and repeated the process until all the nodes were attached. Trees including 50, 100 and 200 mutations were considered, thus, a total of 60 topologies are eventually generated. An example of a tree with 50 mutations is reported in Fig. 3.

We sampled 1000 single cells to generate the ground-truth single-cell genotypes matrix D_{gt} (cells $\times$ mutations) from each topology T_{gt}. In particular, we populated each row of D_{gt} (i.e., cell genotype) by randomly selecting a node. Then, the genotype of a cell (i.e., row of D_{gt}) is populated by assigning 1, if the mutation is included in the shortest path from the selected node to the root, or 0, otherwise. Notice that this path is unique because each node can have only one parent. Since it is unrealistic to observe a high number of clones in a cancer sample, we increased the probability of selecting any of the leaf nodes (i.e., the most recent clones) with respect to that of selecting one of the internal nodes. The former probability is 5 times higher than the latter. As a result, D_{gt} is a binary matrix with 1000 rows and m columns (notice that each clone can be defined by the last mutation accumulated, so the number of clones equals the number of mutations).

In order to include data-specific noise in the simulated datasets, we defined *low*, *middle*, and *high* noise levels by setting the rates of False Positives (α), False Negatives (β), and Missing values (γ) as follows:

- Low noise level: $\alpha = 0.005$, $\beta = 0.05$, and $\gamma = 0$
- Middle noise level: $\alpha = 0.01$, $\beta = 0.1$, and $\gamma = 0.1$
- High noise level: $\alpha = 0.02$, $\beta = 0.2$, and $\gamma = 0.2$

From each D_{gt}, 3 different noisy datasets D are generated, by randomly selecting α of the entries equal to 1, β of 0 entries and changing them into 0 and 1, respectively. Then, a fraction γ of all the entries are replaced with missing values (NA).

Simulation Settings. The procedure described above yields a total of 180 noisy datasets. In this preliminary analyses, we employed SCITE [16] as inference framework, since it is one of the state-of-the-art approaches for single-cell mutational tree inference. In particular, each inference is performed multiple times for each dataset, with distinct values of MCMC iterations. We performed 10 independent SCITE runs (with 10 restarts), with the following MCMC iterations:

- for models with $m = 50 \rightarrow$ [1000, 2000 *(short)*, ..., 6000 *(average)*, ..., 10000 *(long)*] MCMC iterations,

- for models with $m = 100 \rightarrow$ [5000, 10000 *(short)*, ..., 30000 *(average)*, ..., 50000 *(long)*] MCMC iterations,
- for models with $m = 200 \rightarrow$ [50000, 100000 *(short)*, ..., 300000 *(average)*, ..., 500000 *(long)*] MCMC iterations.

Performance Metrics. In order to compare the solution provided by the COB-tree algorithm and the corresponding ML solutions, we considered the differences with respect the ground-truth topologies on simulated data. To this aim, we computed two different metrics (i.e., Parent-Child distance PC and Clonal Genotype errors GC), which assess either the local or the global structure in terms of errors between the obtained COB and ML trees, and the ground-truth topologies.

- Parent-Child distance (PC). This metric is widely used to compare different trees, for example in [1,13]. In brief, the parent-child distance between two trees enumerates the edges unique in either trees. Small values of this metric reflect a correct recovery of the relations between two consecutive nodes, but disregard their position in the topology, so it is considered a local measure. We compute the PC_{ML}, and PC_{COB} for evaluate the goodness of maximum likelihood and optimal branching tree respectively.
- Clonal Genotype errors (CG). As explained above, each clone can be associated with the node representing the last accumulated mutation. So, its genotype includes all the mutations in the path from such node to the root. Thus, clonal genotypes depend on the overall topology. For each inference, we transformed the ground-truth topology, the ML tree, and the COB-tree model into the corresponding clonal genotype matrices. Then, we computed the Hamming distance, i.e., the total number of errors, between the clonal genotype of the ground-truth and either the ML tree (CG_{ML}) or the COB-tree model (CG_{COB}).

Finally, we define $\Delta PC = PC_{ML} - PC_{COB}$ distance and $\Delta CG = CG_{ML} - CG_{COB}$. Positive values indicate an improvement of our approach with respect to the ML tree.

Characterization of the Solution Space. In order to provide a way of characterizing and visualizing the solution space of the inference framework, we decided to plot the distribution of the sampled trees during the MCMC. To do this, we applied Principal Coordinate Analysis (PCA) [14,15] on the distance matrix computed considering the PC distance.

This approach returns a 2-dimensional representation of the tree space, where the relative distance among each point (i.e., sampled trees) is maintained. We added the value of the likelihood L of each tree as a third dimension, by computing the $-\ln(L)$. Notice that, after the transformation, the best likelihood values are the lowest.

Real Data Processing. As a proof of principle, we applied the COB-tree algorithm to a real-world longitudinal scRNA-seq dataset of patient-derived xenografts

(PDXs) of $\mathsf{BRAF}^{\mathsf{V600E/K}}$ mutant melanomas produced in [26]. The authors generated four datasets collected at different time points, before, during, and after therapy administration of a BRAF/MEK-inhibitor with a total of 674 cells. For the aim of the current work, we pre-processed them as independent datasets to defined a set of highly confided SNVs, by using GATK pipeline [10] for alignment and variant calling, and by using filters based on statistical significance (explained in the following). Thus, we selected a set of 55 mutations with a highly significance, well separated from the background noise.

For variant calling from scRNA-seq data, we applied the same steps performed in [22] which are here briefly reported:

1. Considering the expression data, we discarded cells with a high fraction of mitochondrial reads. So, we kept 475 high-quality cells.
2. Via *SRA toolkit* we downloaded the FASTQ files (one for each cell) from the GEO dataset using the accession number GSE116237.
3. Using Trimmomatic (v. 0.39) [4], we removed the nucleotides with a poor quality score.
4. Using 2-pass mode STAR aligner [11], we aligned the reads to the human reference genome (GRCh38 release). This step generate one SAM file for each cell.
5. We added read groups, we sorted, we marked duplicates and we indexed the reads in each BAM file via Picard tools
6. With GATK (v. 3.8.1) we hard clipped intronic regions (via SplitNCigarReads command) and we re-calibrate the base alignment (via BaseRecalibrator command)
7. Finally, we used HaplotypeCaller and VariantFiltration to call Single Nucleotide Variants and to remove the ones with poor quality score (we applied default parameters).

The above steps generated a VCF file for each cell which we merged together and load it into an R environment, as matrix with cells as row and mutations as columns, to perform downstream filtering steps. In particular:

1. Considering each entry of the mutational matrix, we set as NA (i.e., missing data) mutations that fall in a position covered with less than 3 reads
2. We removed mutations (columns) which display a frequency of missing data higher than 0.4 in at least one time point.
3. We compute the frequency of every mutation in each time point and we remove rare mutations by keeping only those with a frequency sum higher than 0.15

We selected 55 high-quality mutations, and we use those mutational profiles for generating the input file to infer the tumour evolution via SCITE. We run one inference using a false discovery rate equal to 0.02, an allele dropout rate of 0.2, and 30 restarts of Markov Chain Monte Carlo with a length of 20000 steps.

SCITE considers the first 25% trees sampled as burning and discards them. Thus, we sampled the remaining 450000 trees from which we removed the ones with a likelihood value less than 1.3× Maximum log-Likelihood (we highlight that the log-likelihood is negative so we are discarding trees with a low likelihood value). Finally, we draw the solution landscape by considering 4740 unique trees using the approach described in the above section.

2 Results

Characterization of the Search Space. We first characterized the search space of the clonal trees inferred via one of the state-of-the-art approaches for mutational tree inference – SCITE [16]. SCITE was selected for the extremely efficient implementation and the good computational costs. The experiments were executed by repeatedly performing the inference on synthetic datasets generated from a number of distinct topologies (the whole simulation settings is described in the Materials and Methods section).

In Fig. 1 we report the results of the PCA applied on the trees sampled from the inference of a selected dataset, generated from a tree topology including $m = 100$ mutations. We show the solution considering three different MCMC lengths i.e., 10000 (short), 30000 (average), and 50000 (long) steps.

As one can see, the 10 independent chains tend to reach the same global minimum but, when the MCMC is short, the trees with better likelihood are far away from each other. We also marked the COB-tree solutions with red dots, and they appear to be placed in a central position among the trees sampled late in the inference. This interesting result suggests that, in this case, applying a consensus approach that does not depend on a clustering step seems to be a reasonable algorithmic choice.

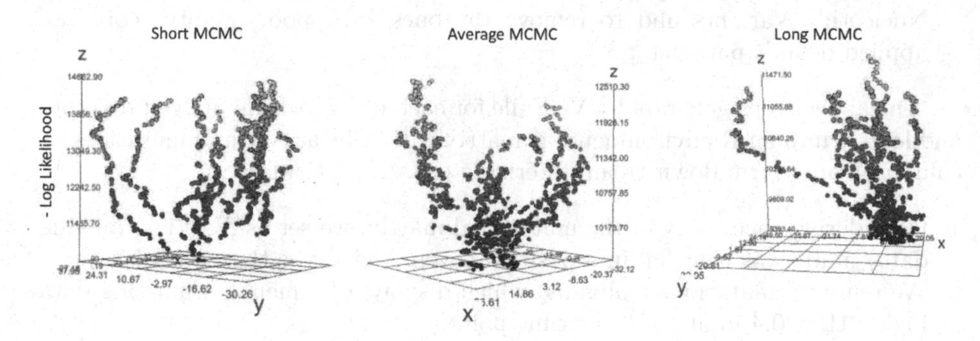

Fig. 1. Visual representation of the solution space explored during the inference of a clonal tree from the same synthetic single-cell dataset (with 100 mutations and 1000 cells), in three independent MCMC runs with 10 restarts (via SCITE). From left to right, the total number of MCMC increases (10000, 30000, and 50000 steps). The solution space is defined by computing a PCA on the distance matrix (using parent-child distance) of the trees sampled during each inference. Z-axis reports the corresponding likelihood value. Red dots indicate the position of the COB-tree solutions. (Color figure online)

Performance Assessment of COB-Tree. We applied the COB-tree algorithm to the synthetic datasets described in the Materials and Methods section. To this end, we considered all the trees sampled during the MCMC. Since SCITE discards the first 25% trees, we only considered the remaining 75% and kept only the trees with a likelihood between L_{best} and $L_{best} \times 1.3$, so to focus on the final part of the MCMC. The Tarjan algorithm was finally applied to retrieve the unique COB-tree model. Notice that we also kept track of the ML tree (L_{best}). We considered three MCMC lengths to evaluate how this affects the performance the COB-tree method. Results are reported in Fig. 2.

It is possible to observe how our approach improves the local phylogenetic structure, by recovering a better ordering of the accumulation of mutations. Instead, the improvement is not that evident when considering the CG metric underlying the global phylogenetic structure. Even though the COB-tree method often improves this metric as well, it sometimes returns trees with a global structure very far from the ground-truth.

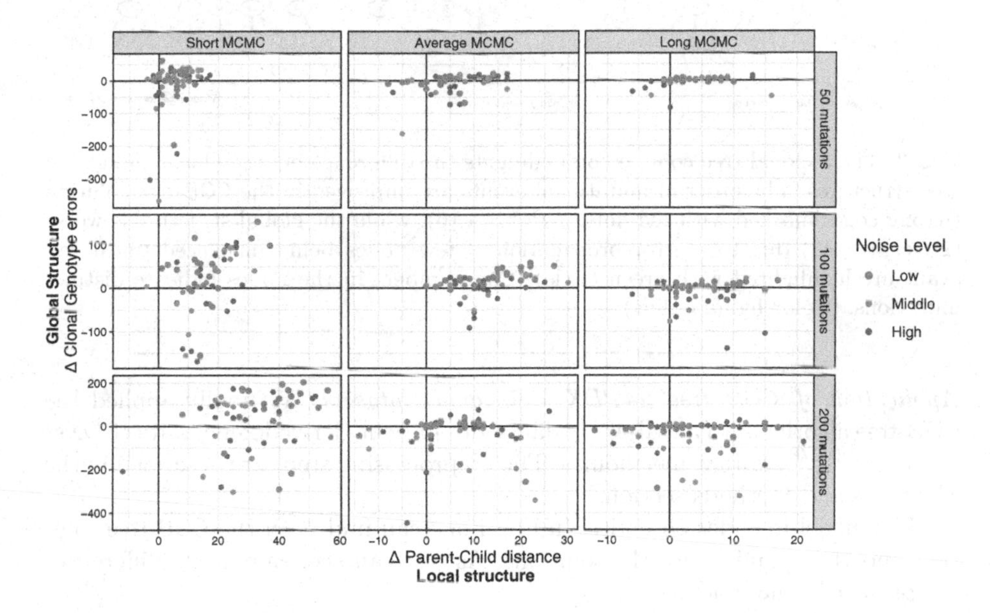

Fig. 2. Differences between Maximum Likelihood and COB trees are reported. Positive values of ΔCG errors or ΔPC distance indicate an improvement for the global structure or the local structure of the COB-tree solutions over the ML ones. Colours indicate the level of noise (i.e., rate of FP events, FN events, and missing values) included in the simulated datasets. (Color figure online)

Note that it is possible to have trees with high PC distance values and low CG metric values. A possible explanation is illustrated in the example depicted in Fig. 3. In the plot, it is possible to observe how COB-tree retrieves a better ordering of mutation pairs. Still, the few errors drastically changed the global topology of the tree by shifting an entire subtree.

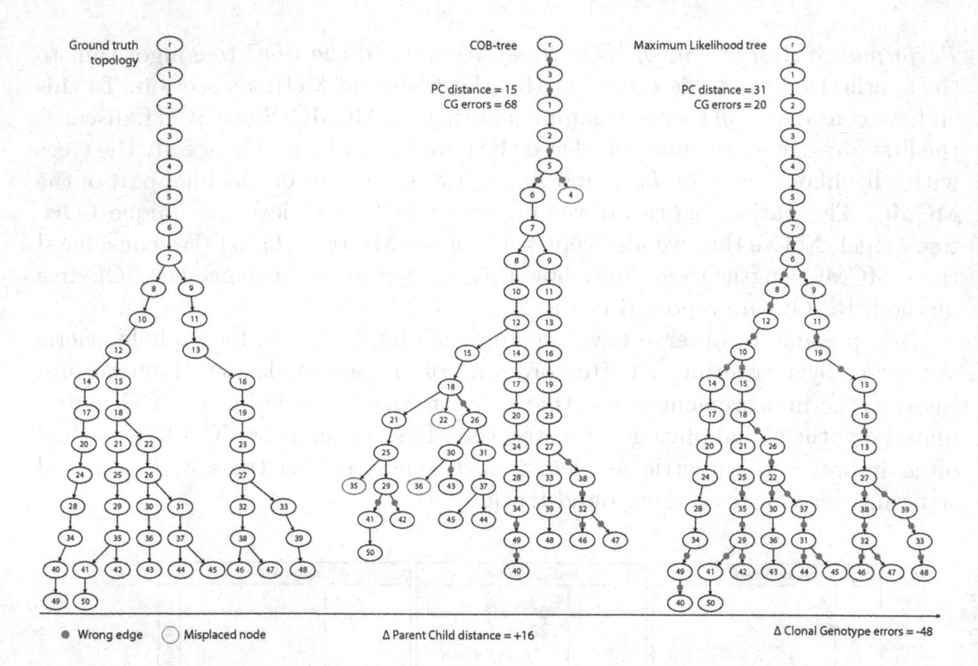

Fig. 3. The 3 clonal tree comparison highlights the difference between local and global tree structures. The order of mutational events are improved in the COB-tree solution (wrong edges marked with red dots, $\Delta PC = +16$), while the global structure is worse ($\Delta CG = -48$) due to a error propagation of few nodes being misplaced (most relevant are highlighted with green circle). The numbers in the nodes indicate distinct mutations. (Color figure online)

Application of COB-Tree to PDX Melanoma Datasets. We finally applied the COB-tree algorithm to a real-world dataset of patient-derived xenografts (PDXs) of $\text{BRAF}^{\text{V600E/K}}$ mutant melanomas. The preprocessing steps are described in the Materials and Methods section.

The model includes 55 nodes. In Fig. 4 one can find both the COB-tree solution and three equivalent ML solutions. As one can see, significant differences are present in the model.

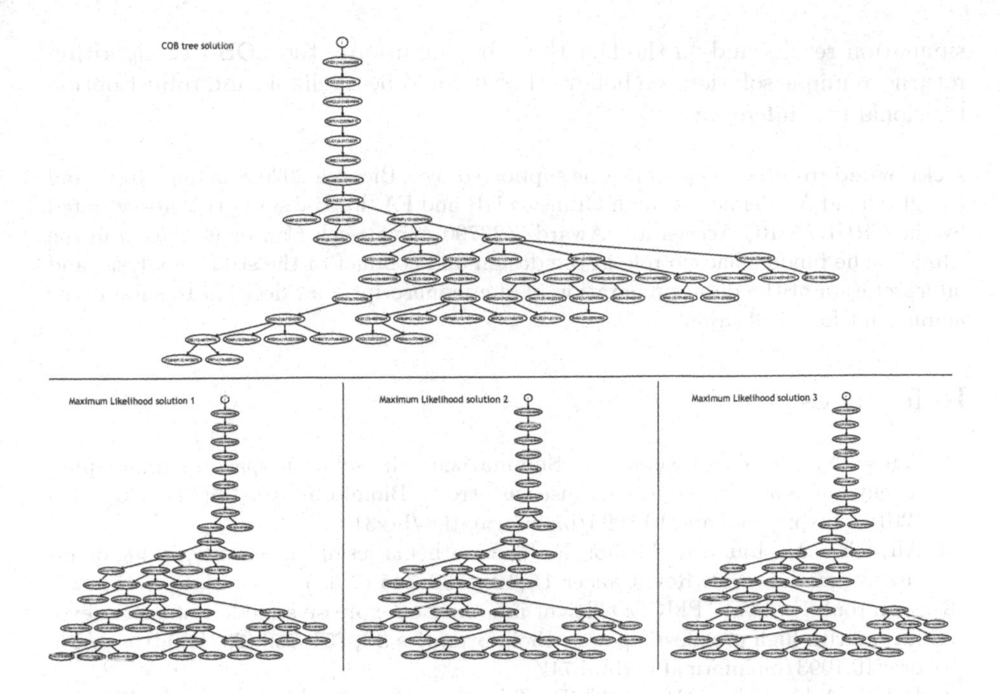

Fig. 4. Clonal tree comparison between the COB-tree solution and three equivalent maximum likelihood solutions returned by SCITE considering the mutational profiles from the melanoma PDX dataset. The node labels define the gene involved and the genome position of the SNVs. The inference was performed with SCITE using a false discovery (fd) rate of 0.2, an allele dropout (ad) rate equal to 0.2, 30 restarts, and a MCMC with 20000 steps.

3 Conclusions

The preliminary analyses illustrated in this work show that in case one is interested in defining the ordering of mutational events the COB-tree algorithm is a reliable and effective option.

This aspect might be of particular relevance, for instance, if someone is interested in detecting possible patterns of repeated cancer evolution across different patients [6], as this might help in identifying possible therapeutic targets, as well as weak points of specific tumor types.

Currently, other methods to generate aggregated trees are available [7]. They are based on specific assumptions and exploit different strategies to summarize the solution space. A comparison between them and COB-tree could be useful to highlight the specific behavior of each method.

Further development of the COB-tree algorithm are underway, aimed, e.g., at better exploiting the properties of the solution space so to improve the performance with respect to the global structure too. However, on the basis of the simulation results and on the fact the – by definition – the COB-tree algorithm returns a unique solution, we believe that it could be a reliable and robust option for clonal tree inference.

Acknowledgments. This work was supported by a Bicocca 2020 Starting Grant and Google Cloud Academic Research Grant to DR and FA. Partial support is also granted by the CRUK/AIRC Accelerator Award #22790 "Single-cell Cancer Evolution in the Clinic". The funders had no role in the design and conduct of the study, analysis, and interpretation of the data, preparation of the manuscript, and decision to submit the manuscript for publication.

References

1. Aguse, N., Qi, Y., El-Kebir, M.: Summarizing the solution space in tumor phylogeny inference by multiple consensus trees. Bioinformatics **35**(14), i408–i416 (2019). https://doi.org/10.1093/bioinformatics/btz312
2. Altrock, P.M., Liu, L.L., Michor, F.: The mathematics of cancer: integrating quantitative models. Nat. Rev. Cancer **15**(12), 730–745 (2015)
3. Angaroni, F., et al.: PMCE: efficient inference of expressive models of cancer evolution with high prognostic power. Bioinformatics **38**, 754–762 (2021). https://doi.org/10.1093/bioinformatics/btab717
4. Bolger, A.M., Lohse, M., Usadel, B.: Trimmomatic: a flexible trimmer for Illumina sequence data. Bioinformatics **30**(15), 2114–2120 (2014)
5. Bouckaert, R., et al.: BEAST 2: a software platform for Bayesian evolutionary analysis. PLOS Comput. Biol. **10**(4), e1003537 (2014). https://doi.org/10.1371/journal.pcbi.1003537
6. Caravagna, G., et al.: Detecting repeated cancer evolution from multi-region tumor sequencing data. Nat. Methods **15**(9), 707–714 (2018). https://doi.org/10.1038/s41592-018-0108-x
7. Christensen, S., et al.: Detecting evolutionary patterns of cancers using consensus trees. Bioinformatics **36**(Suppl. 2), I684–I691 (2020). https://doi.org/10.1093/bioinformatics/btaa801. pmid: 33381820
8. Christensen, S., et al.: Detecting evolutionary patterns of cancers using consensus trees. Bioinformatics **36**(Suppl. 2), i684–i691 (2020)
9. Chu, Y.-J.: On the shortest arborescence of a directed graph. Sci. Sinica **14**, 1396–1400 (1965)
10. DePristo, M.A., et al.: A framework for variation discovery and genotyping using next-generation DNA sequencing data. Nat. Genet. **43**(5), 491 (2011)
11. Dobin, A., et al.: STAR: ultrafast universal RNA-seq aligner. Bioinformatics **29**(1), 15–21 (2013)
12. Edmonds, J.: Optimum branchings. J. Res. Natl. Bureau Stand. B **71**, 233–240 (1967)
13. Govek, K., Sikes, C., Oesper, L.: A consensus approach to infer tumor evolutionary histories. In: Proceedings of the 2018 ACM International Conference on Bioinformatics, Computational Biology, and Health Informatics, BCB 2018, pp. 63–72. Association for Computing Machinery, New York (2018). https://doi.org/10.1145/3233547.3233584. ISBN 978-1-4503-5794-4
14. Gower, J.C.: Adding a point to vector diagrams in multivariate analysis. Biometrika **55**(3), 582–585 (1968). https://doi.org/10.1093/biomet/55.3.582

15. Gower, J.C.: Some distance properties of latent root and vector methods used in multivariate analysis. Biometrika **53**(3–4), 325–338 (1966). https://doi.org/10.1093/biomet/53.3-4.325
16. Jahn, K., Kuipers, J., Beerenwinkel, N.: Tree inference for single-cell data. Genome Biol. **17**(1), 86 (2016). https://doi.org/10.1186/s13059-016-0936-x
17. Kuipers, J., Moffa, G.: Uniform random generation of large acyclic digraphs. Stat. Comput. **25**(2), 227–242 (2013). https://doi.org/10.1007/s11222-013-9428-y
18. Langmead, B., Salzberg, S.L.: Fast gapped-read alignment with Bowtie 2. Nat. Methods **9**(4), 357–359 (2012)
19. O'Reilly, J.E., Donoghue, P.C.J.: The efficacy of consensus tree methods for summarizing phylogenetic relationships from a posterior sample of trees estimated from morphological data. Syst. Biol. **67**(2), 354–362 (2018). https://doi.org/10.1093/sysbio/syx086
20. Patruno, L., et al.: A review of computational strategies for denoising and imputation of single-cell transcriptomic data. Briefings Bioinform. **22**(4), bbaa222 (2021). https://doi.org/10.1093/bib/bbaa222
21. Ramazzotti, D., et al.: CAPRI: efficient inference of cancer progression models from cross-sectional data. Bioinformatics **31**(18), 3016–3026 (2015). https://doi.org/10.1093/bioinformatics/btv296
22. Ramazzotti, D., et al.: LACE: inference of cancer evolution models from longitudinal single-cell sequencing data. J. Comput. Sci. **58**, 101523 (2022). https://doi.org/10.1016/j.jocs.2021.101523. https://www.sciencedirect.com/science/article/pii/S1877750321001848
23. Ramazzotti, D., et al.: Learning mutational graphs of individual tumour evolution from single-cell and multi-region sequencing data. BMC Bioinform. **20**(1), 210 (2019). https://doi.org/10.1186/s12859-019-2795-4
24. Ramazzotti, D., et al.: Variant calling from scRNA-seq data allows the assessment of cellular identity in patient-derived cell lines. Nat. Commun. **13**(1), 1–3 (2022)
25. Ramazzotti, D., et al.: VERSO: a comprehensive framework for the inference of robust phylogenies and the quantification of intra-host genomic diversity of viral samples. Patterns **2**(3), 100212 (2021). https://doi.org/10.1016/j.patter.2021.100212
26. Rambow, F., et al.: Toward minimal residual disease-directed therapy in Melanoma. Cell **174**(4), 843–855.e19 (2018). https://doi.org/10.1016/j.cell.2018.06.025
27. Schwartz, R., Schäffer, A.A.: The evolution of tumour phylogenetics: principles and practice. Nat. Rev. Genet. **18**(4), 213–229 (2017). https://doi.org/10.1038/nrg.2016.170
28. Singer, J., et al.: Bioinformatics for precision oncology. Briefings Bioinform. **20**(3), 778–788 (2019)
29. Tarjan, R.E.: Finding optimum branchings. Networks **7**(1), 25–35 (1977). https://doi.org/10.1002/net.3230070103

Computational Investigation of the Clustering of Droplets in Widening Pipe Geometries

Hans-Georg Matuttis[1](✉), Johannes Josef Schneider[2,3], Jin Li[3], David Anthony Barrow[3], Alessia Faggian[4], Aitor Patiño Diaz[4], Silvia Holler[4], Federica Casiraghi[4], Lorena Cebolla Sanahuja[4], Martin Michael Hanczyc[4,5], Mathias Sebastian Weyland[2], Dandolo Flumini[2], Peter Eggenberger Hotz[2], Pantelitsa Dimitriou[3], William David Jamieson[6], Oliver Castell[6], and Rudolf Marcel Füchslin[2,7]

[1] Department of Mechanical and Intelligent Systems Engineering, The University of Electrocommunications, Chofu Chofugaoka 1-5-1, Tokyo 182-8585, Japan
hg@mce.uec.ac.jp

[2] Institute of Applied Mathematics and Physics, School of Engineering, Zurich University of Applied Sciences, Technikumstr. 9, 8401 Winterthur, Switzerland
johannesjosefschneider@googlemail.com,
{scnj,weyl,flum,eggg,furu,escl}@zhaw.ch

[3] School of Engineering, Cardiff University, Queen's Buildings, 14-17 The Parade, Cardiff CF24 3AA, Wales, UK
{LiJ40,Barrow,dimitrioup}@cardiff.ac.uk

[4] Laboratory for Artificial Biology, Department of Cellular, Computational and Integrative Biology (CIBIO), University of Trento, 38123 Trento, Italy
{alessia.faggian,aitor.patino,silvia.holler,
federica.casiraghi,lorena.cebolla,martin.hanczyc}@unitn.it

[5] Chemical and Biological Engineering, University of New Mexico, MSC01 1120, Albuquerque, NM 87131-0001, USA
{jamiesonw,CastellO}@cardiff.ac.uk

[6] Welsh School of Pharmacy and Pharmaceutical Science, Cardiff University, Redwood Building, King Edward VII Avenue, Cardiff CF10 3NB, Wales, UK

[7] European Centre for Living Technology, S.Marco 2940, 30124 Venice, Italy
http://www2.matuttis.mce.uec.ac.jp

Abstract. Experimentally, periodically released droplets in systems of widening pipes show clustering. This is surprising, as purely hydrodynamic interactions are repulsive so that agglomeration should be prevented. In the main part of this paper, we investigate the clustering of droplets under the influence of phenomenological hydrostatic forces and some hypothetical attraction. In two appendices, we explain why a direct

---

The original version of this chapter was previously published without open access. A correction to this chapter is available at
https://doi.org/10.1007/978-3-031-31183-3_25

C. De Stefano et al. (Eds.): WIVACE 2022, CCIS 1780, pp. 82–93, 2023.
https://doi.org/10.1007/978-3-031-31183-3_7

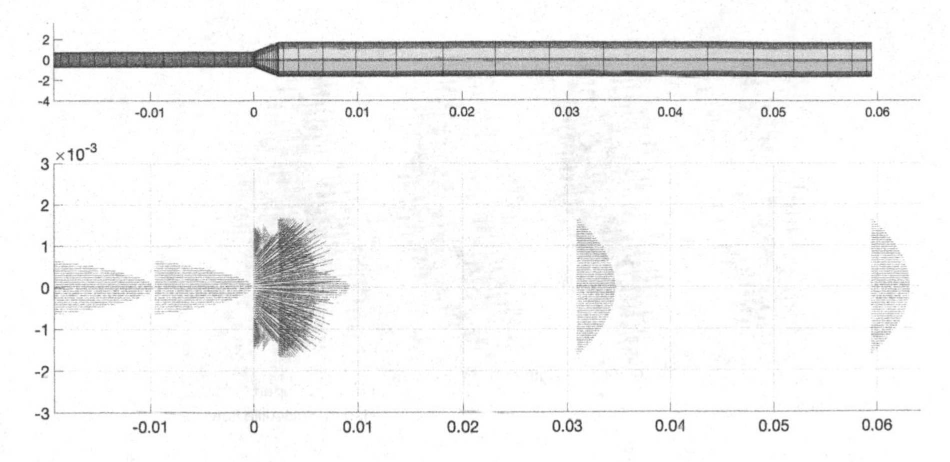

Fig. 1. Cylindrical pipe of the model system (above) with dimensions in meter and Hagen-Poiseuille flow profile (below) with the widening of the pipe to an "O"-shape at the right of $x = 0$, where the fanning out of the flow is computed based on the assumption of continuity, respectively incompressibility.

numerical simulation for this system is rather more difficult (and probably not possible with current methods) than the "simple" geometry would suggest.

Keywords: Micro-fluidics · droplets in fluids · multiphase flow · clustering of droplets

1 Introduction

The purpose of this work is the search for a theoretical understanding of the clustering of oil droplets (density $1.2\,\mathrm{g/cm^3}$) suspended in water (density $1.0\,\mathrm{g/cm^3}$) in an opening pipe as in the experiments by Li and Barrow [7]. Intuitively, one would expect that under the influence of purely hydrodynamic interactions, there should only dominate repulsion between droplets and no clustering should occur. We investigate the basic properties of the interaction between droplets, together with the effect of the simulation geometry with the phenomenological modeling of hydrodynamic forces and interactions. We desist from trying to implement a full-fledged direct numerical simulation (DNS) of droplets in flow, as the state of the art of direct numerical simulation is still far away from modelling particles, let alone droplets, in three dimensional fluid geometries with the exact boundary conditions. Some problems of the standard discretizations for DNS methods with freely moving bodies inside are explained in Appendix A, while the issues of "approximate" boundaries are discussed in Appendix B, to make it plain that we have selected our modeling approach from a higher insight of simulation methods, not due to ignorance of more elaborate methods.

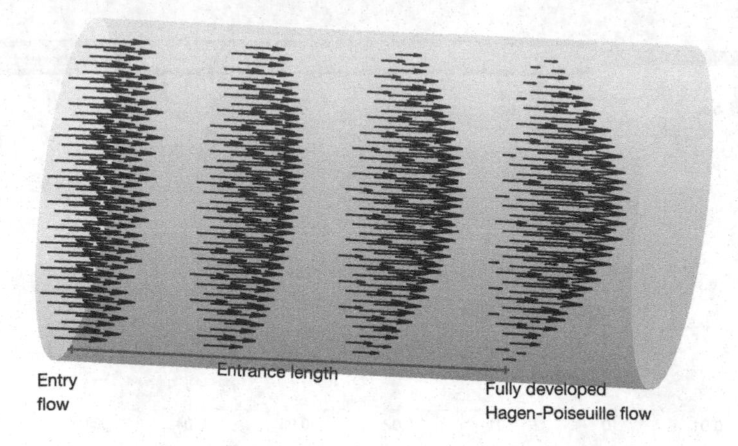

Fig. 2. Flow entering a pipe from a wider vessel with constant flow profile so that only after the "entrance length", a trule parabolic Hagen-Poiseuille flow profile is reached.

2 Our Modeling Approach

For simplification, we treat the experiment by Li and Barrow [7] as a widening pipe (see Fig. 1 above) without the butterfly-pipe mechanism which triggers the separation and release of the droplets.

2.1 Forces on Droplets in Pipes

We want to model small oil droplets with a radius of about 0.6 mm in water, so the surface tension should be large enough so that the deformation can be neglected and the droplets can be treated as rigid. We work with an experimental inside flowrate of 3.8 [ml/h]. Assuming a Hagen-Poiseuille flow with a parabolic profile, the maximal velocity in the center of the narrow inflow pipe (d=1.32 mm) will be $u_{in} = 0.76$ [mm/s], while at the wider O-shaped outflow pipe it will be $u_{out} = 0.34$ [mm/s], see Fig. 1. The Reynolds number Re based on the viscosity of water $\mu = 10^{-3}$ [Pa s] with unit density ρ and the total pipe diameter at inflow of 1.3 mm is

$$Re_{max} = \frac{\rho u_{in} D_{\mathrm{H}}}{\mu} \approx 4.4 \cdot 10^{-4},$$

"deep" in the viscous regime of the Stokes flow ($Re < 1$), and far away for inertia dominated ($Re > 50$), let alone fully developed turbulent flow ($Re > 1000$), so that vortices and turbulence should not play any role. For pipe flow, at inflows from a constant flow profile, it takes a certain distance, the entry (or entrance) length, until the flow profile changes to full developed Hagen-Poiseuille flow, see Fig. 2. For laminar flow, is typically taken "within 5%" [15] of the channel diameter, scaled by the Reynolds number, so for our case at the narrow inflow we have

$$L_{h,n} = 0.05 \cdot ReD = 5.2 \cdot 10^{-7}[m],$$

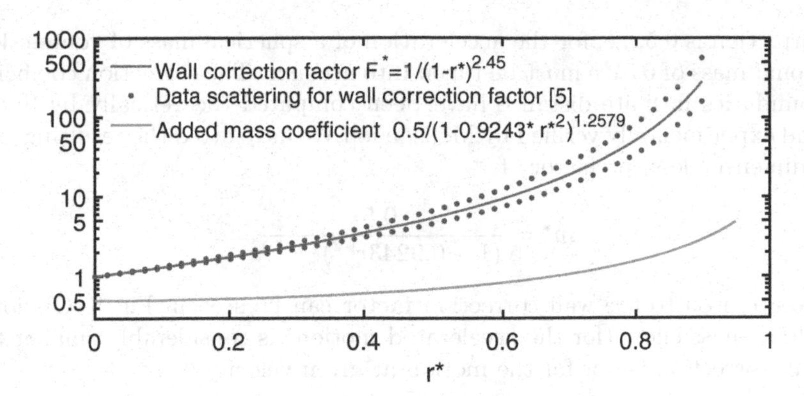

Fig. 3. (Dimensionless) wall correction factor and added mass over the reduced diameter from 0 to 1.

and at the wider outflow, $L_{h,w} = 1.460 \cdot 10^{-6}[m]$, still considerably less than a droplet diameter. This means that for our setup, the flow assumes a parabolic profile practically instantaneously, even between droplets. The entrance length is a confirmation that viscous forces dominate and that the assumption of Hagen-Poiseuille flow is justified. (The assumption of a constant flow speed would reduce the Reynold number and therefore the entrance length, which would again confirm the prevalence of Hagen-Poiseuille flow.) We neglect any "elastic" deformations of the droplet which may lead to deformations of the shape, so for the force on the droplet we use Stokes law

$$F_d = 6\pi\mu R v, \tag{1}$$

where v is the velocity difference between the center of mass of the droplet and the velocity of the flow in the pipe. The Stokes law is only valid for walls far away. For more narrow geometries, a (dimensionless) wall correction factor F^* must be included, which takes the influence of the walls into account. Fitting the experimental data from Iwaoka and Ishii [5] for the dimensionless

$$r^* = \frac{r_{particle}}{r_{pipe}}, \tag{2}$$

we found that at least for the experimental data range $(0 \leq r^* \leq 0.9)$

$$F^* = \frac{1}{(1 - r^*)^{2.45}}. \tag{3}$$

was a good parametrization, see Fig. 3. Another issue of the Stokes law is that it is valid only for equilibrium velocities. When the spheres undergo an acceleration in a dense liquid (with a density comparable to the density of the body), there is an added (or vitual) mass of the fluid around the particle which must be accelerated, too. [14]. For a sphere, that gives an additional inertia term, which in a cylindrical vessel depends on the vessel radius. For the boundaries at infinity, the correction is 0.5, i.e. for the acceleration of a spherical mass of volume V, an additional mass of $0.5V\rho$ must be taken into account. The correction coefficients for boundaries in finite distance have been computed theoretically by Smythe [13] and experimentally verified by Mellsen [8]. We have fitted the resulting graph to a (dimensionless) prefactor of

$$m^* = \frac{0.5}{(1 - 0.9243r^{*2})^{1.2579}}. \tag{4}$$

The comparison to the wall correction factor can be seen in Fig. 3. Obviously, the added mass effect (for the accelerated motion) is considerably smaller than the wall correction factor for the motion at given velocity.

2.2 The Interaction of Droplets

It is possible to expand the repulsive force between two droplets which are approaching each other with velocity v as a correction to the Stokes law [2,3]. For our simulation, we have parameterized the resulting prefactor of the interaction law for equal sized spheres, see Fig. 2a) of Goddard et al. [3]. This means that for distances $d > 10r$ with r being the particle radius, the repulsive force decays so much that it corresponds to Stokes' law alone while for distances $d < 0.1r$, the repulsive force increases as a prefactor to Stokes's law with a $1/d$ dependence, while for the intermediate range $10r > d > 0.1r$, there is a smooth transition between the repulsive (short range) and the neutral (long range) regime.

3 Results: Gravity in X-Direction

For this preliminary investigation, we make the system more symmetric by setting the gravitation in x-direction: We first want to understand the effect of the interactions we have implemented without interference of an asymmetric effect from gravity between an upper and lower boundary. We integrate the equations of motion with MATLAB's adaptive time ode15s-integrator and enforce the use of the BDF2 (Backward differentiation formula of second order) with a relative error tolerance of 0.1% and absolute errors of 1% of the droplet radius and 1% of the maximal velocity in the narrow channel. For this preliminary investigation, we work with a constant augmented mass of 50% only, without the corrections

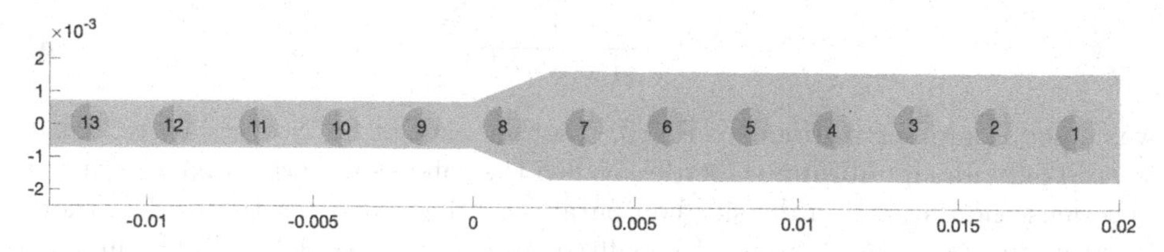

Fig. 4. Release of the droplets with only hydrodynamic interactions and initial disorder.

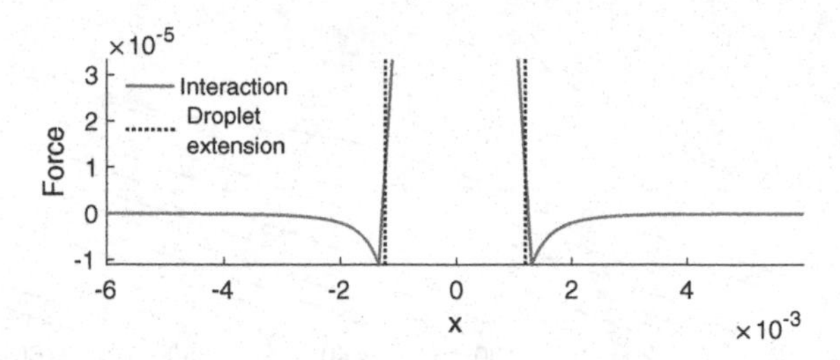

Fig. 5. Force between the droplets from Eq. (5) drawn for $k_{mult} = 1$ with the extension of the droplets.

due to the closeness of the boundary. To take into account that the release of the droplet at the inlet will be affected by an oscillation of the droplet which may shift its center of mass, we add ± 3.75 % equally distributed randomness. Our hope was that this disorder in the initial coordinates would trigger the clustering. The result for hydrodynamic forces only was disappointing and is shown in Fig. 4. The droplets are released like pearls on a string. While the initial disorder is amplified by the hydrodynamic interaction, so that the scattering of the vertical position around the symmetry axis is enhanced, no clustering occurs.

3.1 Introducing an Additional Interaction

We next introduce an additional interaction. From comparison with the clustering in the experiment, we have to assume that there must be an additional attraction. On the other hand, once the droplets are in contact, they feel a kind of "elastic" repulsion, because they do not fuse. We construct the interaction for the droplet radius r_{dr} so that

$$F^{int} = k_{mult} \begin{cases} -2.5 \cdot 10^{-4} \frac{r_{dr}}{r^2} & \text{for } r > 2.15\, r_{dr}, \\ +10^{-4} \frac{(2r_{dr} - r)}{r_{dr}} + 10^{-5} \exp\left(1.2 \frac{1.05 r_{dr}}{r - 2r_{dr}}\right) & \text{else,} \end{cases} \tag{5}$$

so the upper term is a Coulomb-type attraction, while the lower term is a linear repulsion with an added exponential term to guarantee that no fusion (penetration) of droplets can occur. The Coulomb-type has been chosen because $1/r^2$ is the only algebraic form which allows to model a spatially decaying interaction between agglomerations by their centers of mass. While the coefficients in the interaction look a bit strange, they have been chosen so that repulsion changes continuously into attraction when the droplets come into contact, see Fig. 5.

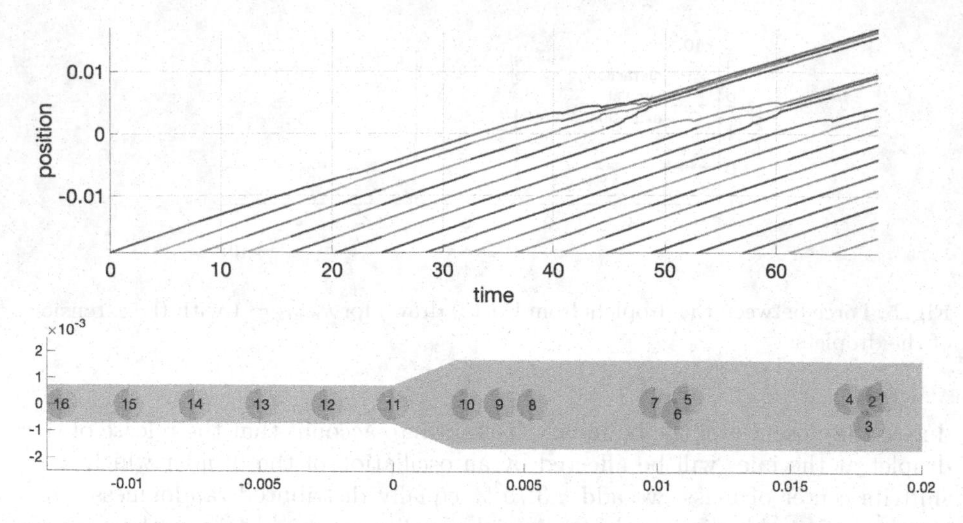

Fig. 6. Wordlines (x-coordinates over time) above and (three-dimensional) clusters for the final timestep of the worldlines below for $k_{mult} = 0.8$. At $t = 107$[s] (not shown), the first two clusters fuse to a seven-droplet cluster, but the three-droplet clusters behind seem to be stable.

3.2 Clustering for Different Values of k_{mult}

With attraction, the dynamics in the simulation develops two timescales: The slow timescale of the advection of the droplets through the pipe and the fast timescale due to the reordering of the droplet positions into clusters at close range under the influence of the attractive force. Depending on the prefactor k_{mult}, we obtain clusters with a different number of droplets. In Fig. 6, we have shown the result for $k_{mult} = 0.8$ with three droplets per clusters in "equilibrium" and only the first cluster with four droplets. The release of three-droplet cluster was sustained also beyond the simulation time shown here. Interestingly, the number of droplets in a cluster is not proportional to the attraction strength, because the force equilibrium between attraction and hydrostatic repulsion is rather subtle: For $k_{mult} = 0.4$, a leading cluster of three droplets develops, which is then followed by pairs of droplets which over time fuse into clusters of four droplets, see Fig. 7. While the first four-droplet cluster fuses with the leading three-droplet cluster, the later four-droplet clusters stay intact when they are flushed along the system. Typically, the number of droplets in a cluster for a given value of k_{mult} is constant, except the first cluster, which may have one droplet less or more and which then fuses with the second cluster.

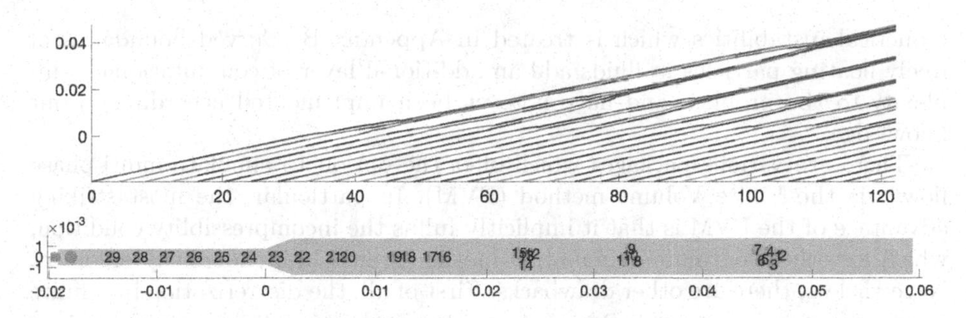

Fig. 7. Wordlines (x-coordinates over time) above and (three-dimensional) clusters for the final timestep of the worldlines below for $k_{mult} = 0.4$. At the widening outlet, first two-droplet clusters form which then fuse into four-droplet clusters.

4 Summary, Conclusions and Future Work

For the current state of the art of fluid simulation techniques, direct numerical simulations of (hydrophobe) droplets in (hydrophile) fluids seem to be unfeasible. Therefore, to understand the experimentally observed clustering of droplets, we have proposed a phenomenological interaction of the droplets based on known hydrodynamic interactions with an additional Coulomb-type attraction. The phenomenology of this model is rather rich, as the number of droplets in a cluster is not proportional to the interaction strength: Due to a subtle balance of attraction and hydrodynamic repulsion, there can be clusters with more droplets for weaker attraction. In the next step, a better physical understanding of the possible attractive interactions and a matching with the experimental data have to be obtained.

Appendix A: Problem of Direct Numerical Simulation

Ideally, one would investigate the flow of spherical droplets in another fluid with a "direct numerical simulation" of two-phase flow, with a geometry of droplets (themselves a continuum) inside a continuum of the outer fluids, with a suitable boundary in between. Unfortunately, such an approach is currently not feasible, neither from the computational nor from the algorithmic aspect. The success of the Finite Element Methods in structural dynamics (that means, problems which are predominantly linear) is so much taken for granted that in engineering tasks, there is a downright reflex to use (or at least recommend) Finite Elements. For flow problems of particles (droplets included) which come closer than several mesh sizes while additional refinement is not possible, there are indeed computational issues [9]. Already for two dimensions, the simulation of polygonal particles in fluids is rather difficult and expensive if the boundary conditions have to be taken into account exactly, see [9] for a suitable simulation method and [10,11] for the resulting problems which have to be dealt with for the grid generation. If the boundary conditions are approximated, there is a danger of numerical instabilities which is treated in Appendix B. Curved boundaries of freely floating particles in fluids add an additional layer of computational complexity to the problem and have not yet been implemented according to our knowledge.

There are various packages which offer the option to simulate multi-phase flows via the Finite Volume method (FVM). In particular, the most striking advantage of the FVM is that it implicitly fulfils the incompressiblity condition, which for other continuum simulation methods must be ponderously enforced. Nevertheless, there are other drawbacks: First of all, the discretization into finite volumes makes a uniform connectivity for the underlying mesh necessary. As a result, in screenshots of simulations, small "steps" can be seen which represent the elementary volumes of the underlying uniformly connected grid. This step-like outlines leads to the problems with the zero-order approximation of boundaries which is discussed in Appendix B. A second drawback is that the non-linear Navier-stokes equations are solved in a linearized form: This gives a fast update speed, but leaves a remaining uncertainty about the validity of the result.

The least suitable method for two-phase flows is the Finite Difference Method: The finite-difference approximation of differential operators demands an underlying square grid, which severely limits the accuracy of the simulation. Worse, in spite of various approaches which claim to "interpolate" boundaries with "immersed"boundaries (interpolated and therefore of rather dubious validity), one can show the congruence of finite-difference methods with zero-order Finite Element methods [4]: This means that the boundaries in finite difference methods are always approximated in zero order, there are no boundary conditions "on" the boundary, just in the middle of the "element" next to the boundary. The issue of zero-order boundaries and the very likely occurence of numerical instabilities is treated in the next Appendix.

Appendix B: Little Known Hitches in the Approximation of Boundary Conditions to Zero Order

B. P. Leonard treated the snags in the application of low order finite difference methods in his highly readable, enligthening and amusing treatise "A survey of finite differences of opinion on numerical muddling of the incomprehensible defective confusion equation" [6]. The title refers to the confusion in the discussion of the error order when diffusion equations are augmented with advective terms, as well as the "remedies" which make the numerical solutions more "stable" but physically implausible. Nevertheless, Leonard's treatise only focused on the discretization of the equations. There are additional issues for the treatment of boundaries, which affect in particular finite difference and finite volume methods, when smooth slanted or curved surfaces are implemented by a "stepped" profile outline. For a partial differential equation

$$\frac{df(t)}{dt} = \hat{L}f(t), \tag{6}$$

(with $\hat{L}$ as the sum of the spatial differential operators) the product

$$\tilde{f}(t) = \underbrace{\exp\left(t(\hat{L})\right)}_{\text{time evolution operator}} \underbrace{\psi_0}_{\text{initial condition}} \tag{7}$$

is the formal solution (including additional spatial dependencies). When we solve the problem numerically, so that there is an additional error ϵ_L in the discretization of the spatial differential operator, our solution becomes

$$\tilde{f} = \exp\left(t(\hat{L} + \epsilon_L)\right) \psi_0. \tag{8}$$

The solution of a numerical simulation is "stable" if we don't generate any exponentially diverging noise via ϵ_L. If we add additional noise ϵ_0 to the initial condition by not using the exact, but stepped we can write the time evolution formally as

$$\begin{aligned} \tilde{f} &= \exp\left(t(\hat{L} + \epsilon_L)\right) (\psi_0 + \epsilon_0) \\ &= \exp\left(t(\hat{L} + \epsilon_L)\right) \psi_0 + \exp\left(t(\hat{L} + \epsilon_L)\right), \epsilon_0 \end{aligned} \tag{9}$$

and we can hope that the additional exponential term will not add much. As a simple test case of a higher-order solution method with zero order boundary, we can try to solve the harmonic oscillator where we replace the trapeze-method with the rectangle midpoint method. For quadrature over intervalls of length Δx, the error for the midpoint method is

$$\epsilon_{mid} = \frac{1}{24} \Delta x^3 f''(x_i), \tag{10}$$

while surprisingly, the error of the trapeze rule is twice as large,

$$\epsilon_{tra} = \frac{1}{12} \Delta x^3 f''(x_i), \tag{11}$$

(both errors from [1]), but which can be understood graphically, see Fig. 8.

We can now shift from spatial coordinates x to time coordinates t and create time integrators for differential equations

$$\frac{dy}{dt} = f(t) \tag{12}$$

from quadrature formulae by integrating Eq. (12) to

$$y(t + \tau) = y(t) + \int_t^{t+\tau} f(t) d\tau, \tag{13}$$

and insert our favorite quadrature scheme, which is the basis of the trapeze rule for time integration. When we replace the trapeze rule with the midpoint rule,

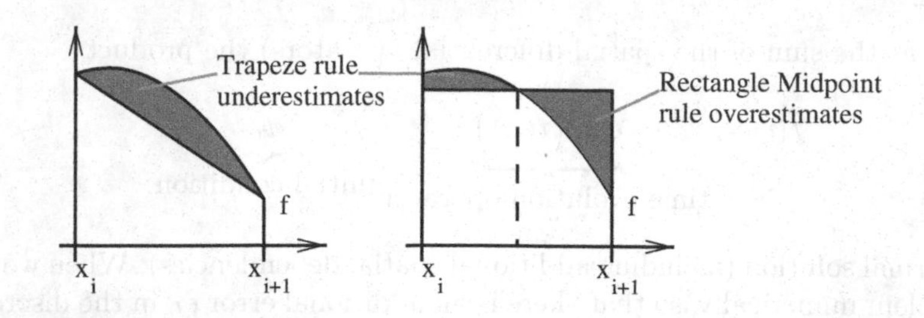

Fig. 8. Approximation of the integral of function f in the interval x_i to $x_i + 1$ with the trapeze rule, which underestimates (red area) the integral more than the midpoint rule overestimates it (error compensation of red and green area). (Color figure online)

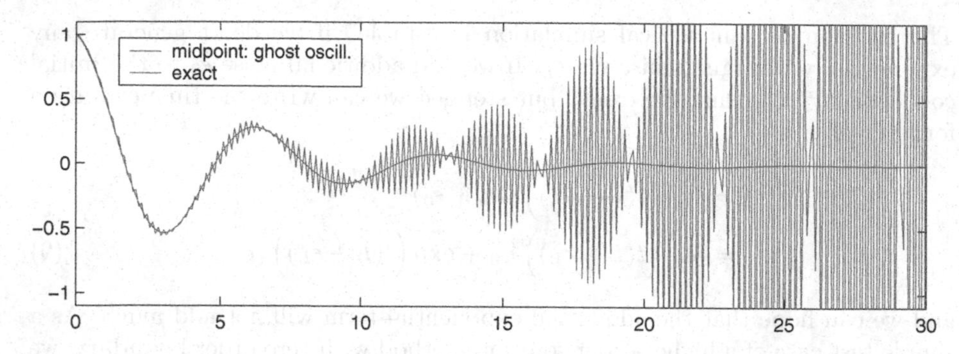

Fig. 9. Solution of a damped harmonic oscillator with the exact solution in black and the ghost oscillations generated by the midpoint time integrator.

we would expect better accuracy, because of the smaller local error Eq. (10) for the midpoint rule, compared to Eq. (11). The actual result can be seen in Fig. 9: While it starts well enough (at least more accurately than the Euler-Integrator would do), the numerical solution starts to develop high-frequency ghost oscillations with exponentially increasing amplitude, even for the (exponentially) decaying exact solution. The culprit is the midpoint value $f((x_i + x_{i+1})/2)$ in the integration, which, when we look at the original quadrature in Fig. 8, clearly is not an initial or final value for the integral in Eq. 13 - quite in contrast to the values at the end of the interval which are used for the Trapeze rule. The random osciallations around the true values at the initial value of each timestep are then amplified to non-decaying oscillations of high frequency. The same can be expected to happen for the random assignments of zero-order boundary conditions in lattice methods. In fact, such oscillations have already been observed for interpolated boundaries of circles moving over square grids [12], and in truely zero-order fashion the noise amplitude did not decay even for increasing area size of the circles: A good reason to keep one's fingers away from DNS simulations without the exact implementation of the boundaries. Such an exact implementation of spherical droplets in a fluid, on the other hand, would necessitate the use of curvilinear meshes near the droplet boundaries, and, as the positions change in every timestep, an adaptive remeshing in every timestep becomes necessary. From the standpoint of the first author of this article, who has experience with such issues for polygonal particles and two dimensions [10,11], the issue of droplets in fluids in three dimensions are at present not yet tractable in direct numerical simulations.

References

1. Acton, F.: Numerical Methods that Work. MAA Spectrum, Mathematical Association of America (1990)
2. Farooq, M.U., Kaneda, Y.: The hydrodynamic interaction of two spheres in a viscous fluid at small non-zero Reynolds number - axisymmetric case. J. Phys. Soc. Jpn. **54**(7), 2477–2484 (1985). https://doi.org/10.1143/JPSJ.54.2477
3. Goddard, B.D., Mills-Williams, R.D., Sun, J.: The singular hydrodynamic interactions between two spheres in stokes flow (2020)
4. Gresho, P.M., Sani, R.L.: Incompressible Flow and the Finite Element Method, Volume 2: Isothermal Laminar Flow. Wiley (2000)
5. Iwaoka, M., Ishii, T.: Experimental wall correction factors of single solid spheres in circular cylinders. J. Chem. Eng. Jpn. **12**, 239–242 (1979)
6. Leonard, B.P.: A survey of finite differences of opinion on numerical muddling of the incomprehensible defective confusion equation. In: Hughes, T. (ed.) Finite Element Methods for Convection Dominated Flows, New York, 2–7 December 1979. American Society of Mechanical Engineers. Applied Mechanics Division (1987)
7. Li, J., Barrow, D.A.: A new droplet-forming fluidic junction for the generation of highly compartmentalised capsules. Lab Chip **17**, 2873–2881 (2017). https://doi.org/10.1039/C7LC00618G
8. Mellsen, S.B.: On the added mass of sphere in a circular cylinder considering real fluid effects. Ph.D. thesis, California Institute of Technology (1966)
9. Mueller, J., Kyotani, A., Matuttis, H.G.: Towards a micromechanical understanding of landslides - aiming at a combination of finite and discrete elements with minimal number of degrees of freedom. J. Appl. Math. Phys. **8**, 1779–1798 (2020)
10. Mueller, J., Kyotani, A., Matuttis, H.G.: Grid-algorithm improvements for dense suspensions of discrete element particles in finite element fluid simulations. In: EPJ Web of Conferences, vol. 249, 09006 (2021)
11. Mueller, J., Matuttis, H.G.: Improved grid relaxation with zero-order integrators. J. Appl. Math. Phys. **9**, 1257–1270 (2021)
12. Ristow, G.H.: Wall correction factor for sinking cylinders in fluids. Phys. Rev. E **55**, 2808–2813 (1997). https://doi.org/10.1103/PhysRevE.55.2808
13. Smythe, W.R.: Flow around a sphere in a circular tube. Phys. Fluids **4**(6), 756–759 (1961)
14. Stokes, G.G.: On the effect of the internal friction of fluids on the motion of pendulums. Trans. Cambridge Philos. Soc. **9**, 8 (1851)
15. Tritton, D.: Physical Fluid Dynamics, 2nd edn. Oxford Science Publication, Clarendon Press (1988)

Network Creation During Agglomeration Processes of Polydisperse and Monodisperse Systems of Droplets

Johannes Josef Schneider[1,2](✉), Alessia Faggian[3], Aitor Patiño Diaz[3], Jin Li[2], Silvia Holler[3], Federica Casiraghi[3], Lorena Cebolla Sanahuja[3], Hans-Georg Matuttis[4], Martin Michael Hanczyc[3,5], David Anthony Barrow[2], Mathias Sebastian Weyland[1], Dandolo Flumini[1], Peter Eggenberger Hotz[1], Pantelitsa Dimitriou[2], William David Jamieson[6], Oliver Castell[6], Patrik Eschle[1], and Rudolf Marcel Füchslin[1,7]

[1] Institute of Applied Mathematics and Physics, School of Engineering, Zurich University of Applied Sciences, Technikumstr. 9, 8401 Winterthur, Switzerland
johannesjosefschneider@googlemail.com, {scnj,weyl,flum,eggg,escl,furu}@zhaw.ch

[2] School of Engineering, Cardiff University, Queen's Buildings, 14-17 The Parade, Cardiff CF24 3AA, Wales, UK
{LiJ40,Barrow,dimitrioup}@cardiff.ac.uk

[3] Laboratory for Artificial Biology, Department of Cellular, Computational and Integrative Biology (CIBIO), University of Trento, 38123 Trento, Italy
{alessia.faggian,aitor.patino,silvia.holler,federica.casiraghi, lorena.cebolla,martin.hanczyc}@unitn.it

[4] Department of Mechanical Engineering and Intelligent Systems, The University of Electrocommunications, Chofu Chofugaoka 1-5-1, Tokyo 182-8585, Japan
hg@mce.uec.ac.jp

[5] Chemical and Biological Engineering, University of New Mexico, MSC01 1120, Albuquerque, NM 87131-0001, USA

[6] Welsh School of Pharmacy and Pharmaceutical Science, Cardiff University, Redwood Building, King Edward VII Avenue, Cardiff CF10 3NB, Wales, UK
{jamiesonw,CastellO}@cardiff.ac.uk

[7] European Centre for Living Technology, S.Marco 2940, 30124 Venice, Italy
https://www.zhaw.ch/en/about-us/person/scnj/

Abstract. We simulate the movement and agglomeration of oil droplets in water under constraints, using a simplified stochastic-hydrodynamic

This work has been partially financially supported by the European Horizon 2020 project *ACDC – Artificial Cells with Distributed Cores to Decipher Protein Function* under project number 824060.

The original version of this chapter was previously published without open access. A correction to this chapter is available at
https://doi.org/10.1007/978-3-031-31183-3_25

C. De Stefano et al. (Eds.): WIVACE 2022, CCIS 1780, pp. 94–106, 2023.
https://doi.org/10.1007/978-3-031-31183-3_8

model. We analyze both local and global properties of the networks formed by the agglomerations of droplets for various system sizes. We focus on the differences of these properties for monodisperse and polydisperse systems of droplets. For the mean degree, we obtain different values for critical exponents.

Keywords: Network analysis · monodisperse · polydisperse · droplets · cluster · agglomeration

1 Introduction

Within the scope of the European Horizon 2020 project *ACDC – Artificial Cells with Distributed Cores to Decipher Protein Function*, we intend to develop a probabilistic compiler [2,20] to aid the three-dimensional agglomeration of particles filled with various chemicals in a specific way in order to e.g. create macromolecules via a gradual chemical reaction scheme [10,11]. Hereby we aim at the creation of some specific macromolecules, but in contrast to the work of [7], we govern the successive reaction process by a specific design of the three-dimensional structure of the agglomeration. Furthermore, we are interested in the question of which molecules could be produced to which extent within randomly created agglomerations in comparison to the production within the primordial soup in order to have a closer look at the origin of life [9].

In such an agglomeration, neighboring droplets can form connections, either by simply touching each other or by getting glued to each other by matching pairs of single-stranded DNA, which are attached to hulls enclosing their surfaces. Chemicals contained within the droplets can move to neighboring droplets either directly, as hydrophobic compounds can be exchanged between adjacent oil droplets at the contact face, or, if the oil droplets are contained in a hull comprised of amphiphilic molecules like phospholipids, through pores within bilayers. Thus, a complex network is created, with the droplets being the nodes of this graph and the existing connections being the edges between the corresponding droplets.

In this paper, we present computational results for basic simulations of simplified agglomeration processes of oil droplets in water to mimick experiments. In the experiments performed by our co-authors at Cardiff University, a microfluidic approach is used to generate droplets: A stream of fluid within another fluid is split up in a series of spherical droplets of (almost) equal size after passing through a t-junction if the ratio of the pressures between the two fluid streams is chosen in an appropriate range [6]. In contrast, the manual emulsion method [3] called Rakka, which is used by our co-authors in experiments at the University of Trento, mechanically sends excitations in a large oil drop lying at the bottom of a container filled with water and containing amphiphilic molecules, thus splitting it in a polydisperse system of droplets with a wide range of radii. The droplets sink to the bottom of the cylinder where they agglomerate, while forming connections. Thus, this scenario of droplets randomly placed in a cylinder defines the starting point for the simulations.

Within the scope of this paper, we focus on the question of how polydispersity influences the agglomeration process of the particles and some specific properties of the networks created. In order to deal only with this question and to exclude effects from other experimental properties, we simulate the droplets as hard spheres and ignore details of the surface structure of the particles, attractive forces as well as adhesion effects. As the extension of the bilayers is very small and as due to their small radii [1], the droplets keep their spherical shape during the experiments, hence, this simplified approach is justified. This paper is organized as follows: In Sect. 2, we describe our simulation technique in detail, before presenting our computational results of a network analysis of the agglomerations both for polydisperse and for monodisperse systems of various system sizes in Sect. 3. Section 4 provides a summary of the results and an outlook to future work.

2 Simulation Details

At the beginning of the simulations, we randomly place N spherical particles in a cylindrical container with radius 1 mm and height 4 mm in a way that they do not overlap with each other and that they do not overlap with the walls of the cylinder, as shown in the left part of Fig. 1. For the polydisperse system, we randomly choose the particle radii r_i uniformly from the interval [10–50] μm, whereas we set all radii $r_i \equiv 30\,\mu$m for the monodisperse system.

Fig. 1. Initial (left) and final (right) configuration generated in a simulation of the agglomeration process of a multidisperse system of 2,000 spherical particles in a cylinder.

After this initialization, we perform the main simulation which is comprised of 10^7 time steps of a duration of $\delta t = 10^{-5}$ s. In each time step, the particles are subjected to various forces:

– They sink in water due to gravity $\boldsymbol{F}_G$ reduced by the buoyant force $\boldsymbol{F}_b$:

$$\boldsymbol{F}_G(i) - \boldsymbol{F}_b(i) = \frac{4\pi}{3} r_i^3 \left(\varrho_{\text{oil}} - \varrho_{\text{water}}\right) g \tag{1}$$

For the oil density, we use the value $\varrho_{\mathrm{oil}} = 1.23\,\mathrm{kg/l}$, just as in the experiments of our co-authors in Trento.

- Secondly, the spatial components $v_{x,y,z}(i)$ of the velocity vectors $\boldsymbol{v}(i)$ are subjected to random velocity changes: They are randomly altered by up to $\pm 5\%$ of their absolute values in order to take at least in this small random way into account that the containers are moved by the experimentalists in the laboratory in Trento.
- The particles are also subjected to the Stokes friction force $\boldsymbol{F}_S$:

$$\boldsymbol{F}_S(i) = -6\pi\eta r_i \boldsymbol{v}(i) \tag{2}$$

The viscosity of water at 25 °C is $\eta = 0.891\,\mathrm{mPas}$.

- As in classic hydrodynamics, the concept of added mass [19] is used. When applying Newton's second law, we have to consider an effective mass of the particle, i.e., $\boldsymbol{F}(i) = m_{\mathrm{eff}}(i)\boldsymbol{a}(i)$. This effective mass is composed of the mass $m(i)$ of particle i and of the added mass $m_{\mathrm{added}}(i)$. This added mass is caused by the inertia of the surrounding fluid, which needs to be deflected or attracted if the particle itself is accelerated or decelerated in the water, and can be determined to being half of the mass of the water displaced by oil particle i. Thus, we use for the deceleration caused by the Stokes friction the equation

$$\boldsymbol{a}_S(i) = \frac{\boldsymbol{F}_S(i)}{m(i) + m_{\mathrm{added}}(i)}. \tag{3}$$

When working with such a set of second order differential equations governing the laws of motion for the particles, the question arises as to which integrator to use. Due to the stochastic nature of random velocity changes, only an Euler scheme with very small time intervals is suitable for the determination of new velocities and positions [4]. In the case of collisions between pairs of particles or between particles and walls, a mostly elastic collision dynamics is imposed. Overlaps occurring at the end of each time step are resolved as in [8,12].

When simulating a polydisperse and a monodisperse system of droplets, generating movies from these simulations, and watching them, one sees at first sight a striking difference between polydisperse and monodisperse systems: While the droplets in a monodisperse system sink to the bottom of the cylinder with almost equal speed, in the case of the polydisperse system the largest droplets rush down fastest, whereas the smallest droplets sink comparatively very slowly towards the bottom of the cylinder. Also the agglomeration processes at the bottom of the cylinder look different as the final agglomerations themselves. Thus, we evaluate these agglomerations and their time evolutions with some standard measures from network analysis in order to quantify these differences.

For this purpose, we performed 100 simulation runs each both for the polydisperse and for the monodisperse system for the system sizes N = 10, 20, 30, 40, 50, 60, 70, 80, 90, 100, 110, 120, 130, 140, 150, 160, 170, 180, 190, 200, 250, 300, 350, 400, 450, 500, 550, 600, 650, 700, 750, 800, 850, 900, 950, 1000, 1100, 1200, 1300, 1400, 1500, 1600, 1700, 1800, 1900, 2000, 2500, 3000, 3500, 4000, 4500, and 5000. From each simulation run, we store each 1000th configuration, such that we have a set of 10^4 configurations per simulation run, with a time interval of 0.01 s between successive configurations. The results presented for the network analysis in the next section are averaged over the measurements taken from the 100 simulations. Thus, for each curve in Figs. 2, 4, 6, and 8 in total 10^6 configurations had to be evaluated.

3 Network Analysis

For network analysis, first of all a network related to the problem to be considered has to be defined. As we are interested mainly in structures resulting from neighborhood relationships, we have a look at the adjacency matrix η with

$$\eta(i,j) = \begin{cases} 1 & \text{if droplets } i \text{ and } j \text{ are neighbors of each other} \\ 0 & \text{otherwise.} \end{cases} \tag{4}$$

We consider a pair (i, j) of droplets as neighboring each other, meaning (almost) touching each other, if the condition

$$\sqrt{(x_i - x_j)^2 + (y_i - y_j)^2 + (z_i - z_j)^2} \leq r_i + r_j + 0.1\,\mu\text{m} \tag{5}$$

is fulfilled, i.e., if the distance between their midpoints is smaller or equal to the sum of their radii plus some small offset. Please note that one usually sets $\eta(i, i) \equiv 0$ for all nodes i. The matrix η contains all the information about the network.

When analyzing a network, one mostly takes either an atomistic view, looking at the various nodes and determining their network related properties, or a global view, determining clusters of nodes and dealing with questions like whether the network is percolating [18]. We present basic results both for the atomistic and for the global view.

3.1 Degrees of the Particles

As an example for local network analysis, we consider the degrees of the particles. A degree $k(i)$ of a node i is defined as the number of nodes it is attached to via edges in the network. It can be easily calculated by

$$k(i) = \sum_{j=1}^{N} \eta(i,j). \tag{6}$$

We are interested in the average $\langle k \rangle$ of all degrees, its time evolution, and its final value for various system sizes, both for the polydisperse and for the monodisperse systems.

Figure 2 shows the increase of $\langle k \rangle$ over time t for various system sizes N. The increases are steeper for the monodisperse system and approach their final values less smoothly than in the case of the polydisperse system. Furthermore,

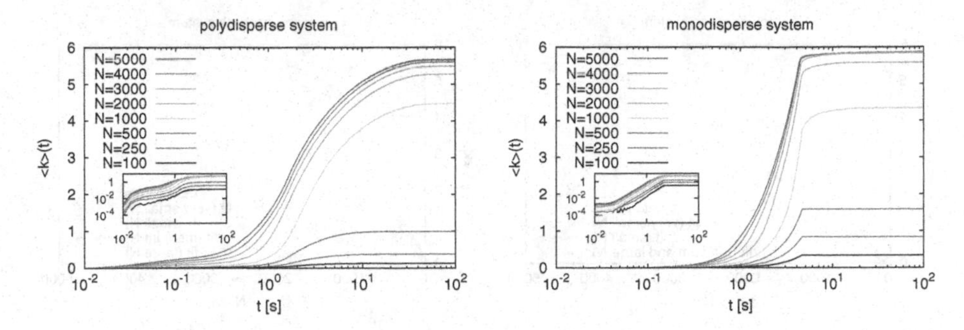

Fig. 2. Time evolution of the average value $\langle k \rangle$ of the degrees for various system sizes: On the left, results for the polydisperse system are presented, on the right, results for the monodisperse system are shown. In the insets, the curves are redrawn in a double-logarithmic way to put more emphasis on short time scales.

the insets reveal that $\langle k \rangle$ increases in a double-sigmoidal way for the polydisperse system, while this is not the case for the monodisperse system, for which only one sigmoidal increase can be observed. The final values for $\langle k \rangle$ are larger for small and large system sizes for the monodisperse system, such that we now have a closer look at the final values also for other system sizes.

Figure 3 depicts the final values for various system sizes, both for the polydisperse system and for the monodisperse systems. In both cases, we find that there are regimes of system sizes in which we can fit the data points to various specific fit functions well known from the studies of phase transitions and order parameters in statistical physics [17]. Besides the linear or quadratic increase for small N, we find an interesting critical behavior. When investigating the data points for the polydisperse system in Fig. 3, we get the overall behavior [15]

$$\langle k_f \rangle = \begin{cases} c_1 N & \text{for } N \leq 40 \\ c_1 N + c_2 N^2 & \text{for } N \leq 80 \\ c_3 \dfrac{N^\alpha}{(N_{\text{crit},1} - N)^\gamma} & \text{for } 40 \leq N \leq 900 \\ c_4 \tanh\left(c_5 \left(N - N_{\text{crit},2}\right)^\beta\right) & \text{for } 850 \leq N \end{cases} \tag{7}$$

with the prefactors $c_1, \ldots, c_5$, the critical exponents α, β, and γ, and the critical numbers $N_{\text{crit},1}$ and $N_{\text{crit},2}$ of particles. Various fits of the functions in Eq. 7 with similar fitting qualities result in $\alpha = 1 \ldots 1.1$, $\gamma = 0.4 \ldots 0.5$, $\beta = 0.1 \ldots 0.18$, $N_{\text{crit},1} = 940 \ldots 1000$, and $N_{\text{crit},2} = 780 \ldots 845$. Please note that the prefactor $c_4 = 6 \ldots 7$ provides an estimate for $\langle k_f \rangle$ in the limit $N \to \infty$. While the values for prefactors and critical numbers of particles will change if altering the simulation parameters, the theory of critical phenomena decrees that the critical exponents α, β, and γ and the form of the functions stay identical [17].

For the monodisperse system, we find striking similarities, but also differences to the polydisperse system. Here we get a linear increase of $\langle k_f \rangle$ for $N \leq 160$, a critical increase for $200 \leq N \leq 900$ according to the same law as for the

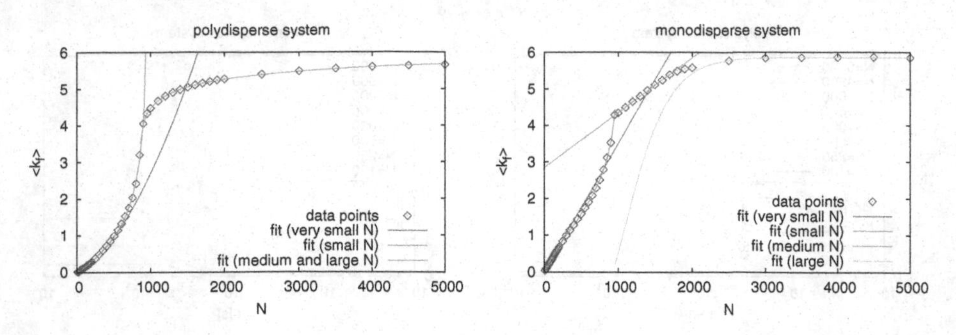

Fig. 3. Final values $\langle k_f \rangle$ for the averages of the degrees for various system sizes N: Both for the polydisperse (left) and for the monodisperse system (right), we find ranges of N for which we can fit the data points to specific fit functions: for the polydisperse system, the data points are fitted to $1.2 \times 10^{-3} N + 1.5 \times 10^{-6} N^2$ for $N \leq 80$, to $2.222848282 \times 10^{-2} \times N^{1.064929}/(982.95 - N)^{0.4597383}$ for $40 \leq N \leq 900$, and to $6.4592928303905470 \tanh(0.41066892597996169 \times (N - 843.037477864008)^{0.14512724870611132})$ for $850 \leq N \leq 5000$. For the monodisperse system, we display the fit functions $3.5 \times 10^{-3} N$ for $N \leq 160$, $4.9962568900578735 \times 10^{-2} \times N^{0.78025269698399657}/(1008.0838515966243 - N)^{0.22394426252398117}$ for $200 \leq N \leq 900$, $2.85 + 1.5 \times 10^{-3} \times N$ for $950 \leq N \leq 1700$, and $5.8606051850230925 \tanh(7.4870316562044323 \times 10^{-4} \times (N - 923.86956752140270)^{1.1002364522169625})$ for $1900 \leq N \leq 5000$. The more exactly given fit parameters were determined with an altered conjugate gradients method [16].

polydisperse system, but then in contrast to the polydisperse system a linear increase for $950 \leq N \leq 1700$, and finally we tried to fit a tangens hyperbolicus also to the data points in the regime of large $N \geq 1900$. However, this last fit to a tangens hyperbolicus is of a worse quality, the data points are not sufficient to prove that we have this same behavior here. The small value for c_4 here suggests that the assumption of a tangens hyperbolicus might be wrong in the case of a monodisperse system. For the largest values $N \geq 3000$, the measured averages of $\langle k_f \rangle$ might simply fluctuate in the range $[5.84 - 5.87]$. These results do not prove but seem to hint that in the limit $N \to \infty$, the final average value for $\langle k_f \rangle$ is significantly larger for the polydisperse system than for the monodisperse system.

In the next step, we have a look at the time evolutions of the maximum degree $k_{\max} = \max\{k(i)\}$ for various system sizes, which are shown in Fig. 4. For the polydisperse system, we find a sigmoidal increase. In the case of the monodisperse system, the increase is steeper, but is then stopped rather abruptly when reaching the final values $k_{\max,f}$. These final values are shown in Fig. 5. We find that $k_{\max,f}$ is slightly larger for the monodisperse system for small system sizes, but then the curves cross in the range $500 \leq N \leq 700$. From then on, $k_{\max,f}$ is larger for the polydisperse system. For the monodisperse system, $k_{\max,f}$

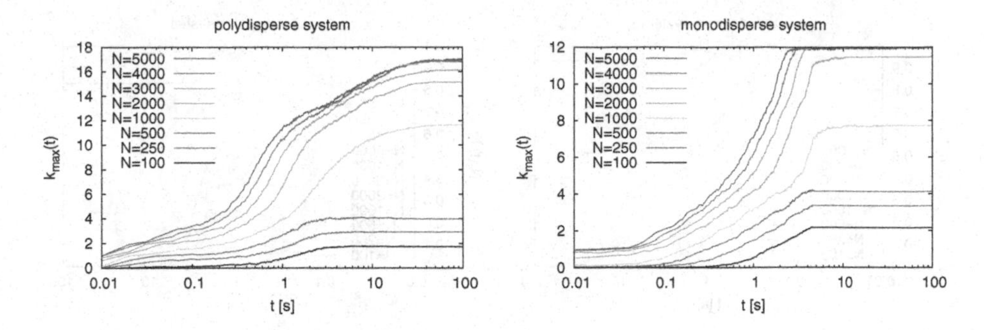

Fig. 4. Time evolution of the maximum $k_{\max}$ of the degrees for various system sizes, both for the polydisperse system (left) and for the monodisperse systems (right).

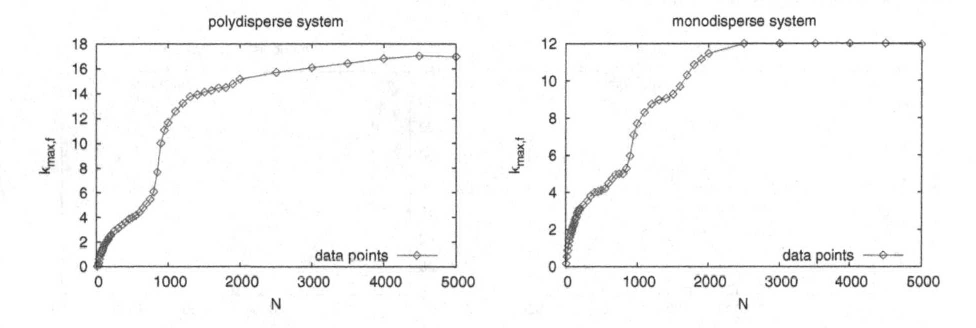

Fig. 5. Final values $k_{\max,\mathrm{f}}$ of the maximum of the degrees for various system sizes, both for the polydisperse system (left) and for the monodisperse systems (right).

is bounded by the kissing number of 12 in three dimensions [5,13]. Also the polydisperse system contains such a bound, depending on the ratio between the radii of the largest and of the smallest particles, but this bound is considerably larger in our case, in which this ratio is 1:5 [13], such that it does not play a role here as $k_{\max,\mathrm{f}}$ is much smaller for the system sizes we consider here.

3.2 Clusters of Particles

As an example for global network analysis, we present results for the decrease of the number of clusters and the increase of the maximum cluster size. A cluster is defined as a subset C of nodes in the network for which the condition holds that for each arbitrarily chosen pair (i, j) of nodes with $i, j \in C$, a path from i to j exists on which one walks only over edges starting at a node in C and ending at a node in C, and for which no node is left out which fulfills this condition.

We are interested in the number N_c of clusters and its time evolution for various system sizes both for the polydisperse and for the monodisperse systems. As the droplets are placed in the cylinder at the beginning of the simulation without touching each other, we generally get $N_c = N$ at the beginning of the

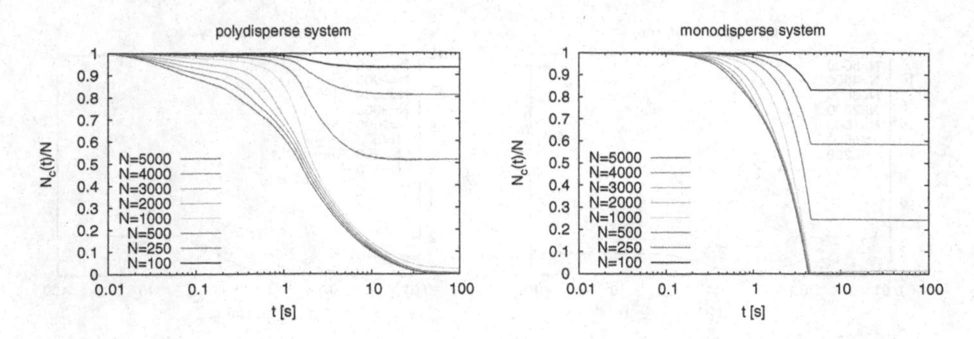

Fig. 6. Time evolution of the number N_c of clusters, renormalized by the number N of particles for various system sizes, both for the polydisperse system (left) and for the monodisperse systems (right).

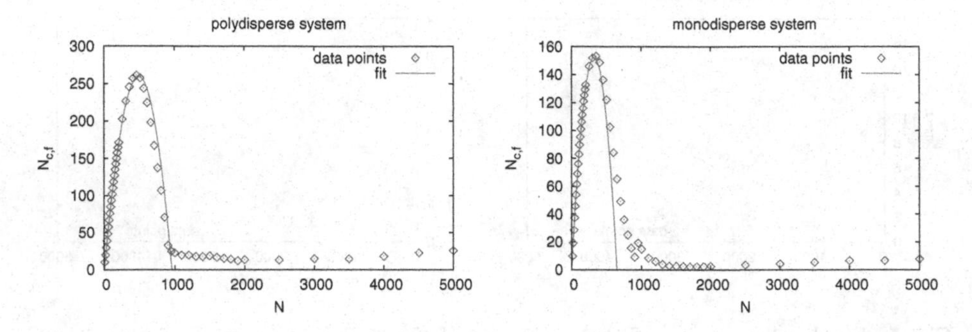

Fig. 7. Final values $N_{c,f}$ of the number of clusters for various system sizes, both for the polydisperse system (left) and for the monodisperse systems (right): The fit function for the polydisperse system is given by $260 - 0.00115 \times (N - 480)^2$, for the monodisperse system by $153 - 0.00145 \times (N - 321)^2$.

simulation, as each node forms a cluster of its own at the beginning. (Please note that we also count these "one-node clusters".) With an increasing number of edges, clusters unite and N_c decreases. In order to better compare the results for various system sizes, we plot the number N_c renormalized by the system size N in Fig. 6. We get a sigmoidal decrease for the polydisperse system, but again we find an abrupt end of the decrease for the monodisperse system when the final value $N_{c,f}$ is reached, which is plotted in Fig. 7. Both for the polydisperse and for the monodisperse system, we find that the final number of clusters can be fitted almost quadratically for small system sizes N. For large system sizes, the number of clusters does not decrease to 1, but a small number of clusters remains.

Thus, the question arises whether the system is split in various clusters of almost equal size or in various clusters with strongly differing sizes or whether the system is dominated by one large cluster. Therefore, we have a look at the size $C_{\max}$ of the largest cluster, which is the number of nodes contained in this

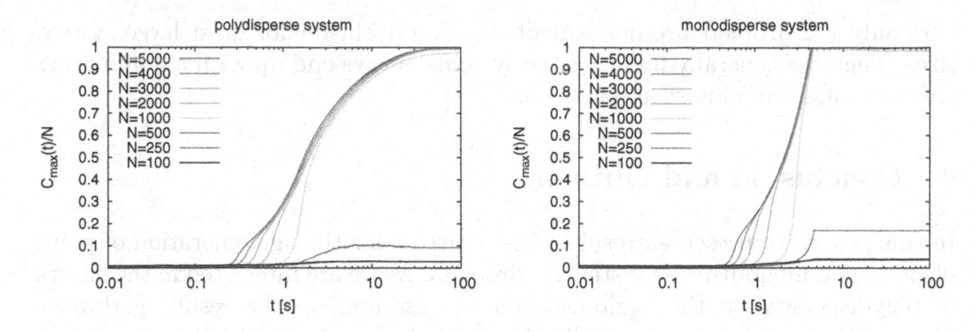

Fig. 8. Time evolution of the size $C_{\max}$ of the largest cluster, renormalized by the number N of particles for various system sizes, both for the polydisperse system (left) and for the monodisperse system (right).

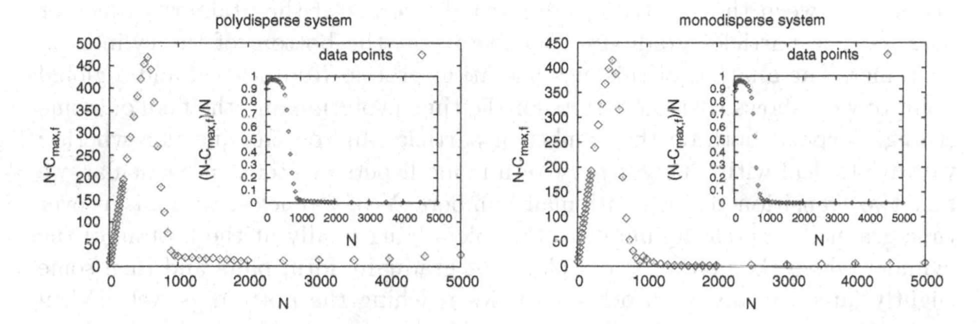

Fig. 9. Deviation of the final values of the size $C_{\max,f}$ of the largest cluster from the system size, for various system sizes, both for the polydisperse system (left) and for the monodisperse system (right). In the insets, the data points are replotted to display the relative deviation of $C_{\max}$ from the system size.

cluster. In order to again better compare the results for the time evolution of this observable, we renormalize it with the system size and plot $C_{\max}/N$ in Fig. 8. This ratio increases sigmoidally over time, reaching a value of (almost) 1 for large system sizes. Again, the increase stops abruptly for monodisperse systems when the final value $C_{\max,f}$ is reached. When plotting $C_{\max,f}$ versus the system size N, one only sees a linear increase of $C_{\max,f}$ for medium and large N [15]. Thus, we here have a look at the absolute deviations $N - C_{\max,f}$ of these final values from the corresponding system sizes and at the relative deviations $(N - C_{\max,f})/N$, which are plotted in Fig. 9. Both for the polydisperse and for the monodisperse system, we see an almost linear increase of the absolute deviation at small system sizes, which is reflected by a virtually constant value of the relative deviation, which is slightly smaller than 1. Then the absolute and the relative deviation decrease to relatively small values till $N = 1000$. For larger system sizes, only 12–28 droplets are on average not part of the largest cluster for the polydisperse system. In the case of the monodisperse system, this number is even smaller, here only 1–2 droplets are not part of the largest cluster for most large system sizes. Thus, we generally find that the systems always end up with a dominating cluster containing almost all particles.

4 Conclusion and Outlook

In this paper, we presented results of simulations for the agglomeration of polydisperse and monodisperse systems of droplets. As we are interested in the effects of polydispersity on the agglomeration process and on the resulting droplet networks, we study a very simplified system, in which the droplets are represented as hard spheres (There are no other attractive or repulsive forces implemented.), subjected to gravity reduced by buoyancy, as well as Stokes friction, added mass effect, random velocity changes, and almost-elastic impacts. Connections between these particles are virtually formed if they (almost) touch or overlap. The particles gradually agglomerate at the bottom of the cylindrical container. The analysis of this agglomeration process from a local and a global point of view shows that the results for the time evolution and the final outcome strongly depend both on the number of particles and on the question whether we have to deal with a polydisperse or a monodisperse system. In particular, we find two transition regimes: at small numbers N of particles, we find an over time gradually increasing number of droplets lying finally at the bottom of the cylinder where they either stay isolated or gradually form pairs and then some slightly larger groups with other particles reaching the bottom as well. When increasing N even further, a network of droplets is created at a bottom layer of the cylinder. During this regime, we find first an increase of the number of clusters and then a decrease again due to the formation of one gradually more dominating cluster containing most droplets.

We intend to continue our investigations by measuring clustering coefficients, fractal dimensions, the locations of droplets with differing radii, and the importance some droplets might have for the overall network. Furthermore, we will perform simulations also with other simulation parameters, like another size of the cylinder and with other polydisperse radius distributions in order to better understand the values for the critical exponents and for the critical numbers and their dependencies. We also plan to extend our investigations first to binary systems, in which two particle types A and B are present and connections can only be forged between pairs of $A-B$ but not $A-A$ or $B-B$ [14] and then to ternary systems, in which there are three particle types A, B, and C with connections between adjacent pairs of $A-B$ particles but in which the additional C-particles are unable to form any connections. Hereby we want to study the breakdown of the size of the largest cluster with increasing density of C-particles. Furthermore, we want to add gluing forces between particles to find out how they change the results.

References

1. Aprin, L., Heymes, F., Laureta, P., Slangen, P., Le Floch, S.: Experimental characterization of the influence of dispersant addition on rising oil droplets in water column. Chem. Eng. Trans. **43**, 2287–2292 (2015)
2. Flumini, D., Weyland, M.S., Schneider, J.J., Fellermann, H., Füchslin, R.M.: Towards programmable chemistries. In: Cicirelli, F., Guerrieri, A., Pizzuti, C., Socievole, A., Spezzano, G., Vinci, A. (eds.) WIVACE 2019. CCIS, vol. 1200, pp. 145–157. Springer, Cham (2020). https://doi.org/10.1007/978-3-030-45016-8_15
3. Hadorn, M., Boenzli, E., Eggenberger Hotz, P., Hanczyc, M.M.: Hierarchical unilamellar vesicles of controlled compositional heterogeneity. PLoS ONE **7**, e50156 (2012)
4. Kloeden, P., Platen, E.: Numerical Solution of Stochastic Differential Equations. SMAP, Springer, Heidelberg (2013). https://doi.org/10.1007/978-3-662-12616-5
5. Kucherenko, S., Belotti, P., Liberti, L., Maculan, N.: New formulations for the kissing number problem. Discret. Appl. Math. **155**, 1837–1841 (2007)
6. Li, J., Barrow, D.A.: A new droplet-forming fluidic junction for the generation of highly compartmentalised capsules. Lab Chip **17**, 2873–2881 (2017)
7. Marshall, S.M., et al.: Identifying molecules as biosignatures with assembly theory and mass spectrometry. Nat. Commun. **12**, 3033 (2021)
8. Müller, A., Schneider, J.J., Schömer, E.: Packing a multidisperse system of hard disks in a circular environment. Phys. Rev. E **79**, 021102 (2009)
9. Oparin, A.: The Origin of Life on the Earth, 3rd edn. Academic Press, New York (1957)
10. Schneider, J.J., Weyland, M.S., Flumini, D., Füchslin, R.M.: Investigating three-dimensional arrangements of droplets. In: Cicirelli, F., Guerrieri, A., Pizzuti, C., Socievole, A., Spezzano, G., Vinci, A. (eds.) WIVACE 2019. CCIS, vol. 1200, pp. 171–184. Springer, Cham (2020). https://doi.org/10.1007/978-3-030-45016-8_17
11. Schneider, J.J., Weyland, M.S., Flumini, D., Matuttis, H.-G., Morgenstern, I., Füchslin, R.M.: Studying and simulating the three-dimensional arrangement of droplets. In: Cicirelli, F., Guerrieri, A., Pizzuti, C., Socievole, A., Spezzano, G., Vinci, A. (eds.) WIVACE 2019. CCIS, vol. 1200, pp. 158–170. Springer, Cham (2020). https://doi.org/10.1007/978-3-030-45016-8_16
12. Schneider, J.J., Müller, A., Schömer, E.: Ultrametricity property of energy landscapes of multidisperse packing problems. Phys. Rev. E **79**, 031122 (2009)
13. Schneider, J.J., et al.: Geometric restrictions to the agglomeration of spherical particles. In: Schneider, J.J., Weyland, M.S., Flumini, D., Füchslin, R.M. (eds.) WIVACE 2021. CCIS, vol. 1722, pp. 72–84. Springer, Cham (2022). https://doi.org/10.1007/978-3-031-23929-8_7
14. Schneider, J.J., et al.: Influence of the geometry on the agglomeration of a polydisperse binary system of spherical particles. In: The 2021 Conference on Artificial Life, ALIFE 2021 (2021). https://doi.org/10.1162/isal_a_00392
15. Schneider, J.J., et al.: Paths in a network of polydisperse spherical droplets. In: The 2022 Conference on Artificial Life, ALIFE 2022 (2022). https://doi.org/10.1162/isal_a_00502
16. Schneider, J.J., Kirkpatrick, S.: Stochastic Optimization. Springer, Heidelberg (2006)
17. Stanley, H.E.: Introduction to Phase Transitions and Critical Phenomena. Oxford University Press, New York (1972)

18. Stauffer, D., Aharony, A.: Introduction to Percolation Theory. Taylor & Francis (1992)
19. Stokes, G.G.: On the effect of the internal friction of fluids on the motion of pendulums. Trans. Cambridge Philos. Soc. **9**, 8–106 (1851)
20. Weyland, M.S., Flumini, D., Schneider, J.J., Füchslin, R.M.: A compiler framework to derive microfluidic platforms for manufacturing hierarchical, compartmentalized structures that maximize yield of chemical reactions. In: The 2020 Conference on Artificial Life, ALIFE 2020, pp. 602–604 (2020). https://doi.org/10.1162/isal_a_00303

Artificial Chemistry Performed in an Agglomeration of Droplets with Restricted Molecule Transfer

Johannes Josef Schneider[1,2(✉)], Alessia Faggian[3], William David Jamieson[2], Mathias Sebastian Weyland[1], Jin Li[2], Oliver Castell[2], Hans-Georg Matuttis[4], David Anthony Barrow[2], Aitor Patiño Diaz[3], Lorena Cebolla Sanahuja[3], Silvia Holler[3], Federica Casiraghi[3], Martin Michael Hanczyc[3,5], Dandolo Flumini[1], Peter Eggenberger Hotz[1], and Rudolf Marcel Füchslin[1,6]

[1] Institute of Applied Mathematics and Physics, School of Engineering, Zurich University of Applied Sciences, Technikumstr. 9, 8401 Winterthur, Switzerland
johannesjosefschneider@googlemail.com, {scnj,weyl,flum,eggg,furu}@zhaw.ch

[2] School of Engineering, Cardiff University, Queen's Buildings, 14-17 The Parade, Cardiff CF24 3AA, Wales, UK
{jamiesonw,LiJ40,CastellO,Barrow}@cardiff.ac.uk

[3] Laboratory for Artificial Biology, Department of Cellular, Computational and Integrative Biology (CIBIO), University of Trento, 38123 Trento, Italy
{alessia.faggian,aitor.patino,lorena.cebolla,silvia.holler, federica.casiraghi,martin.hanczyc}@unitn.it

[4] Department of Mechanical Engineering and Intelligent Systems, The University of Electrocommunications, Chofu Chofugaoka 1-5-1, Tokyo 182-8585, Japan
hg@mce.uec.ac.jp

[5] Chemical and Biological Engineering, University of New Mexico, MSC01 1120, Albuquerque, NM 87131-0001, USA

[6] European Centre for Living Technology, S.Marco 2940, 30124 Venice, Italy
https://www.zhaw.ch/en/about-us/person/scnj/

Abstract. Within the scope of the European Horizon 2020 project *ACDC – Artificial Cells with Distributed Cores to Decipher Protein Function*, we aim at the development of a chemical compiler governing the three-dimensional arrangement of droplets, which are filled with various chemicals. Neighboring droplets form bilayers with pores which allow chemicals to move from one droplet to its neighbors. With an appropriate three-dimensional configuration of droplets, we can thus enable gradual biochemical reaction schemes for various purposes, e.g., for the production of macromolecules for pharmaceutical purposes. In this paper, we

This work has been partially financially supported by the European Horizon 2020 project *ACDC – Artificial Cells with Distributed Cores to Decipher Protein Function* under project number 824060.

The original version of this chapter was previously published without open access. A correction to this chapter is available at
https://doi.org/10.1007/978-3-031-31183-3_25

C. De Stefano et al. (Eds.): WIVACE 2022, CCIS 1780, pp. 107–118, 2023.
https://doi.org/10.1007/978-3-031-31183-3_9

demonstrate with artificial chemistry simulations that the ACDC technology is excellently suitable to maximize the yield of desired reaction products or to minimize the relative output of unwanted side products.

Keywords: artificial chemistry · agglomeration · droplet · preference

1 Introduction

1.1 Background

Fig. 1. Agglomeration of 3,000 spherical droplets with radius 30 µm at the bottom of a cylinder, created in a computer simulation as described in [9]. The configurations in both pictures are identical with respect to the geometric locations of the droplets, but the left graphics shows a bipartite system with two particle types, marked as red and green, whereas the right graphics shows a tripartite system with three particle types, marked as red, green, and blue. These two configurations serve as the common underlying geometric structure and, with respect to particle types, as starting points for our artificial chemistry simulations.

In organic chemistry, one usually deals with molecules containing functional groups, like the hydroxyl `-OH` group of alcohols, the carbonyl `-C=O` group of ketones and aldehydes, and the carboxyl `-COOH` group of organic acids. These and other functional groups allow specific reactions, e.g., an alcohol and a carboxylic acid can form an ester, while splitting off a water molecule. Besides these well-known natural functional groups, additionally corresponding pairs of artificial functional groups have been created in the last decades in the field of click chemistry [3,5] to better govern reaction processes. However, the number of different active groups which can be used simultaneously in a reaction environment as intended is usually rather limited [13], which limits this approach.

Therefore, we apply another approach: Instead of having one well-stirred pot containing all educts for the chemical reaction process, we use droplets within a hull of amphiphilic molecules like phospholipids. The droplets are filled with only one chemical each at the beginning and are put in a cylinder where they

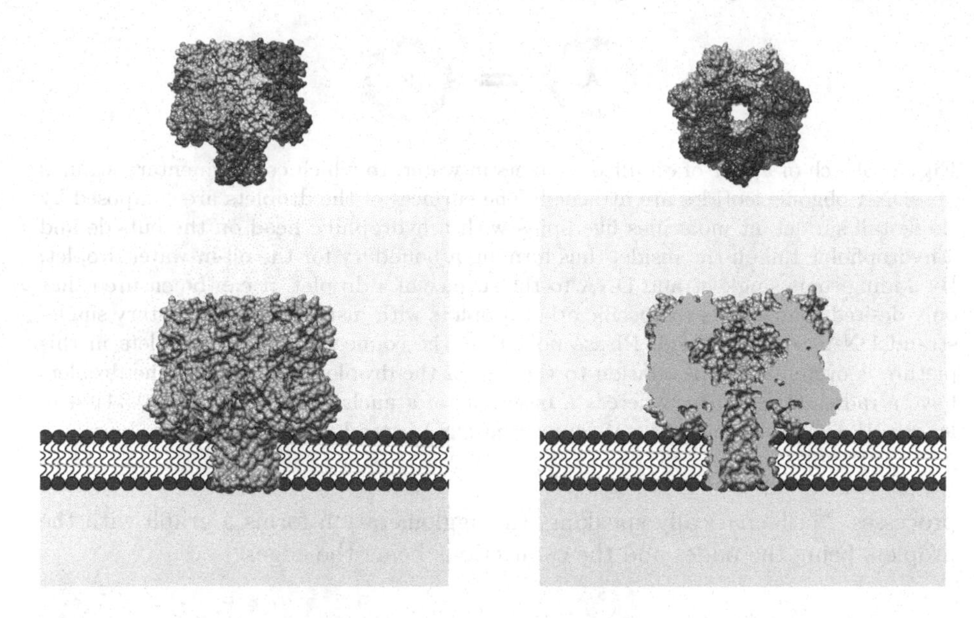

Fig. 2. Alpha hemolysin (aHL) pore: The top row shows the aHL, which is comprised of seven macromolecules, from the side view (left), displaying its trunk, and from the bottom up view (right), displaying the narrow channel in its center [12]. The bottom row reveals how the aHL sticks its trunk through the bilayer between two droplets (left) and presents in detail a cut through the pore formed (right).

sink and agglomerate at the bottom. Such an agglomeration is shown in Fig. 1. In the computer simulation, 3,000 spherical droplets were first placed randomly in a cylinder and sink to the bottom under the influence of gravity reduced by the buoyant force, Stokes friction, added mass effect, small random velocity changes, and quasi-elastic collisions among each other and with the walls of the cylinder [9,11]. Droplets touching each other can form a small bilayer and are then able to exchange small molecules through pores in these bilayers, as shown at the example of an alpha hemolysin (aHL) pore in Fig. 2. The creation of such connections can also be restricted so that neighboring pairs can only form connections if additional conditions are fulfilled. Such additional conditions can be realized by e.g. placing the seven constituent molecules of aHL only in some of the droplets. Covering the hulls of the droplets with single strands of DNA, as shown in Fig. 3, is another way of imposing such restrictions. Only if neighboring droplets are covered with complementary strands of DNA, a connection for exchanging molecules can be formed. Such restrictions offer a wide variety of altering the time evolution of gradual chemical reaction schemes, which will be useful, as we will show later. Furthermore, this overall approach with an agglomeration of droplets with bilayers which allow the exchange of molecules mimicks living organisms with their spatial structuring. The droplets correspond to the cells and their hulls to the membranes. One can also use a chemist's language and call the droplets the containers and the exchange of molecules transport

Fig. 3. Sketch of a pair of oil-filled droplets in water, to which complementary strands of ssDNA oligonucleotides are attached: The surfaces of the droplets are composed by single-tail surfactant molecules like lipids with a hydrophilic head on the outside and a hydrophobic tail on the inside, thus forming a boundary for the oil-in-water droplet. By adding some single-strand DNA to the surface of a droplet, it can be ensured that only desired connections to specific other droplets with just the complementary single-strand DNA can be formed. Please note that the connection of the droplets in this picture is overenlarged in relation to the size of the droplets. In realiter, the droplets have a radius of 1–50 µm, whereas a base pair of a nucleic acid is around 0.34 nm in length [1], such that the sticks of connecting DNA strands are roughly 5 nm long.

processes. Mathematically speaking, the agglomeration forms a graph with the droplets being the nodes and the connections being the edges.

1.2 Gillespie Algorithm

While reaction kinetics is often modeled with ordinary differential equations, in solutions in organic chemistry the number of molecules is too small to justify the assumption of continuous concentrations. Therefore, we use Gillespie's stochastic simulation method [4], in a way called the 'direct method' [2], which we want to briefly describe here: Gillespie's stochastic simulation method deals with the scenario in which there are various chemical reactions possible in the system. For each possible reaction i, a so-called propensity $p(i)$ has to be calculated. The simplest possible reaction is the unimolecular reaction with one educt A only:

$$A \longrightarrow \text{product(s)}$$

In order to calculate the propensity of this reaction i, one needs the number $N(A)$ of molecules of type A and its reaction constant $k(i)$. Then the propensity of this unimolecular reaction is given by

$$p(i) = N(A) \times k(i). \tag{1}$$

For a bimolecular reaction j with two different educts A and B, i.e.

$$A + B \longrightarrow \text{product(s)},$$

the propensity is given by

$$p(j) = N(A) \times N(B) \times k(j)/V. \tag{2}$$

Here the number of possible combinations of an A and a B molecule is multiplied with the reaction constant $k(j)$, which has to be renormalized by the volume V of the container [2]: The larger the volume, the less likely it is that an A molecule comes near a B molecule. To simplify matters, we set all $k(i) \equiv k = 1$ and, as we work with a monodisperse system in which each droplet has the same volume $V(i) \equiv V$, we set $V = 1$. According to [2], the transport of a molecule can be treated like a unimolecular reaction. In the next step, one needs to sum up all n propensities into an overall propensity P, $P = \sum_{i=1}^{n} p(i)$. Then one needs to choose two independent random numbers r and z, which are uniformly chosen from the interval $(0; 1)$. From z, one derives an exponentially distributed random number e by setting $e = -\ln(z)$. The time interval between the previous reaction and the current reaction is given by e/P. While z and e are used to determine the time in which the new reaction takes place, the other random number r is used to determine which of the reactions is to be chosen. That reaction i is chosen for which the condition

$$\sum_{j=1}^{i-1} p(i) < r \times P \leq \sum_{j=1}^{i} p(i) \tag{3}$$

holds. This way, a series of reactions can be created with the Gillespie algorithm, taking care of the correct time scale and the different propensities of the various reactions. The algorithm stops if the overall propensity P is zero, meaning that there is no further reaction possible, or if a predefined maximum amount of time has evolved [2].

This Gillespie algorithm can be easily extended to a system like ours with various containers, so that all possible reactions within each container and all possible transport processes between each pair of containers are represented in the $p(i)$, which are summed up in P. With the random number r, one selects the respective $p(i)$ to determine not only the reaction but also the container or the transport process between a pair of containers.

In the next sections, we study various scenarios for connections between the droplets in the spatial arrangement shown in Fig. 1 for simple reaction schemes.

2 Reaction with Two Educts

Table 1. Properties of the networks, as described in scenarios 1 and 2 in Sect. 2

	scenario 1	scenario 2
number of edges	8983	4495
maximum degree of a node	12	10
mean degree of a node	5.98	2.99
number of clusters	1	114
maximum cluster size	3000	2876

The simplest bimolecular reaction is

$$A + B \longrightarrow AB.$$

This reaction, which shall be studied in this section, has two educt molecules A and B and one product molecule $A - B$, shortly written as AB. In this section, we do not allow any further reactions. Additionally to this bimolecular reaction, we have transport processes of the three different molecules A, B, and AB. The overall number of transport processes for a droplet i is thus three times the number of droplets droplet i is attached to. In order to get a well-stirred agglomeration of droplets, we might want to create connections between every pair of neighboring droplets. This scenario, to which we want to refer to as scenario 1 in this section, should lead to a fast stirring and a fast reaction process. But one might wonder whether it would not be better to allow only connections between droplets which initially only contain A molecules and those which initially only contain B molecules. Then the molecules would not need to jump first to other molecules of their own kind but can directly move into those containers which initially contain only their desired reaction partners for the reaction process $A + B \rightarrow AB$. Thus, this scenario 2 might lead to an even faster reaction process.

Using the droplet agglomeration comprised of 1,500 droplets filled with 1,000 molecules A and 1,500 droplets filled with 1,000 molecules B in Fig. 1, we create a network each for these two scenarios and perform a network analysis, with the most important results given in Table 1. We find that at least one droplet has indeed the maximum possible number of neighboring droplets in scenario 1, which equals the kissing number in three dimensions [6,10], but the mean number of droplets a droplet is attached to is only roughly 6 in scenario 1 and 3 in scenario 2. In scenario 1, all 3,000 droplets form one cluster, whereas there is a large dominating cluster containing most of the droplets, some unconnected droplets, and a few small separate clusters in scenario 2. (Please note that for other agglomerations, also for scenario 1, the largest cluster might not contain all droplets.)

In the next step, we have a look at the time evolution of the reaction $A+B \longrightarrow AB$ in these two scenarios, which is shown in Fig. 4. For comparison, we also show the reaction dynamics, if we would throw all molecules into a single pot with a volume of 3,000 times the volume of one droplet. In order to study the dynamics also at short time scales, all diagrams are drawn with a logarithmic time axis. We find sigmoidal decreases of $N(A)$ and $N(B)$ (the curves for the time evolutions of $N(A)$ and $N(B)$ coincide), while the number $N(AB)$ of the product molecule increases correspondingly. The pot scenario depicts the fastest dynamics by far. The more elaborate scenario 2, which we hoped would generate a faster dynamics than scenario 1, turns out to be equally fast only at short time scales, for medium time scales it slows down. And even worse, as some of the droplets are not connected to the largest cluster, not all educts are able to react, such that the yield is also smaller when using this scenario. Summarizing, we have to state that according to the results presented so far, the ACDC technology seems to be disadvantageous compared to a well stirred pot and trying to improve the method with a seemingly clever ansatz even leads to a significant deterioration.

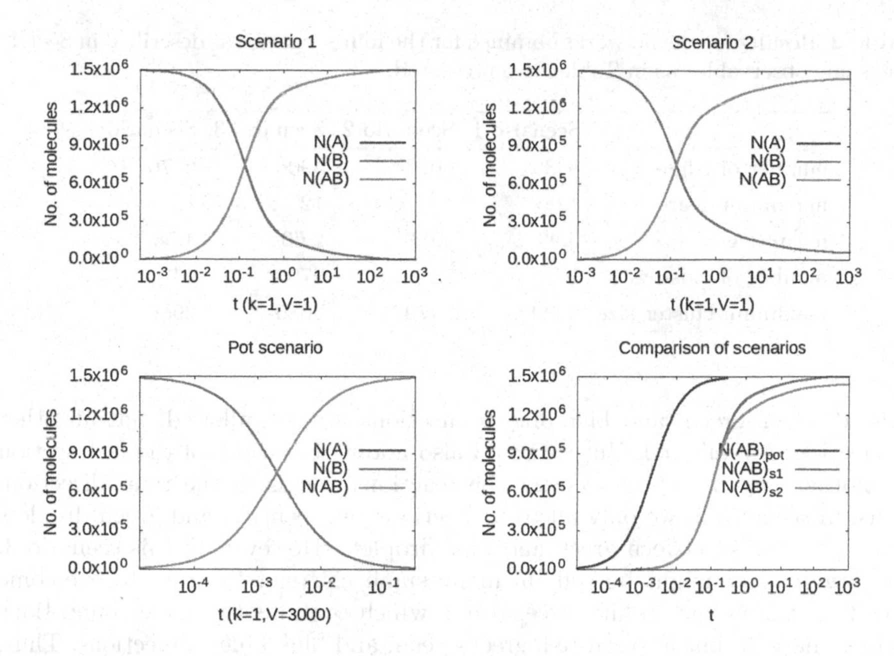

Fig. 4. Results for the time evolution of the reaction process $A+B \longrightarrow AB$, for scenario 1 (top left), scenario 2 (top right), and the pot scenario (bottom left), as described in the text. The picture at the bottom right replots the increases of the numbers of AB molecules for the three scenarios for a better comparison.

3 Reaction with Three Educts and One Desired Product

The next more complex example for a reaction scheme would of course involve three different types of educts A, B, and C to create a molecule $A - B - C$ (short ABC) via a gradual reaction scheme in two steps. Hereby we allow four reactions as follows:

$$A + B \longrightarrow AB$$
$$B + C \longrightarrow BC$$
$$AB + C \longrightarrow ABC$$
$$A + BC \longrightarrow ABC$$

As a starting point, we use the right graphic in Fig. 1, with 1,000 red droplets initially filled with 1,000 molecules of type A, 1,000 green droplets initially filled with 1,000 molecules of type B, and 1,000 blue droplets initially filled with 1,000 molecules of type C. Also here we investigate various scenarios: In scenario 1, each droplet is connected with each of its neighbors as in the previous section. For scenario 2, we want to generalize the scenario 2 of the last section. Thus, in scenario 2, each droplet is connected to neighbors with a different color, i.e.,

Table 2. Results for the networks obtained for the four scenarios as described in Sect. 3. The same observables as in Table 1 are presented.

	Scenario 1	Scenario 2	Scenario 3	Scenario 4
number of edges	8983	6013	3906	6876
maximum degree	12	12	12	12
mean degree	5.98	4.00	2.60	4.58
number of clusters	1	22	270	14
maximum cluster size	3000	2979	2680	2986

red-red, green-green, and blue-blue connections are not allowed, but all other connections are allowed. But we could also adopt scenario 2 of the last section in another way by trying to steer the reaction process in the right direction. Thus, in scenario 3, we only allow connections between red and green droplets and connections between green and blue droplets. However, in this scenario 3, the large cluster might be split in many small clusters. In order to overcome this disadvantage, we define a scenario 4, which contains the same connections as in scenario 3, but also red-red, green-green, and blue-blue connections. Thus, in scenario 4, only red-blue connections are forbidden. Again we perform a network analysis and show the most important results in Table 2. We find that the seemingly cleverest scenario 3 contains the smallest number of edges, but also the largest number of smaller sized clusters.

Figure 5 displays the time evolution of the numbers of molecules for the three educts A, B, and C, for the intermediary products AB and BC, and for the final product ABC, for the scenarios described above and also, for comparison, for the pot scenario again. We generally find a sigmoidal decrease for all three educts, with the decrease of educt B being much faster than those of educts A and C, which coincide. This deviation can be easily explained, as molecule B is part of both reaction processes leading to intermediary products. The two intermediary products show an almost identical rise and decrease again. When comparing the sigmoidal increase of the number of ABC molecules for the various scenarios, we find that of course again the well-stirred pot is much faster than any of the ACDC technology scenarios. Furthermore, we find for short time scales that scenarios 1 and 2 as well as scenarios 3 and 4 exhibit the same dynamics, while at medium time scales, scenario 4 approaches the dynamics of scenarios 1 and 2. The optimum yield is of course only obtained with scenario 1, but it takes a long time for the last educts to move around in the labyrinth of the agglomeration until a partner molecule for a reaction can be found. The worst yield is obtained with scenario 3, which again was originally designed to help and accelerate the gradual reaction processes. Summarizing, also here we have to state that the application of the ACDC technology seems to be disadvantageous when compared to a well-stirred pot and the attempt to improve it leads to a significant deterioration again.

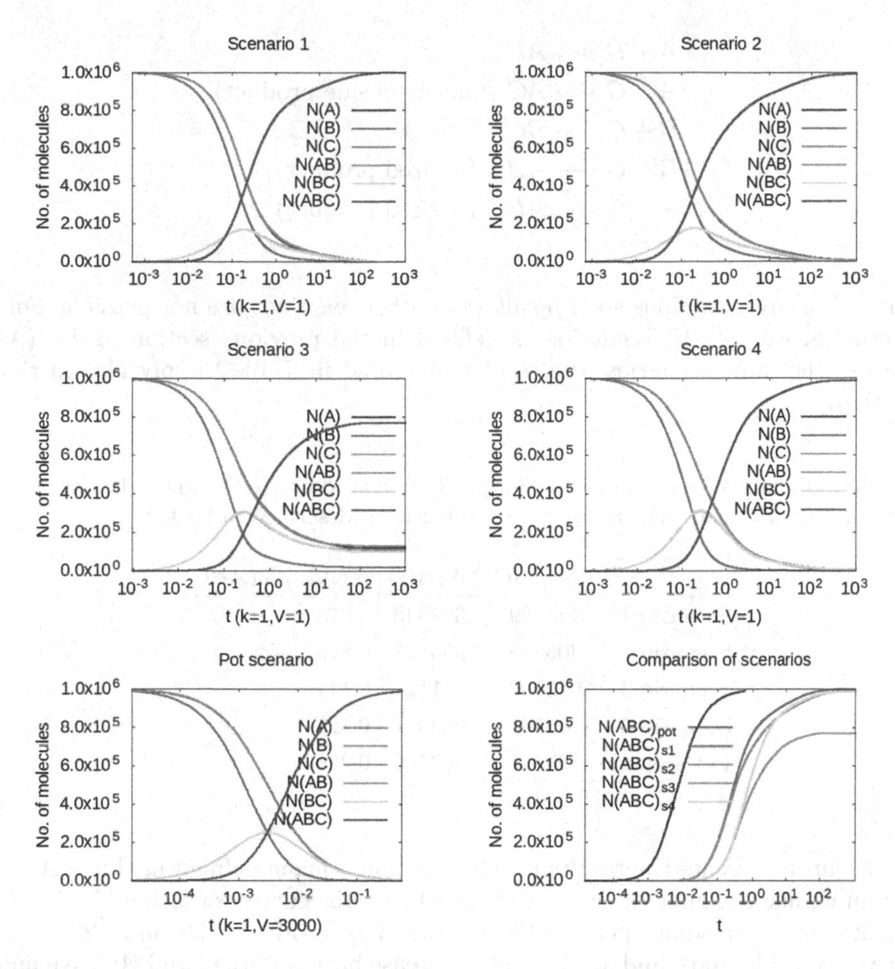

Fig. 5. Results for the time evolution of the gradual chemical reaction scheme for the scenarios described in Sect. 3. The overall numbers of the educt molecules A, B, and C, of the intermediary products AB and BC, and of the final product ABC exhibit a rich behavior. Like in Fig. 4, the graphic at the bottom right displays a comparison of the dynamics of the various scenarios by replotting the sigmoidal increase of the overall number of product molecules.

4 Reaction Scheme with Three Educts, One Desired Product, and One Undesired Side Product

In the next step, we extend slightly the reaction scheme presented in the last section with a further reaction, leading to an undesired side product AC. Thus, we have:

$$\begin{aligned}
A + B &\longrightarrow AB \\
A + C &\longrightarrow AC \text{ (undesired side product)} \\
B + C &\longrightarrow BC \\
AB + C &\longrightarrow ABC \text{ (desired product)} \\
A + BC &\longrightarrow ABC \text{ (desired product)}
\end{aligned}$$

Only these five reactions shall be allowed, other reactions are not possible. Furthermore, we use the scenarios as defined in the previous section again. (As we use the same scenarios, the results presented in Table 2 apply also in this section.)

Table 3. Final numbers of molecules for the desired product ABC and for the undesired side product AC and their ratio for the scenarios as described in Sect. 4.

	$N(ABC)$	$N(AC)$	$N(AC)/N(ABC)$
Scenario 1	408769	357343	0.874
Scenario 2	406865	355197	0.873
Scenario 3	640700	71112	0.111
Scenario 4	798561	96804	0.121
Pot scenario	399811	367796	0.920

Figure 6 presents the results for the reaction scheme defined in this section. Again we find a sigmoidal decrease of the educts (the curves for A and C coincide again), an intermediate peak of the intermediary products AB and BC (their curves coincide, too), and an sigmoidal increase both for ABC and AC. We find that the introduction of an undesired side product changes the final outcomes for the scenarios dramatically. Still, the pot scenario leads to the fastest dynamics. But all scenarios using the ACDC technology lead to a larger yield of the desired product ABC. When we aim at maximizing the number of ABC molecules, the best results are obtained with scenario 4, followed by scenario 3. The exact final values for the numbers of molecules of the desired product ABC and the undesired side product AC are provided in Table 3, together with the ratio between them, as instead of maximizing $N(ABC)$, one might want to aim at minimizing the ratio between the undesired side product and the desired product. While scenario 4 leads to the by far largest yield of the desired product ABC, scenario 3 provides a slightly better ratio. The question which of these two scenarios shall be chosen depends on the question how important the condition to minimize the ratio is.

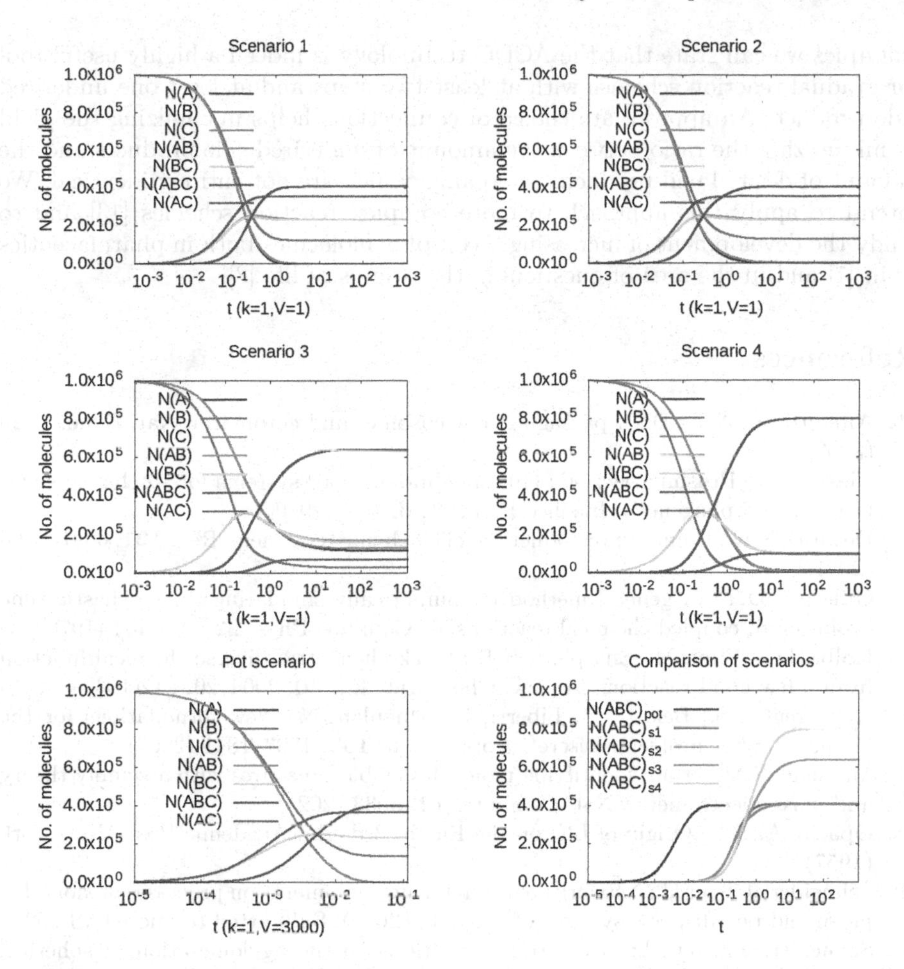

Fig. 6. Results for the reaction processes defined in Sect. 4. The results are shown in the same way as in Fig. 5 for the reaction processes of Sect. 3, but additionally with curves for the side product AC.

5 Conclusion and Outlook

In this paper, we applied some small toy instances of artificial chemistries to a network of droplets, which are initially filled with only one chemical each. In comparison to a well-stirred pot, in which one would usually put all the ingredients for the desired gradual reaction scheme, such an agglomeration of droplets allows for some additional steering possibilities for the reaction scheme by restricting the types of connections in the network via which molecules can move from one droplet to another one. While this network approach leads to a much slower dynamics than the well-stirred pot, one can show that an appropriate choice of connections can lead to much better results, if undesired side products occur in the reaction scheme. Summarizing the results for these simple examples we can state that the ACDC technology is indeed a highly useful tool for gradual reaction schemes with at least two steps and at least one undesired side product. An appropriate choice of connections helps maximizing the yield or minimizing the ratio between the amount of undesired side products and the amount of desired end products – two aims which are not entirely identical. We intend to apply this approach to more complex reaction schemes [13] and to study the development of increasingly complex molecules both in pharmaceutics as in [7] and in the area of questioning the origins of life [8].

References

1. Annunziato, A.T.: DNA packaging: nucleosomes and chromatin. Nat. Educ. **1**, 26 (2008)
2. Cieslak, M., Prusinkiewicz, P.: Gillespie-lindenmayer systems for stochastic simulation of morphogenesis. In silico Plants 1, diz009 (2019)
3. Devaraj, N.K., Finn, M.: Introduction: click chemistry. Chem. Rev. **121**, 6697–6698 (2021)
4. Gillespie, D.T.: A general method for numerically simulating the stochastic time evolution of coupled chemical reactions. J. Comput. Phys. **22**, 403–434 (1976)
5. Kolb, H.C., Finn, M., Sharpless, K.B.: Click chemistry: diverse chemical function from a few good reactions. Angew. Chem. Int. Ed. **40**, 2004–2021 (2001)
6. Kucherenko, S., Belotti, P., Liberti, L., Maculan, N.: New formulations for the kissing number problem. Discret. Appl. Math. **155**, 1837–1841 (2007)
7. Marshall, S.M., et al.: Identifying molecules as biosignatures with assembly theory and mass spectrometry. Nat. Commun. **12**, 3033 (2021)
8. Oparin, A.: The Origin of Life on the Earth, 3rd edn. Academic Press, New York (1957)
9. Schneider, J.J., et al.: Network creation during agglomeration processes of monodisperse and polydisperse systems of droplets (2022). Submitted to microTAS 2022
10. Schneider, J.J., et al.: Geometric restrictions to the agglomeration of spherical particles. In: Schneider, J.J., Weyland, M.S., Flumini, D., Füchslin, R.M. (eds.) WIVACE 2021. CCIS, vol. 1722, pp. 72–84. Springer, Cham (2022). https://doi.org/10.1007/978-3-031-23929-8_7
11. Schneider, J.J., et al.: Influence of the geometry on the agglomeration of a polydisperse binary system of spherical particles. In: The 2021 Conference on Artificial Life, ALIFE 2021 (2021). https://doi.org/10.1162/isal_a_00392
12. Song, L., Hobaugh, M.R., Shustak, C., Cheley, S., Bayley, H., Gouaux, J.E.: Structure of staphylococcal α-hemolysin, a heptameric transmembrane pore. Science **274**(5294), 1859–1865 (1996). Structure was uploaded by the authors on https://www.rcsb.org/structure/7ahl and was visualized using Chimera https://www.cgl.ucsf.edu/chimera
13. Weyland, M.S., et al.: The MATCHIT automaton: exploiting compartmentalization for the synthesis of branched polymers. Comput. Math. Methods Med. **2013**, 467428 (2013)

Modelling Wet-Dry Cycles in the Binary Polymer Model

Federica Senatore[1], Roberto Serra[1,2,3], and Marco Villani[1,2(✉)]

[1] Department of Physics, Informatics and Mathematics, Modena and Reggio Emilia University, Modena, Italy
marco.villani@unimore.it
[2] European Centre for Living Technology, Venice, Italy
[3] Institute of Advanced Studies, University of Amsterdam, Amsterdam, The Netherlands

Abstract. A key question concerning the origin of life is whether polymers, such as nucleic acids and proteins, can spontaneously form in prebiotic conditions. Several studies have shown that, by alternating (i) a phase in which a system is in a water-rich condition and (ii) one in which there is a relatively small amount of water, it is possible to achieve polymerization. It can be argued that such "wet-dry" cycles might have actually taken place in the primordial Earth, for example in volcanic lakes. In this paper, using a version of the binary polymer model without catalysis, we have simulated wet and dry cycles to determine the effectiveness of polymerization under these conditions. By observing the behavior of some key variables (e.g., the number of different chemical species which appeared at least once and the maximum length of the species currently present in the system) it is possible to see that the alternation of wet and dry conditions can indeed allow a wider exploration of different chemical species when compared to constant conditions.

Keywords: Protocells · Origin of life · Gillespie algorithm · Semipermeable membrane

1 Introduction

Polymers play key roles in life, and it suffices to recall the ubiquitous presence of nucleic acids and proteins in living beings to be sure about their importance. So a key question concerning the origin of life (OoL) [10, 16, 18, 31] is whether such polymers can spontaneously form in prebiotic conditions. Studies spanning several decades have shown that there are plausible routes to the formation of key monomers (the basic building blocks of polymers) as aminoacids and nucleobases, but their polymerization is not a straightforward consequence of the presence of monomers [16, 19–21], although it is possible [3]. For example, the formation of a peptide bond linking two aminoacids implies the release of a water molecule, so it is unfavored in water solutions.

In this paper we will be mainly concerned with polypeptides, which can be qualitatively classified as oligopeptides, with a few monomers, and long polypeptides such as

C. De Stefano et al. (Eds.): WIVACE 2022, CCIS 1780, pp. 119–129, 2023.
https://doi.org/10.1007/978-3-031-31183-3_10

proteins. Several empirical tests have shown that an effective way to achieve polymerization is by putting the system through wet-dry cycles, where a water-rich (i.e. "wet") condition is followed by a drier one ("dry"), which is then followed by another wet period, a.s.o. [2, 23]. One may observe that a totally dry system is likely to be inert, but it suffices that in the "dry" phase the water content be small to allow some reactions to take place (so, the dry phase is really a dampish one; however, we will continue to use the usual "wet-dry" phrasing). Wet-dry cycles might easily have taken place on the primeval earth in warm ponds, where the energy for drying was supplied by the sun, or in thermal pools and near volcanic lakes, where the energy came from terrestrial sources [2, 24].

The effectiveness of polymerization can be analyzed by looking at different variables; the more relevant for the OoL are (i) the number of different molecular species that are "discovered" and (ii) the length of the longer polymers obtained. Indeed, the most important property of polypeptides is their catalytic activity (enzymes) and, while a few oligopeptides can catalyze some reactions, most catalysts are long peptides – therefore the ability to synthesize long polypeptides matters. The importance of generating a high diversity of molecular types, for the development of self-replicating sets of molecules, is also evident.

In order to study the effects of various choices of parameter values and cycling strategies on the effectiveness of polymerization, we have applied a well-known model, i.e. the binary polymer model (BPM) introduced by Stuart Kauffman [14] several years ago. The model, described in detail in Sect. 2, concerns the formation of "informational" polymers, whose identity is specified by the succession of their monomers. For the sake of simplicity, only two types of monomers are considered (when thinking of real systems, the two types might be identified with some classes of aminoacids, e.g. polar or apolar ones, rather than with the 20 present-day aminoacids). The polymers in BPM are strings of monomers, and they interact according to the law of mass action: two types of reactions can take place, i.e. cleavage, where a molecular chain is broken into two smaller ones, and condensation, where two molecules are chained to produce a longer one. The original molecules are usually at least partially random, and the evolution of the population of molecules is followed in time.

In the original BPM, condensations and cleavages can take place only when catalyzed by an already existing polymer. While some alternatives have been explored, in most cases it is assumed that each polymer (longer than a minimum threshold) has a chance to catalyze any reaction. The set of pairs {catalyst, catalyzed reaction} defines a particular "artificial chemistry" and the behaviour of the system is analyzed with statistical methods, either varying parameters and initial composition in a specific chemistry or averaging over different chemistries. Several studies of the BPM have provided important results, concerning e.g. the role of collectively autocatalytic sets [4, 5, 13] and RAF sets [11, 28].

However, the fact that catalyzed reactions only are allowed may be a problem here, since we are interested in the appearance of long polymers, while shorter ones are supposed not to have catalytic properties. Therefore we have introduced a non-catalyzed form of the model, which we call the NCBPM, where condensations and cleavages can take place without any support from a catalytic molecule. When applications to

OoL scenarios is concerned, we may suppose that there is a widespread low-level of catalysis, which might be provided e.g. by surfaces, clays, metal ions or membranes. This is however not explicitly dealt with in the model, where it shows up in nonvanishing reaction rates in the absence of polymer catalysts. One key question then concerns under which conditions long polymers can form, and how they are affected by wet-dry cycling.

It is necessary to define the environment where the reactions take place. While a simple case like that of a closed container can provide some clues about the behavior of the system, the most interesting case within the OoL scenario is that of a protocell with semipermeable barriers immersed in a large container, think e.g. of a lake or a pond, which can vary from wet to dry (dampish). These changes can be modelled by assigning different kinetic constants in the two cases; for simplicity, in this paper we will assume that in each condition all condensations have the same constant, and all cleavages have the same constant. Of course, in order to determine the reaction rates, in the former case the constant is multiplied times the product of the concentrations of the two condensing molecules, while in the latter case the constant is multiplied times the concentration of the cleaved molecule.

So, wet-dry cycles are straightforwardly modelled by different values for the constants of all the cleavages and the constants of all the condensations. In this way it is possible to test whether wet-dry cycles can provide an advantage with respect to a uniform situation, and under which conditions.

The semipermeable protocell model, and the way in which wet-dry conditions are simulated, are described in Sect. 3, while in Sect. 4 the main results are presented. As usual, some comments and suggestions for further work are found in the final section.

2 The NCBPM Model

The model is inspired by the "binary polymer model" (BPM) introduced in [14, 15]. The entities of the model are the different chemical species. Species are oriented linear chains composed by symbols taken from the alphabet A = {A, B} ("bases" in the following). We call "monomers" the species formed by a single symbol, while "polymers" those formed by two or more symbols. For simplicity, we assume below that monomers A and B have the same properties and that the energy required to produce AAB is the same as to produce BBA (there are no differences between chemical bonds linking different types of monomers).

In this model, two different kinds of reaction (which occur without catalysis) are defined:

1. condensation, i.e. two reactants are concatenated together into a longer product (e.g. $A + B \rightarrow AB$);
2. cleavage, i.e. a single species is cut into two shorter ones (e.g. $AB \rightarrow A + B$);

Note that different reactions can create the same chemical species (e.g., $B + BA \rightarrow BBA$ and $ABBA \rightarrow A + BBA$). It can also be noted that in the overall system condensations and cleavages have both constructive and destructive functions. A cleavage obviously eliminates a molecule of the reacting species, but produces two new molecules (which cause an increase in the total concentration of species). A condensation produces a new molecule of a species different from those of the reactants, but at the cost of a decrease in the reactants' concentration. Consequently, the overall dynamics of heterogeneous mixtures is not obvious.

We assume that all possible reactions can occur; the reaction rates depend on the concentration of the reagent (i.e., cleavages), or in the case of two reagents (i.e. condensations), on the product of the concentrations of the reactants.

If $\boldsymbol{N}$ is the total number of different types of chemical species present in the system at a given time t and L_i is the length of species i, then the total number $\boldsymbol{R}$ of conceivable reactions at a given time t is given by:

$$R = \sum\nolimits_{i=1}^{N} (L_i - 1) + N^2 \tag{1}$$

Therefore, as the number N of existing species increases, the number of all possible reactions increases more than linearly, as observed also in [6, 7, 15].

Since we are interested in small-sized systems characterized by low concentrations (and therefore a small number of molecules) – a situation typical of OoL scenarios -, the model is simulated using the well-known Gillespie algorithm [8, 9] according to which, at each simulation step, a single reaction occurs and so there is an asynchronous stochastic update of the concentrations of the species. For simplicity, in this work we assume that all condensations have the same kinetic coefficient K_{cond}, and all cleavages have the same kinetic coefficient K_{cl}. The values of these coefficients will define the wet and dry conditions used in this work. The choice of parameters is largely arbitrary: we have used some sets of parameters and verified that the qualitative trends described here were also present in different sets of parameters.

Given the reaction constants and the number of molecules of each reactant, the algorithm computes:

- the time at which the next reaction will occur;
- the probability of occurrence in the next time interval of each reaction, so that the choice of the reaction which actually takes place is made proportionally to these probabilities.

After having chosen the reaction to be performed, the algorithm updates both the time and the number of molecules according to it.

3 The Semipermeable Protocell Model

Small flow reactors have also been proposed as protocell models [12, 29], but they are indeed not well adequate in modelling some interesting phenomena, since (i) they require a constant inflow that has no physical analogue in a vesicle, where the flows are affected by internal chemical phenomena, and (ii) they usually allow the outflow of all the solutes. In this work, as already done in [26, 27, 30], we introduce a semipermeable membrane that separates the reaction network from the external environment. We emphasize the importance of this case in the OoL scenario, as an essential basis for modeling protocellular systems [2, 22, 25, 26].

The semi-permeable membrane is here modelled by allowing only some species to enter and leave the protocell, namely those that are shorter than an arbitrary length L_{perm}. All the species that are longer than L_{perm} either remain entrapped within the protocell or never enter it from the outside (in the example of Fig. 1, only the species shorter than three bases can cross the membrane). Species having different structure or length can have different permeability properties; however, these differences involve only quantitative and not qualitative changes, and will therefore be ignored in the following.

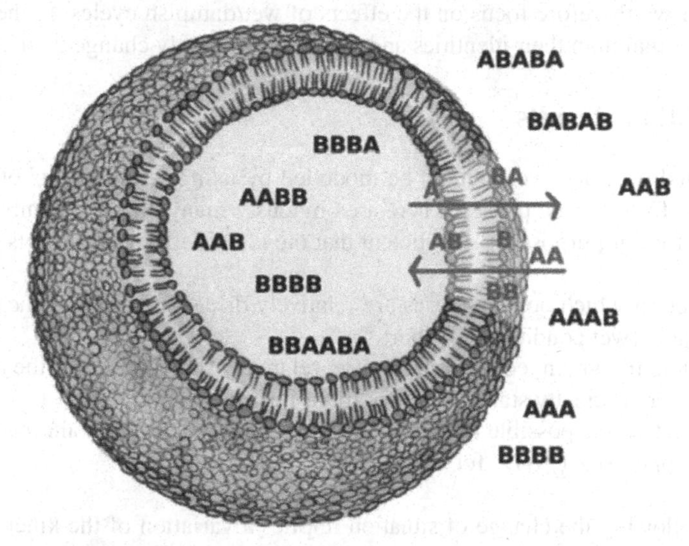

Fig. 1. Schematic representation of the semi-permeable membrane (from [27]). The membrane is here represented as a lipid bilayer that shapes a spherical vesicle. In this example, only the species shorter than three bases can cross the membrane.

The concentration of the permeable molecules is homogeneous both inside and outside the protocell, i.e., we assume infinitely fast diffusion in both aqueous phases. Transmembrane diffusion takes place at a finite rate according to Fick's law, i.e., in a way proportional to the concentration difference of the permeable species [1]. In this work the volume of the protocell is assumed to be constant, so the contribution of the surface to the transmembrane solute fluxes can be incorporated into the membrane permeability coefficient.

Therefore:

$$\frac{dC_i}{dt} = K_{diff}\left(\left[C_i^{out}\right] - \left[C_i^{in}\right]\right) \tag{2}$$

where dC_i/dt is the rate of intake of the chemical i (number of molecules), K_{diff} is proportional to its transmembrane diffusion coefficient (which for simplicity is supposed to have the same value for all chemical species) divided by the (constant) membrane thickness, and $[C_i^{out}]$ and $[C_i^{in}]$ are the concentrations of the chemical i outside and inside the protocell, respectively.

In Fig. 1 the shape of the protocell is spherical; let us remember that we assume a constant volume. These conditions are not verified in extreme scenarios such as those present in [2], where there are situations in which the membranes rupture. However, it is possible to hypothesize less extreme scenarios, in which the protocells are able to withstand osmotic stress without breaking [2, 17] and therefore are able to maintain their own identity. On the other hand, it is possible to modify our current model by introducing the possibility of ruptures and subsequent recompositions: yet, this activity requires the study of populations of objects, and we therefore postpone these studies in future works. In this paper we therefore focus on the effects of wet/dampish cycles, in the case that the protocells maintain their identities and do not significantly change their volume.

4 Simulation Results

As anticipated, wet-dry cycles could be modelled by using different sets of values of the constants for different phases. There are obviously many possible combinations of parameters; for our purposes it is sufficient that the identified parameter sets represent:

1. a situation in which condensations are relatively disadvantaged and the molecules are unstable (wet condition, for short "W").
2. a situation in which condensations are relatively favoured and the molecules produced are typically stable (dry condition, for short "D")
3. consequently, it is possible to realize situations where there is an alternation of wet and dry conditions ("WD" for short)

In the following the change of situation implies a variation of the kinetic constant of three orders of magnitude for both groups of reactions (condensation and cleavages), but we have verified that lower or higher variations maintain the trends that we will comment on below. For simplicity, in all cases the transmembrane diffusion coefficient is kept constant.

The starting situation consists of a protocell with only monomers inside, in equilibrium with the external environment (in which the concentration of these species is supposed to be constant); only monomers can cross the membrane. All conceivable reactions are feasible.

At each step, the Gillespie algorithm chooses the next reaction and quantifies the time interval to the next step: if the species produced is not already present in the system, a molecule of this species is produced and the list of existing species is updated; if the species produced is already present, the quantity of its molecules is updated. The quantities of the reagents are decreased, and it is possible that some species be completely consumed: in this case their quantity becomes zero. They will come back into play only if some reaction produces a molecule of theirs again.

It is therefore interesting to study, among others, the following variables:

- the number of existing species
- the number of species that appeared at least once during the dynamics; this variable provides a measure of the breadth of the portion of the chemical space explored by the system

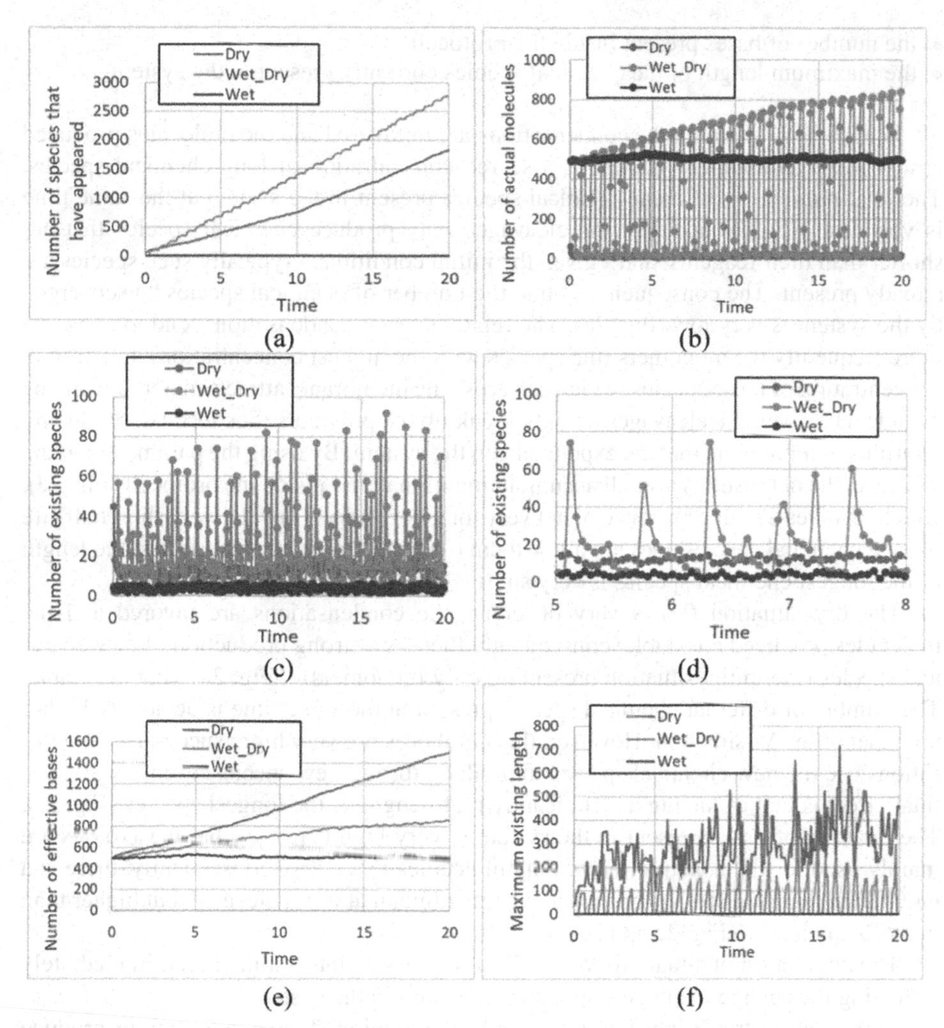

Fig. 2. Behavior of protocells in different phases (wet, dry and alternance WD in each image - time: arbitrary units). (a) Number of different chemical species discovered inside the container during the evolution of the system. (b) Number of molecules (the sum of the molecules of all the chemical species) present inside a protocell. The volume of a protocell does not change, so rather than the concentration we prefer to directly indicate the number of molecules (c) Number of different chemical species present at the same time within a protocell. (d) An enlargement of the previous image. (e) Number of bases present inside a protocell. (f) Length of the longest chemical species presents within the protocell. The Gillespie coefficients of the condensations and of the cleavages in the W phase are respectively $K_{cond} = 3.16 \cdot 10^{-4}$ and $K_{cl} = 3.16 \cdot 10^{1}$, in the D phase $K_{cond} = 3.16 \cdot 10^{-1}$ and $K_{cl} = 3.16 \cdot 10^{-2}$. The change of situation therefore implies a variation of three orders of magnitude for both groups of reactions. In all cases the transmembrane diffusion coefficient is kept constant ($K_{diff} = 0.1$); initially the protocells have inside only monomers (250 A and 250 B), in equilibrium with the external environment.

- the number of bases present inside the protocell
- the maximum length of the chemical species currently present in the system

In a wet (W) situation the condensations are unfavored and the molecules produced easily break: consequently, there is a scarce concentration of long chemical species. The total number of different chemical species present in the system at the same time is very low (Fig. 2c and 2d): the cleavages only produce chemical species that are shorter than their reagents, and - given the initial conditions - typically such species are already present. The consequence is that the number of chemical species "discovered" by the system is very low (Fig. 2a). The relatively rare condensations tend to consume more frequently the monomers (the species with the highest concentration): the drop in concentration of these species, which can cross the membrane, attracts material from the outside. However, the cleavages quickly break up the polymers thus formed, producing a surplus of monomers that are expelled from the system. By using the parameters of the wet case, the net result is a small accumulation of material inside the protocell (Fig. 2e), which involves a scarce presence of not very long polymers. The relatively short half-life of long chemical species does not allow them to produce even longer species: the length of the longest chemical species is very short (Fig. 2f).

The dry situation (D) is very different: the condensations are favored and the molecules produced are stable: consequently there is a strong production of new chemical species (the initial situation presenting only monomers) – Fig. 2a - that are stable. The number of different chemical species present at the same time is decidedly higher than that of the W situation. However, this number is not very high: successive condensations use the new chemical species to produce further new species, even longer. The final effect is that of having a significantly high length of the longest species (Fig. 2f). The quantity of bases present in the system is very high (Fig. 2a), but these bases are mainly used for the construction of long molecules (Fig. 2b) - consequently, there is a number of chemical species present at the same time that is not elevated, but higher than that of situation W (Fig. 2c and 2d).

The alternation of situations W and D introduces an interesting effect: immediately following the change many new species are created in the system.

In the case of transition from situation W to situation D the condensations produce a large amount of new chemical species. The new species are used to produce further (and longer) new species, but the intermediate lengths are not immediately depleted, because within the protocell there is an abundance of short materials that allow the creation of further species. There is therefore a relatively long phase in which there is a coexistence of a large number of different chemical species - a number by far greater than the number of species simultaneously present during the dry phase alone (Fig. 2c and 2d). Continuing phase D would lead in the long run to a "normal dry" situation. But then there is a new alternation.

The transition from situation D to situation W breaks the equilibrium again: immediately following the change the cleavages break the long molecules, generating a large quantity of molecules of shorter lengths. The starting species are few and very long, so initially the probability of generating already present chemical species is very low: there is therefore a phase in which a coexistence of a large number of different chemical species is observed - a number by far greater than the number of species simultaneously present during the dry phase alone (Fig. 2c and 2d). The passage from phase D to phase W tends over time to generate a large number of molecules of species already present (Fig. 2b), thus decreasing the number of chemical species present at the same time (the typical situation W). But then there is a new alternation – and so on.

The alternation of situations W and D therefore presents several aspects at the same time. With the parameters used in Fig. 2, the final overall effect is to allow the system a much more effective exploration compared to the single W and D situations (a higher number of chemical species discovered - Fig. 2a) despite a number of bases lower with respect the situation D (Fig. 2). The maximum length reached, although less than that of situation D, is also considerable (Fig. 2f). So, a continuous alternation seems to be an important discovery factor.

The frequency of the alternation could be also an important element. Indeed, as the frequency changes, so does the effectiveness of the just described effects. The higher the frequency, the greater the number of species discovered (Fig. 3a) and the shorter the species' maximum length (Fig. 3c). It is interesting to note that the number of bases

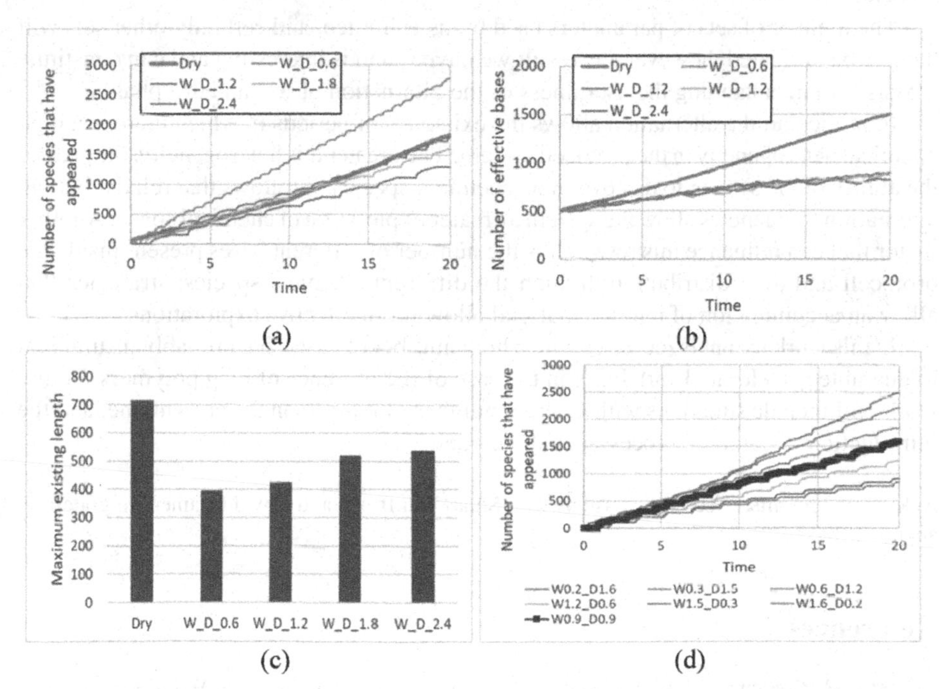

Fig. 3. WD alternation effects. (a) Variation of the alternation frequency, from 0.6 to 2.4 time units (equally divided between phases W and D): effect on the number of different chemical species discovered during the evolution of the protocell. (b) Number of bases present inside the protocell, under the same conditions. (c) Length of the longest chemical species in each situation, compared with the maximum length that appeared using phase D alone. The pure W scenario is not shown, as the maximum length does not exceed a few units. (d) Number of different chemical species discovered during the evolution of the protocell, as the relative length of the W and D phases varies (with the same length of a complete cycle, equal to 1.8 time units). The symmetrical case is highlighted in bold (1.8 time units equally divided between phases W and D).

stored during the alternation (less than the number stored by the D situation alone) does not seem to depend on the frequency of the alternation (Fig. 3b). The relative length of phases W and D (keeping constant the total duration) is also important: increasing the length of phase D with respect to that of phase W implies a greater accumulation of bases within the protocell (data not shown), and a correspondent greater effectiveness in generating new chemical species (Fig. 3d).

5 Discussion

In this work we focused on an important aspect regarding the problem of the Origin of Life (OoL), that is, the effectiveness of polymerization in the period preceding the emergence of the first living forms. Some of the most relevant aspects in this stage are (i) the number of different molecular species that are "discovered" and (ii) the length of the longest polymers that are present. Both aspects are important for the creation of molecules with catalytic activity - which in turn are essential elements for building autocatalytic systems.

The number of sets of parameters used by us is limited, and certainly other sets will have to be evaluated: however, the result we have found is interesting and can constitute a basis for understanding the usefulness of the alternation of wet and dry phases.

In particular, the alternation allows the existence of periods in which there is a high chemical heterogeneity in the protocell system. The greater this heterogeneity, the greater the ability of the system to discover new chemical species: it follows that relatively high alternation frequencies allow the system a greater exploration of chemical space. Another factor that can influence this research is the number of different bases present inside the protocell and their distribution through the different chemical species: strategies that allow an accumulation of internal material allow more effective explorations.

In this work we analyzed cases with a high number of bases, but probably insufficient to guarantee an elevated variability in the case of the presence of long polymers. Future work will include situations with a higher number of bases than the current one, and the introduction of catalytic processes in the model.

Acknowledgments. Useful discussions with Stuart Kauffman and David Deamer are gratefully acknowledged.

References

1. Bird, R.B., Stewart, W.E., Lightfoot, E.N.: Transport Phenomena. John Wiley & Sons, New York, NY, USA (1976)
2. Damer, B., Deamer, D.: The hot spring hypothesis for an origin of life. Astrobiology **20**(4), 429–452 (2020). https://doi.org/10.1089/ast.2019.2045
3. Deamer, D.: Where did life begin? Testing ideas in prebiotic analogue conditions. Life **11**, 134 (2021). https://doi.org/10.3390/life11020134
4. Eigen, M., Schuster, P.: The hypercycle: part B. Naturwissenschaften **65**, 7–41 (1978)
5. Eigen, M., Schuster, P.: The hypercycle. A principle of natural self-organization. Part A: emergence of the hypercycle. Naturwissenschaften **64**(11), 541–65 (1977). PMID: 593400. https://doi.org/10.1007/BF00450633

6. Filisetti, A., Graudenzi, A., Serra, R., et al.: A stochastic model of the emergence of autocatalytic cycles. J. Syst. Chem. **2**, 2 (2011). https://doi.org/10.1186/1759-2208-2-2
7. Filisetti, A., Graudenzi, A., Serra, R., et al.: A stochastic model of autocatalytic reaction networks. Theory Biosci. **131**, 85–93 (2012). https://doi.org/10.1007/s12064-011-0136-x
8. Gillespie, D.T.: A general method for numerically simulating the stochastic time evolution of coupled chemical reactions. J. Comput. Phys. **22**, 403–434 (1976)
9. Gillespie, D.T.: Exact stochastic simulation of coupled chemical reactions. J. Phys. Chem. **81**, 2340–2361 (1977)
10. Haldane, J.B.S.: Origin of life. Ration. Annu. **148**, 3–10 (1929)
11. Hordijk, W., Hein, J., Steel, M.: Autocatalytic sets and the origin of life. Entropy **12**, 1733–1742 (2010)
12. Hordijk, W., Steel, M., Kauffman, S.: The structure of autocatalytic sets: evolvability, enablement, and emergence. Acta Biotheor. **60**(4), 379–392 (2012)
13. Jain, S., Krishna, S.: Autocatalytic sets and the growth of complexity in an evolutionary model. Phys. Rev. Lett. **81**, 5684–5687 (1998)
14. Kauffman, S.A.: Autocatalytic sets of proteins. J. Theor. Biol. **119**(1), 1–24 (1986). https://doi.org/10.1016/s0022-5193(86)80047-9
15. Kauffman, S.A.: At Home in the Universe: The Search for Laws of Self-Organization and Complexity. Oxford University Press, Oxford (1995)
16. Maynard Smith, J., Szathmáry, E.: The Major Transitions in Evolution. Oxford Universi-ty Press, New York NY USA (1995)
17. Mavelli, F., Ruiz-Mirazo, K.: Theoretical conditions for the stationary reproduction of model protocells. Integr. Biol. **5**(2), 324–341 (2013). https://doi.org/10.1039/c2ib20222k
18. Miller, S.L., Orgel, L.E.: The Origins of Life on the Earth. Prentice Hall, New Jersey, USA (1974)
19. Pascal, R., Pross, A., Sutherland, J.D.: Towards an evolutionary theory of the origin of life based on kinetics and thermodynamics. Open Biol. **3**, 130156 (2013). https://doi.org/10.1098/rsob.130156
20. Oparin, A.I.: Proiskhozhdenie zhizni. Izd. Moskovskii Rabochii, Moscow (1924)
21. Oparin, A.I.: Origin of Life. Macmillan, New York (1938)
22. Rasmussen, S., et al. (eds.): Protocells. MIT Press, Cambridge (2008)
23. Ross, D.S., Deamer, D.: Dry/wet cycling and the thermodynamics and kinetics of prebiotic polymer synthesis. Life **6**, 28 (2016). https://doi.org/10.3390/life6030028
24. Ross, D., Deamer, D.: Prebiotic oligomer assembly: what was the energy source? Astrobiology **19**(4), 517–521 (2019). Epub 2019 Feb 1. PMID: 30707599. https://doi.org/10.1089/ast.2018.1918
25. Ruiz-Mirazo, K., Briones, C., de la Escosura, A.: Prebiotic systems chemistry: new perspectives for the origins of life. Chem. Rev. **114**, 285–366 (2014)
26. Serra, R., Villani, M.: Modelling Protocells—The Emergent Synchronization of Reproduction and Molecular Replication; Understanding Complex Systems Series. Springer, Heidelberg (2017). https://doi.org/10.1007/978-94-024-1160-7
27. Serra, R., Villani, M.: Sustainable growth and synchronization in protocell models. Life **9**, 68 (2019). https://doi.org/10.3390/life9030068
28. Steel, M., Hordijk, W., Xavier, J.C.: Autocatalytic networks in biology: structural theory and algorithms. J. R. Soc. Interface **16**, 20180808 (2019)
29. Vasas, V., Fernando, C., Santos, M., Kauffman, S.: Szathmary E evolution before genes. Biol. Direct **7**(1), 1 (2012)
30. Villani, M., Filisetti, A., Graudenzi, A., Damiani, C., Carletti, T., Serra, R.: Growth and division in a dynamic protocell model. Life **4**, 837–864 (2014)
31. Walker, S.I., Davies, P.C.W., Ellis, G.F.R.: From Matter to Life. Cambridge University Press, Cambridge, UK (2017)

On the Growth of Chemical Diversity

Federica Senatore[1], Marco Villani[1,2](✉), and Roberto Serra[1,2,3]

[1] Department of Physics, Informatics and Mathematics, Modena and Reggio Emilia University, Modena, Italy
marco.villani@unimore.it
[2] European Centre for Living Technology, Venice, Italy
[3] Institute of Advanced Studies, University of Amsterdam, Amsterdam, The Netherlands

Abstract. In complex systems that host evolutionary processes, in which entirely new entities may enter the scene, some variables can sometimes show a "hockey-stick" behavior, that is a long period of slow growth followed by an "explosive" increase. The TAP equation was proposed with the aim of describing the growth of the number of different types of entities in systems where new entities (e.g., artifacts) can be created, supposing that they derive from transformations of pre-existing ones. It shows a very interesting divergence in finite times, different from the usual exponential growth where divergence takes place in the infinite time limit. The TAP equation does not deal with the growth of the number of actual types, but rather with the number of the possible ones (the members of the so-called set of Adjacent Possible), and it can therefore overestimate the actual rate of growth. In this paper, we introduce a model (called BPSM, focused on systems that may be relevant for the origin of life) that takes into account the difference between the Adjacent Possible and the set of types that are actually created. Using simulations, it has been observed that the growth of the number of chemical species in the system resembles that of the corresponding TAP equation. Since in this case only combinations of at most two entities can be considered at each time, the TAP equation can be analytically integrated. Its behavior can be then compared to the (necessarily finite) behavior of model simulations; their behaviors turn out to be quite similar, and proper tests are introduced, which show that they differ from the familiar exponential growth. Therefore, the BPSM model provides a description of the rapid increase of diversity which resembles TAP, while based upon the growth of the actual entities rather than on the Adjacent Possible.

Keywords: Innovation · Binary polymer model · Origin of life · Adjacent possible · TAP equation

1 Introduction

Predicting the future is a great intellectual challenge with terrific applications, and science has provided many important results about it. However, precise predictions turn often out to be impossible, because of different reasons which may include incomplete knowledge of the system, random noise, sensitive dependence upon initial conditions, coupling with an unknown external environment, and others. Of course, the impossibility to precisely

C. De Stefano et al. (Eds.): WIVACE 2022, CCIS 1780, pp. 130–140, 2023.
https://doi.org/10.1007/978-3-031-31183-3_11

predict the future states of a system does not prevent knowledge of some of its aspects, like e.g. energy conservation in Hamiltonian systems.

In this paper we will focus on a class of systems which pose formidable challenges to every attempt to prediction, since they host evolutionary processes where entirely new entities may enter the scene. In these cases, changes do not affect only the values of the state variables, but also the variables themselves and their interactions. Good examples include the appearance of new species in an ecosystem [3, 26] or of new artifacts in an economy [2, 4, 14, 16, 17]. Without losing sight of different application domains, we will mainly focus on evolutionary systems which may be relevant for the origin of life (OOL) [19–21, 29], and in particular for the phase of chemical evolution which took place before that of biological evolution [10, 20, 23–25].

While changes in complex systems often take place slowly, interesting phenomena are sometimes observed, where a long-lasting almost stationary state, or one of slow growth, is followed by an "explosive" growth [1, 14], giving rise to a "hockey stick" behavior in the plot of the value of some relevant variable vs. time. Note that the usual exponential growth curve may provide at most an approximate description of this fast phenomenon: exponentials diverge only in the infinite time limit, while the observed behavior resembles more closely a divergence in finite time.

While finite systems never show true divergence to infinity, a model which shows divergence in finite time would be a good candidate to approximately describe the phenomena which are observed, before the limits are reached. In particular, they may give rise to the steep hockey-stick curves which are otherwise quite hard to describe. Let us observe that divergence in finite time is observed in more than linear growth rate equations (e.g. $dx/dt = kx^2$), while the usual exponential growth behaviour ($x \cong \exp(kt)$), the outcome of a linearly growing rate equation like $dx/dt = kx$, never diverges at finite times.

But how is it possible that the growth rate be superlinear? An interesting proposal is that of the TAP (an acronym for Theory of the Adjacent Possible) equation, proposed by [14] with the purpose of illuminating the possible growth of diversity in systems that are able to generate new entities. The first application has been to the increase of artifact diversity (see also [15], but the TAP equation can shed light also on the growth of the number of different living species which is relevant, inter alia, for the OOL [18]. The key idea is that the birth of a new artifact or species is the outcome of some "combination" of already existing ones.

While the interested reader is referred to the original papers for a detailed discussion, the TAP equation is briefly described in Sect. 2, where it is shown that it can actually account for the sudden jump in the number of different types of entities, after a period of slow increase.

In its original form, the TAP equation has some limitations which can be easily cured by suitable modifications. For example, in the original proposal new types of artifacts or species never die, but death terms can be easily added [27] without affecting the main qualitative behaviors.

A more profound aspect needs more attention: the TAP equation does not deal with the way in which the number of actual artifacts changes in time, but rather with the number of possible types of artifacts – no matter whether they are indeed produced or not. So, assuming that time is discrete, according to TAP the number of artifact types at time $t + 1$ does not depend upon the number of actual artifacts at time t (except in the initial $t = 0$ case) but rather on the number of possible artifacts at time t. This is related to the notion of the space of the Adjacent Possible (AP), which comprises all those states which can be reached by the system at hand in one move, since they are "close" to the present state. This notion is in general qualitative, but in the case of the TAP it acquires a precise quantitative meaning: if the novelties at time $t + 1$ are combinations of existing entities at time t, then all the possible novelties can be computed using combinatorial reasoning (one usually imposes a cut-off on the number of different artifacts which can be combined to generate a new one). The TAP equation is obtained by chaining these different APs, and it actually diverges in finite time.

But this provides an overestimation of the number of possible artifacts at times >1, since the combinations which are computed refer to the AP of the previous time step, not to the actual artifacts which do exist (since they have been built) at that time. In order to get a stronger physical grounding, the TAP approach should be complemented by a selection rule to identify the (number of new) entities which are actually produced. If the selection rule were that only a fixed fraction of the possible entities are built, then the overall behavior would still be the same, with just some slow-down effects. But such a simplistic rule does not look convincing in general.

After reviewing the main features of TAP, in this paper we will consider a growth rule for the number of types of entities, which is loosely inspired by the kinetics of chemical species which are supposed to have been present when life appeared on earth (or somewhere else). The entities of interest are (idealized representations of) polymers - indeed most interesting molecules in living beings belong to this class, including proteins and nucleic acids. Our abstract "informational polymers" are linear chains of "monomers", which can be of two different types, say A and B. Starting from an initial random population, new polymers are created by applying to some individuals one of the following two operators:

1. *cleavage*: applied to a single polymer, it generates two new ones, by cutting the parent at a randomly chosen point
2. *condensation*: applied to a couple of existing polymers, it generates a single longer one, by chaining the second to the end of the first one.

The application of cleavage or condensation is called a reaction; in the case of cleavage, there is a reactant and two products, in the case of condensation two reactants and one product. A reaction can generate new molecular types if at least one product is a polymer which was not yet a member of the population.

Like in the TAP equation, our model is concerned with molecular types, without any reference to quantities. A species either exists or not, therefore concentrations are Boolean. The two types of reactions (condensations and cleavages) take place with some probability; for the sake of simplicity, in this paper we choose a single probability for all the condensations and a single probability for all the cleavages. The updating is asynchronous, like in the well-known Gillespie algorithm for the simulation of chemical kinetics, and at each time step a single reaction is chosen among the possible ones, according to its probability, and executed. If one or two new species are generated, they are added to the set of polymers.

The model resembles in many ways the random "binary polymer" model (in the following, BPM) introduced by Stuart Kauffman [12], but departs from it in some key aspects. The BPM deals with both species (i.e., polymer strings) and their quantities, while here, as stated above, we are interested in types only (like in the TAP case) and quantities are either 0 or 1. A modification of the kinetic equations is therefore required: while in chemical kinetics reactants disappear, using binary densities we can either keep all the reactants alive, or make them disappear. Since this latter choice would induce strong instabilities, we prefer the first alternative. A possible physical reason in support of this choice could be inspired by the idea that there are several exemplars of each species, and that just one or a few of them are involved in the reaction which discovers new species. A different possible reason for this choice might be provided by assuming that there is a mechanism which quickly makes copies all the existing polymers.

Note that the action of the cleavage and condensation operators fits perfectly the "combinatorial" hypotheses at the basis of the TAP equation, provided that we restrict the number of "parents" of a novelty to at most two.

In the following Sect. 2 we will summarize the TAP equation, and we will provide an approximate analytical solution (in the continuous time limit) for the case with at most two parents per novelty, showing that it actually leads to divergence in finite time. In Sect. 3 we will introduce our model of interacting polymers, which is inspired by TAP, but distinguishes the space of the actual species from that of the AP. In Sect. 4 we will compare the kinetics of the two cases, and in Sect. 5 we will discuss the results.

2 The TAP Equation

As in [14], let M be the number of different types of artifacts, or of different chemical species. If time increases in discrete steps:

$$M_{t+1} = M_t + P\sum_{i=1}^{M_t} \alpha_i \binom{M_t}{i} \tag{1}$$

with $\alpha_i \leq 1$ and $\alpha_{i+1} \leq \alpha_i$ (combining fewer "parents" is easier).

In the continuous time limit (slightly changing the meaning of the parameter P):

$$\frac{dM}{dt} = P\sum_{i=1}^{M} \alpha_i \binom{M}{i} \tag{2}$$

As discussed in Sect. 1, we are interested in the case where at most two parents can be recombined, so we have terms linear or quadratic in M:

$$\alpha_1 M + \frac{1}{2}\alpha_2 M^2 - \frac{1}{2}\alpha_2 M = \left(\alpha_1 - \frac{1}{2}\alpha_2\right)M + \frac{1}{2}\alpha_2 M^2 = bM + aM^2$$

where $b = \alpha_1 - \frac{1}{2}\alpha_2$ and $a = \frac{1}{2}\alpha_2$. The α_i's are non negative and decreasing, therefore $a, b > 0$. The equation is therefore:

$$\frac{dM}{dt} = P\left(bM + aM^2\right) \tag{3}$$

variables can be separated, so:

$$\frac{dM}{bM + aM^2} = Pdt$$

Integrating both sides, we obtain:

$$M(t) = \frac{bK_0}{e^{-bPt} - aK_0} \tag{4}$$

with $K_0 \equiv \frac{M_0}{aM_0+b}$.

The initial value is positive, its derivative is also positive, therefore M is positive and non decreasing. The numerator in Eq. 4 is positive, the denominator starts from the value 1-aK_0, but the value of the time dependent part decreases towards zero. So M grows until the denominator of Eq. (4) approaches 0, i.e. when:

$$t = t^* = -\frac{1}{bP} ln aK_0 = \frac{1}{bP} ln \frac{aM_0 + b}{aM_0} \tag{5}$$

So M grows to infinity in finite time.

The original Eq. 1 or Eq. 2 can be modified, e.g. adding a linear death term. If the b coefficient is larger than the coefficient of this term, then the behavior is essentially the same, while the process is somewhat slower, and divergence is achieved at a later t^* value.

The drawback of Eq. 1 (and of Eq. 2 as well) which was mentioned in the introduction is that M_1 is obtained by M_0 applying the combinatorial argument which is at the heart of the TAP approach. And M_0 is the set of actual artifacts or species which are initially available. So M_1 is the cardinality of the AP of the system at time 0. But M_2 is then obtained from M_1, not from the number of elements of the set of actual artifacts or species at $t = 1$, which should be a proper subset of the possible ones. Here, one might take the simplifying assumption that the number of actual artifacts at time t is just a fixed fraction f of those which are possible. In this case the equation would remain very similar to Eq. 1 – but the hypothesis that a fixed fraction of the possible new artifacts are actually built when the numbers explode seems highly unrealistic, therefore it is interesting to explore a different approach.

In the case of different approaches, and in particular of simulations, the problem arises of distinguishing between an exponential growth (fast, but which tends to infinity in the limit), and a faster growth that tends to infinity in finite times. So, let us now introduce a method with this aim.

Starting from Eq. 3, it is possible to find the following expression

$$\frac{M}{aM + b} = \frac{M_0}{aM_0 + b} e^{Pbt} = K_0 e^{Pbt} \tag{6}$$

We therefore introduce a function y defined as

$$y \equiv \frac{M}{aM + b} \tag{7}$$

So, in case of TAP growth, y results in the following exponential function

$$y = K_0 e^{Pbt} \tag{8}$$

Otherwise, if it is M that has an exponential growth, i.e. $M(t) = M_0 e^{kt}$, then

$$y = \frac{M_0 e^{kt}}{aM_0 e^{kt} + b} = \frac{M_0}{aM_0 + be^{-kt}} \tag{9}$$

and so y is a logistic function such that

$$\lim_{t \to +\infty} y(t) = \frac{1}{a}$$

Therefore, it is possible to distinguish the two types of growth observing the behavior of the graph of $y(t)$.

3 The Polymer Model

As discussed in the Introduction, we are mainly concerned with the phase of chemical evolution which preceded biological evolution, and during this phase new molecular types are generated, and interact with other chemical species. Many studies have dealt with the cooperation and competition between different species, and the notion of autocatalytic sets has emerged as a condition for sustained replication of a set of molecules [5, 6, 11, 28].

However, catalysis typically requires the presence of relatively large molecules, capable of folding and assuming chemically active forms [Luisi, 2006]. The models concerning the emergence of such a kind of systems must therefore allow the formation of chemical species that are progressively longer, take into account the effects of the presence of an ever-increasing number of new species, and consider the structure of the species that have gradually emerged. Indeed, a striking aspect of modern life is the key role of polymers like polypeptides and nucleic acids: hence, it is necessary to study how they can be generated, and how longer polymers can appear.

In our model, the entities are the different types of chemical species. Each species is an oriented linear chain formed by symbols from the binary alphabet $A = \{A, B\}$: so, this means that AB $\neq$ BA (an organization present for example also in [7, 8, 12, 13]). Species formed by a single symbol are called "monomers", while those composed of more symbols are called "polymers".

As described in Sect. 1, there are two different types of reactions, namely cleavage and condensation. A chemical species can be created by different reactions (e.g. B + BA $\rightarrow$ BBA and ABBA $\rightarrow$ A + BBA). The reactions do not require catalysis, so we assume that all possible reactions can indeed occur.

As anticipated, we are interested in the presence or absence of each substance: the model therefore has a binary basis, in which "0" stands for "absent species" and "1" for "present species". In this work, once a chemical species has been created it remains present in the system, no matter whether it is used as a reactant in some reaction (as discussed in the introduction).

Now, suppose that N is the total number of different types of chemical species present in the system at a given time t, then each species i with length L_i can be a reactant of L_{i-1} cleavages and 2N condensations, so the total number R_c of conceivable reactions at a given time t is given by:

$$R_c = \sum_{i=1}^{N} (L_i - 1) + N^2 \tag{10}$$

It should be noted that as the number of existing species increases, the number of all possible reactions increases more than linearly, as observed in [7, 13].

The model is simulated using the Gillespie algorithm [9] according to which, at each step of the simulation, a single reaction occurs: thus, there is an asynchronous stochastic update of the existing species. The equations are based on the law of mass action: the probability of occurrence of each individual reaction is therefore proportional to the concentration of the reactant (or to the product of the concentrations in the case of several reactants) and to a kinetic coefficient. Concentrations do not only affect the chance that a reaction be chosen, but they also affect the computation of the physical time difference between successive reactions, so that more reactions can take place in a (physical) time unit if there are more interacting entities, *ceteris paribus*. In our model the concentrations of the existing chemical species are all equal to 1: therefore, using the same reaction coefficients for both condensations and cleavages is equivalent to inducing Gillespie's algorithm to choose randomly and uniformly among all possible reactions - a choice which is coherent with the idea of focusing on simple combinatorics of types. As consequence, the more types of chemical species there are in the system, the more reactions will occur since there is a higher chance of encounter between two types of species. In the following we will refer to the model with the initials BPSM (Binary Polymer Species Model).

4 Simulation Results

We simulated the discovery of new chemical species, in the following starting from a pool composed of monomers, dimers and trimers (14 species in all). We will show here the results of a typical simulation; since the system is stochastic, other simulations can provide slightly different results, but the relevant properties are the same (although the time when the number of different types starts to grow fast can be different).

The effect of the condensations is that of assembling longer and longer chemical species, while the cleavages tend to "fill" the gaps left. Both actions can lead to the rediscovery of already known species: however, this situation is more frequent for cleavages, while the condensation products inhabit increasingly large areas of phase space, with a smaller and smaller overlapping probability. The final combined effect is that of obtaining longer and longer single chemical species, and to gradually fill the space of possibilities at shorter lengths (Fig. 1).

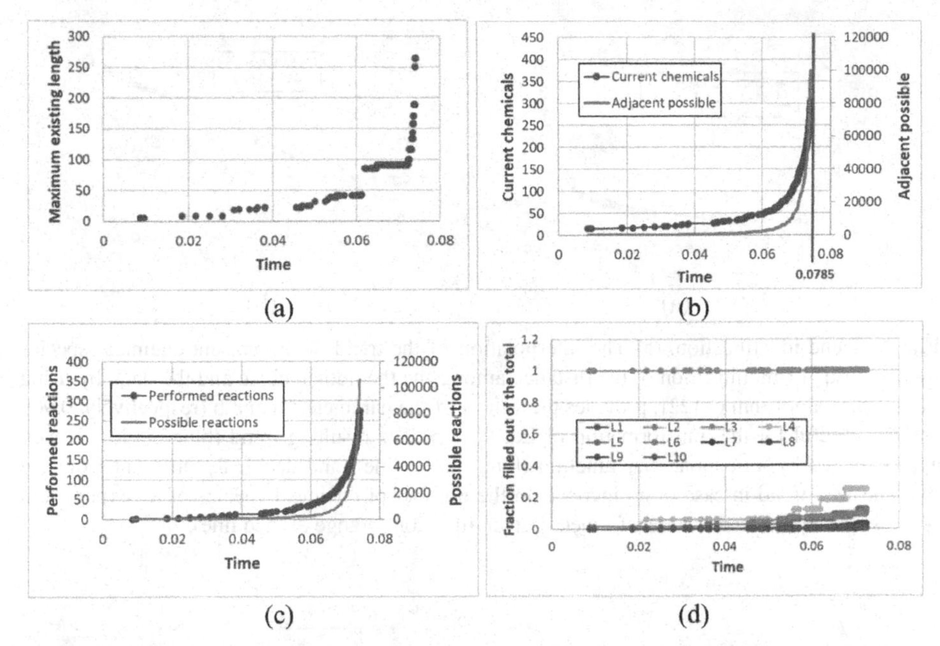

Fig. 1. The model's behavior. (a) Maximum length existing in the pool of chemical species: it can be noted that the process can have jumps, when a condensation uses long molecules as substrates. (b) Number of existing species, and of those possible at that time (the "adjacent possible"). The divergence point of the Eq. 4 is also shown. Note that the ratio "current chemicals"/"adjacent possible" is constantly decreasing. (c) Number of the reactions carried out, and of those possible at that time. After a brief transient, the ratio "performed reactions"/"possible reactions" is decreasing. (d) The fraction of chemical species actually made, out of the total of possible ones, for each length.

All the relevant variables of the system tend to increase as one approaches the time t^* calculated in Eq. (4) (Fig. 1b). We are interested in the type of trend of the curves approaching this limit: in particular, we are interested in the trend of the number of chemical species appeared in the system. For this purpose, it is possible to use the transformation of Eq. (7): the resulting trend is decidedly closer to an exponential than to a logistic, an indication that appears in favor of a divergence in finite times (Fig. 2). It is interesting to note that a similar divergence is also obtained using condensations alone (as shown in Fig. 3).

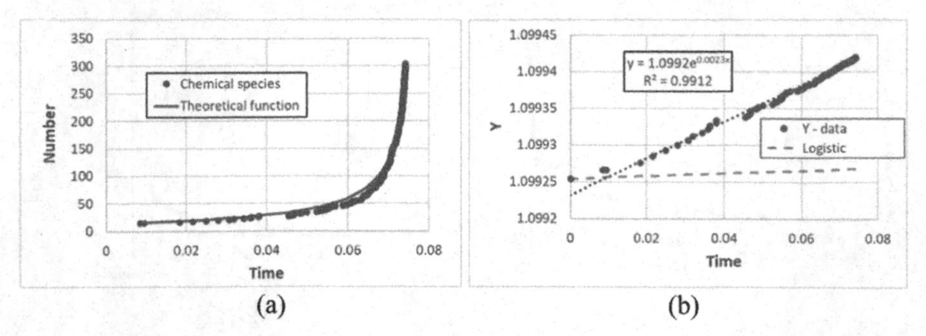

Fig. 2. Trend identification. (a) The interpolation of the trend of the present chemical species by using Eq. 4 (minimization of the distance among the theoretical curse and the data, by using the simplex algorithm of [22]) provides the values of the parameters a and b (respectively, 0.909 and 0.2) useful for the transformation of Eq. (7). (b) The resulting trend (blue circles) is well approximated by an exponential function (blue dotted line), and decidedly different from the hypothesized trend in case of an increase in the number of chemical species of an exponential type, which would lead to a transformed logistic function (orange dashed line).

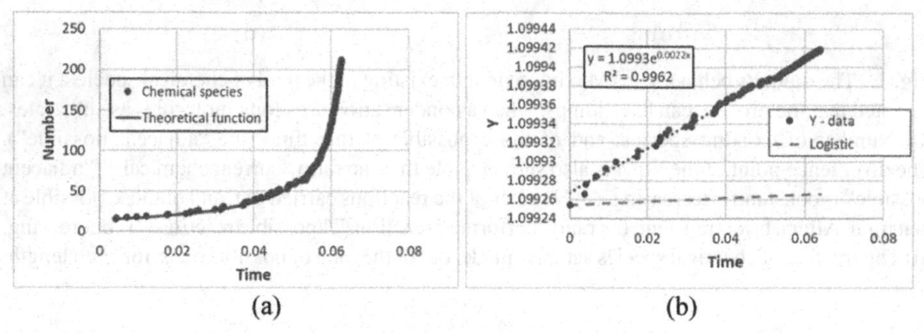

Fig. 3. Trend identification in case of use of condensations only. (a) The interpolation of the trend of the present chemical species by using Eq. (4) ($a = 0.997$, $b = 1.79$). (b) The resulting trend (blue circles) is well approximated by an exponential function (blue dotted line), and decidedly different from the hypothesized trend in case of an exponential increase in the number of chemical species, which would lead to a transformed logistic function (orange dashed line).

5 Discussion

In this paper we have reviewed the TAP equation, which seems able to describe phenomena where the number of different entities (e.g. artifacts, chemical species) grows slowly for a long time, and at a certain time starts to grow at an extremely high rate, giving rise to a curve with almost vertical slope, which reminds us of a "hockey-stick". We have analytically solved the equation in the case where a new entity is obtained changing a single existing one, or at most by combining two of them. The solution actually shows divergence at finite times, thus departing from the exponential growth which diverges in the limit $t \rightarrow \infty$. We have also introduced a way to distinguish between the two different growth curves (TAP and exponential).

We have also observed that the original TAP equation propagates in time combinations of all the elements of the set of the Adjacent Possible, and we have criticized this choice, proposing to take into account only combinations of those artifacts that are actually produced. This requires a criterion for identifying which artifacts in the AP are actually produced, and we have introduced the BPSM model for this purpose. Interestingly, the growth of the number of molecular types resembles the TAP growth rather than the exponential one (although the analysis of BPSM is limited to computer simulations, which necessarily are bound to be finite).

It is worth observing that the model is not limited to aspects related to the origin of life, although that was our initial motivation.

A questionable aspect of BPSM is the fact that "types never die". This can be justified e.g. by assuming that many exemplars of each type are present (or, equivalently, that there is a mechanism which continually copies existing polymers, without producing new ones). In any case, our model is not one of population dynamics. It would be interesting to consider also the dynamics of the number of molecules of each type, thus turning to a model which truly belongs to the population dynamics type; however, this would require the simulation of very large numbers of entities, a nontrivial task using e.g. the original BPM model.

We plan to explore in the future the robustness of the TAP-like growth by studying different selection rules to define the set of actual entities from the AP.

Acknowledgments. Useful discussions with Stuart Kauffman are gratefully acknowledged.

References

1. Bettencourt, L.M.A., Lobo, J., Helbing, D., Kühnert, C., West, G.B.: Growth, innovation, scaling, and the pace of life in cities. Proc. Natl. Acad. Sci. U. S. A. **104**(17), 7301–6 (2007). Epub 2007 Apr 16. https://doi.org/10.1073/pnas.0610172104
2. Bonifati, G., Villani, M.: Exaptation in innovation processes: theory and models. In: Grandori, A. (ed.), Handbook of economic organization. Integrating Economic and Organization Theory Cheltenham, UK and Northampton, MA, USA, Edward Elgar, pp. 172–192 (2013).
3. van Doorn, S., Edelaar, P., Weissing, F.J.: On the origin of species by natural and sexual selection. Science **326**, 1704–1707 (2009)
4. Dopfer, K., Potts, J.: The General Theory of Economic Evolution. Routledge, London (2008)
5. Eigen, M., Schuster, P.: The hypercycle: part B. Naturwissenschaften **65**, 7–41 (1978)
6. Eigen, M., Schuster, P.: The Hypercycle: A Principle of Natural Self-Organization. Springer, Berlin, Germany (1979). https://doi.org/10.1007/978-3-642-67247-7
7. Filisetti, A., Graudenzi, A., Serra, R., et al.: A stochastic model of the emergence of autocatalytic cycles. J. Syst. Chem. **2**, 2 (2011). https://doi.org/10.1186/1759-2208-2-2
8. Filisetti, A., Graudenzi, A., Serra, R., et al.: A stochastic model of autocatalytic reaction networks. Theory Biosci. **131**, 85–93 (2012). https://doi.org/10.1007/s12064-011-0136-x
9. Gillespie, D.T.: Exact stochastic simulation of coupled chemical reactions. J. Phys. Chem. **81**(25), 2340–2361 (1977). https://doi.org/10.1021/j100540a008
10. Haldane, J.B.S.: Origin of life. Ration. Annu. **148**, 3–10 (1929)
11. Jain, S., Krishna, S.: Autocatalytic sets and the growth of complexity in an evolutionary model. Phys. Rev. Lett. **81**, 5684–5687 (1998)

12. Kauffman, S.A.: Autocatalytic sets of proteins. J. Theor. Biol. **119**(1), 1–24 (1986). https://doi.org/10.1016/s0022-5193(86)80047-9
13. Kauffman, S.A.: At Home in the Universe: The Search for Laws of Self-Organization and Complexity. Oxford University Press, Oxford (1995)
14. Koppl, R., Devereaux, A., Herriot, J., Kauffman, S.A.: A Simple Combinatorial Model of World Economic History, Papers 1811.04502, arXiv.org. (2018)
15. Koppl, R., Devereaux, A., Valverde, S., Solè, R., Kauffman, S.A., Herriot, J.: Explaining technology. SRN Electron. J. (2021).https://doi.org/10.2139/ssrn.3856338
16. Lane, D.A., Pumain D., Van der Leeuw, S.E., West, G. (eds.): Complexity Perspectives in Innovation and Social Change. Science+Business Media B.V, Springer, Dordrecht (2009). https://doi.org/10.1007/978-1-4020-9663-1
17. Lane, D.A.: Complexity and innovation dynamics. In: Cristiano, A. (ed.), Handbook on the Economic Complexity of Technological Change, Cheltenham, UK and Northampton, MA, USA, Edward Elgar, pp. 63–80 (2011)
18. Lehman, N.E., Kauffman, S.A.: Constraint closure drove major transitions in the origins of life. Entropy **23**(1), 105 (2021). https://doi.org/10.3390/e23010105
19. Luisi, P.: The Emergence of Life: From Chemical Origins to Synthetic Biology. Cambridge University Press, Cambridge (2006). https://doi.org/10.1017/CBO9780511817540
20. Maynard Smith, J., Szathmáry, E.: The Major Transitions in Evolution. Oxford University Press, New York, NY, USA (1995)
21. Miller, S.L., Orgel, L.E.: The Origins of Life on the Earth. Prentice Hall, New Jersey, USA (1974)
22. Nelder, J.A., Mead, R.: A simplex method for function minimization. Comput. J. **7**(4), 308–313. (1965). https://doi.org/10.1093/comjnl/7.4.308
23. Oparin, A.I.: Proiskhozhdenie zhizni. Izd. Moskovskii Rabochii, Moscow (1924)
24. Oparin, A.I.: Origin of Life. Macmillan, New York (1938)
25. Pascal, R., Pross, A., Sutherland, J.D.: Towards an evolutionary theory of the origin of life based on kinetics and thermodynamics. Open Biol. **3**, 130156 (2013). https://doi.org/10.1098/rsob.130156
26. Schluter, D.: Ecology and the origin of species. Trends Ecol. Evol. **16**, 372–380 (2001)
27. Steel, M., Hordijk, W., Kauffman, S.A.: Dynamics of a birth-death process based on combinatorial innovation. J. Theor. Biol. **491**, 110187 (2020)
28. Steel, M., Hordijk, W., Xavier, J.C.: Autocatalytic networks in biology: structural theory and algorithms. J. R. Soc. Interface **16**, 20180808 (2019)
29. Walker, S.I., Davies, P.C.W., Ellis, G.F.R.: From Matter to Life. Cambridge University Press, Cambridge, UK (2017)

Kernel-based Early Fusion of Structure and Attribute Information for Detecting Communities in Attributed Networks

Annalisa Socievole(✉) and Clara Pizzuti

National Research Council of Italy (CNR), Institute for High Performance Computing and Networking (ICAR), Via Pietro Bucci, 8-9C, 87036 Rende, CS, Italy
{annalisa.socievole,clara.pizzuti}@icar.cnr.it

Abstract. *Early fusion methods* are a category of community detection methods for attributed networks that merge attributes and structure before the method is executed. Typically, a weighted network where the edge between two nodes includes both the structure information (i.e. the weight of the existing link) and the similarity of the attributes is obtained and then, classical community detection algorithms can be applied. In this paper, we investigate the role of kernels on a method based on a Genetic Algoritm (GA). When measuring similarity, kernels are able to provide a more suitable and meaningful form for the similarity matrix in order to facilitate data analysis. Through simulations on both synthetic and real-world attributed networks, we first apply different kernels to @NetGA, a genetic algorithm we proposed for attributed networks which embeds into edges both the structure and the attribute information through a unified distance measure. We compare this novel approach named K@NetGA to @NetGA showing that the Free-Energy kernel is able to sensibly improve the community detection performance. We finally compare our method to other kernel-based community detection methods on a wider set of networks finding that K@NetGA is the best-performing.

Keywords: Community detection · Attributed Networks · Genetic Algorithms · Kernel on graphs

1 Introduction

Attributed graphs model real networks by enriching the nodes with attributes accounting for some properties. Real-world graphs such as the World Wide Web and social networks are more than just their topology. Facebook, for example, forms a social network system where each user is connected through friendship relationships to other users and is characterized by a profile specifying attributes like name, age, birthday, education, email, hometown, and so on.

Clustering attributed networks consists in partitioning them into disjoint communities of nodes that are both well-connected and similar with respect

C. De Stefano et al. (Eds.): WIVACE 2022, CCIS 1780, pp. 141–151, 2023.
https://doi.org/10.1007/978-3-031-31183-3_12

to their attributes. Chunaev [7] classifies the community detection methods for attributed networks in *early fusion methods*, *simultaneous fusion methods*, and *late fusion methods*. The first category merges attributes and structure before the method is executed. Generally, a weighted network is obtained and classical community detection algorithms can be applied. *Simultaneous fusion methods* consider structure and attributes in a joint process, thus specialized implementations must be realized. *Late fusion methods*, on the contrary, first separately perform community detection on structure and attributes, and then fuse the results by, for example, a consensus method.

Inside the first class of approaches, metaheuristic-based methods apply strategies to obtain the optimal value of a fitness function evaluated over a *weighted graph* of the original embedding structure and attributes. In this paper, we extend a metaheuristic-based method proposed in [15] with the concept of *kernels on graphs* to improve the quality of the communities obtained by the method. So far, kernel-based clustering has been mainly considered for networks without attributes with results showing that kernel similarity computed through kernels helps in better grouping nodes (e.g. [17] and [3]). This is due to two important properties of kernels: first, they allow to implicitly compute similarities in a high-dimensional space where data are more likely to be well-separated, and second, they allow to compute similarities between objects that cannot be naturally represented by a simple set of features.

Recently, kernel-based clustering has been extended to attributed networks and evaluated in [4]. The authors combine structure and attribute similarity giving a weight that tries to balance the contribution of the two similarity measures and then apply the classic clustering methods. Similarly to [4], this paper proposes an extension with kernels of the early fusion community detection method for attributed networks based on a genetic algorithm called @NetGA and proposed in [15]. The algorithm is based on a measure that jointly takes into account the similarity between nodes in terms of attributes and the links between them: the *unified distance measure* introduced by Papadopoulos et al. [13]. More specifically, the method first weights each network edge with this distance, then, it optimizes a fitness function based the unified distance measure in order to obtain highly dense communities characterized also by homogeneous attributes. @NetGA has the advantage of finding accurate divisions in most of the synthetic network structure and attributes settings. In particular, when the underlying community structure is not so clearly defined, @NetGA is even able to perfectly find the underlying communities outperforming not only a structure-only benchmark like Louvain [5] but also the attribute-only method k-means [10]. However, in network settings where the attribute noise was high and the community structure very clear, @NetGA was not able to exactly match the ground truth.

In this paper, we apply kernels to investigate their ability in improving the community detection performance of @NetGA. We call this new kernel-based approach, K@NetGA. According to the results of our experiments over synthetically generated attributed networks and two real-world citation networks, the

Free Energy kernel is the most effective in improving community detection performance. In synthetic networks, in particular, the Free Energy kernel allows to match the ground-truth in the network settings of community structure and attribute noise considered. Moreover, when comparing it to other kernel-based methods, K@NetGA achieves the best community detection performance.

The paper is organized as follows. Section 2 describes the related work in the areas of kernel-based clustering and community detection in attributed networks. Section 3 gives some preliminary definitions. Section 4 introduces K@NetGA. Section 5 describes the experimental setup: datasets, algorithms in comparison, and performance indexes. Section 6 describes the results. Finally, Sect. 7 concludes the paper.

2 Related Work

In this section, we discuss the state-of-the art by first focusing on kernel-based clustering and then, on the more specific research area of attributed network clustering.

2.1 Kernel-based Clustering

In [17], Sommer et al. provide an experimental evaluation of a kernel-based k-means clustering algorithm and the Louvain method [5]. The methods are evaluated with six different kernels over a total of 15 graphs, the three well-known small networks Zackary, Polblogs, and Football, 9 newsgroup graphs, and three 600-nodes synthetic networks generated with the LFR benchmark [11]. The results show that the Free Energy and Randomized Shortest-path kernels are the best performing.

Aynulin [2] presents a study for evaluating three clustering algorithms, k-means, Ward, and spectral clustering, when kernels and transformations are applied. Four kernels are analyzed, Walk, Communicability, Forest, and Heat, and eight transformations, two of which with two different parameter values. The results on Zackary, Football, Polbooks, and Newsgroups datasets show that overall the transformations of proximity measures help in improving the quality of the clustering. Between the tested transfomations, the most performing functions are the logarithmic and the power function.

In [4], Aynulin and Chebotarev consider Communicability, Heat, PageRank, Free Energy and Sigmoid Corrected Commute-time kernels to find communities in attributed networks by running k-means and spectral clustering. The kernels are applied on a graph matrix where each entry contains an *extended node similarity* measure which combines structure and attribute similarity as $a^s_{ij} = \beta a_{ij} + (1 - \beta)s_{ij}$, where a_{ij} are the entries of the adjacency matrix, $\beta \in [0, 1]$ is a parameter that tries to balance the weights of the adjacency matrix and s_{ij} is the attribute similarity. s_{ij} is obtained by applying known measures like matching coefficient, cosine similarity, extended Jaccard similarity, Manhattan similarity, and Euclidean similarity. Experimenting on WebKD,

Cora and CiteSeer datasets, the spectral clustering outperforms k-means and the best results are obtained when the cosine similarity is combined with the Free Energy kernel.

2.2 Clustering Attributed Networks

Classically, communities are detected through the network topology. However, when some information is missing or noise is present, enriching the structure data of the algorithm with node attributes can help in detecting more meaningful communities. Detailed surveys of the state-of the art algorithms for community detection in attributed networks can be found in [6] and [7].

According to [7], the community detection methods can be categorized by analyzing how the structure and the attributes are combined. Among these categories, the *early fusion* methods combine attribute and structure similarity and then run the algorithm for community detection.

The algorithm by Tombe at al. [8] defines a unique similarity measure between couples of nodes which sums attribute distance (e.g. Euclidean), and structural distance in terms of shortest path. Then, hierarchical clustering is applied to the resulting weighted adjacency matrix.

Papadopoulos et al. [13] defined the *unified distance measure.* The authors apply a fuzzy clustering method with an objective function based on this measure which assign different weight scores to edges and attribute similarity during the clustering process, by computing iteratively such weights with the gradient descent technique.

In our previous work [15], we exploited the unified distance measure for clustering attributed graphs through a genetic algorithm named @NetGA. To avoid providing a score for attributes and structure beforehand, in a later work [16], we proposed MOGA-@Net, a multi-objective optimization and local merge clustering method for attributed graphs. This approach jointly optimizes the structural quality of the communities as the first objective by testing different topological measures (modularity, community score, conductance), and the attribute similarity as the second objective by trying different similarity measures (Jaccard, Cosine, Euclidean) depending on the type of the input attribute data. A post-processing local merge technique is then applied to generate better solutions.

Recently, Aynulin and Chebotarev [4], presented a performance evaluation of a combined metric given by attribute similarity and edge weight, subsequently transformed with different kernels, to cluster nodes through k-means and spectral clustering. Analogously to [4], in the following, we investigate the use of kernels to detect communities in attributed networks. However, our approach does not need to fix the number of communities in advance.

3 Preliminaries

Before describing our proposal, we provide some preliminary definitions and briefly introduce kernels.

3.1 Definitions

We consider an unweighted undirected *attributed graph* G defined as a tuple $G = (V, E, T, F)$ where V is the set of vertices with $|V| = n$, E is the set of edges with symmetric weights a_{ij} described by the *adjacency matrix* A and where $|E| = m$, $T = \{t_1, t_2, ..., t_d\}$ is the set of attributes which characterizes each node and $F = \{f_1, f_2, ..., f_d\}$ is a set of functions. Each node v_i is described by a feature vector $T_{v_i} = \{f_1(v_i), f_2(v_i), ..., f_d(v_i)\}$, obtained by the functions $f_t : V \rightarrow Z_t$, with $1 \leq t \leq d$ and Z_t the domain of attribute t.

The node *degree* is the sum of all the edges connected to the node. The corresponding *degree matrix* $D = diag(A \cdot 1)$, where $1 = (1, ..., 1)^T$ is the all ones vector, is a diagonal matrix containing the degrees of all nodes on the main diagonal. Given A and D, $L = D - A$ is the *Laplacian matrix*, and the *Markov matrix* $P = D^{-1}A$.

3.2 Kernels

A *kernel* on G is a measure between the vertices of G that can be represented through a positive semidefinite matrix named Gram matrix K, with real entries k_{ij} showing the *similarity* between nodes i and j [1]. Distance measures that are computed through a kernel matrix have the property of being Euclidean, that is, the nodes of the graph can be mapped in a Euclidean space where the distances between nodes are preserved. This alternative representation of points in space often makes data analysis more meaningful than the original one.

In this paper, we take into account a set of kernels that have proven to have good performances for the task of community detection [1,4,17]. The difference between these kernels relies on the matrix on which the kernel is applied which can be the adjacency matrix A of G, the Laplacian L of G, the Markov matrix P, and the parameters used.

- Communicability. $K^{Comm}(t) = \exp(tA)$, $t > 0$.
- Heat. $K^{Heat}(t) = \exp(-tL)$, $t > 0$.
- Personalized Page Rank. $K^{PPR}(\alpha) = [I - \alpha P]^{-1}$, $0 < \alpha < 1$.
- Free Energy. Given P, $W = \exp(-\alpha A) \circ P$, $Z = (I - W)^{-1}$, the Free Energy kernel is given by the transformation $K^{FE} = -\frac{1}{2} H \Delta^{FE} H$, where $H = I - \frac{1}{n} 1 \cdot 1^T$, and $\Delta^{FE} = \frac{\Phi + \Phi^T}{2}$ with $\Phi = \frac{log(Z)}{\alpha}$ and $\alpha > 0$.

4 K@NetGA

Given an attributed graph, the objective of a community detection method for such a graph is partitioning $\mathcal{C}$ of the nodes of V in k communities $\mathcal{C} = \{C_1, ..., C_k\}$ such that (1) the internal structure of communities is dense while the communities are loosely connected between them, and (2) the nodes of a community are highly similar.

In this paper we propose K@NetGA, a method based on our previous work @NetGA, a genetic algorithm that optimizes a fitness function based on the *unified distance measure* (*udm*), introduced by Papadopoulos et al. [13]. This metric embeds network structure and attribute similarity in a single edge weight by combining *similar connectivity* (*SC*) and *attribute distance* (*AD*) between two nodes i and j as

$$d(i,j) = W_{attr} \cdot AD(i,j) + W_{links} \cdot SC(i,j) \tag{1}$$

where W_{attr} and W_{links} are weights representing the importance of attributes and edges, respectively, and AD and SC are measured as

$$AD(i,j) = \sum_{\alpha \in A} W_\alpha \cdot \delta_\alpha(i,j), \quad \sum_{\alpha \in A} W_\alpha = 1 \tag{2}$$

with W_α weight of attribute α and $\delta_\alpha(i,j)$ the attribute distance between nodes i and j for attribute α (see [13] for its expression for numerical or categorical attributes), and

$$SC(i,j) = \frac{1}{N} \sum_{k=1}^{N} [w(i,k) - w(j,k)]^2 \tag{3}$$

where $w_{i,j}$ is 1 if there is a link between nodes i and j, 0 otherwise. Differently from our previous work @NetGA, once computed the distance matrix $Dist = (d_{ij})$ containing all the distances $d(i,j)$, we apply kernels to $Dist$. Note that $Dist$ is considered an input weighted adjacency matrix for our problem since the kernels are applied to this matrix. We call the resulting kernel-based unified distance measure matrix $K^{Dist} = (k^d_{ij})$. From this kernel-based matrix, the fitness function optimized by K@NetGA is defined as follows.

Given a network division $\mathcal{C} = \{C_1, \ldots, C_k\}$ for the attributed graph G composed by k communities, we define the *kernel-based clustering unified distance measure* $kcudm(\mathcal{C})$ of the partition $\mathcal{C}$ by computing for each community C, the kernel-based *udm* between pairs of nodes i and j belonging to C, and then averaging the results with respect to the number of communities k:

$$kcudm(\mathcal{C}) = \frac{1}{k} \sum_{C \in \mathcal{C}} \sum_{\{i,j\} \in C \, i \neq j} k^d(i,j) \tag{4}$$

where $k^d(i,j)$ is the kernel-based *udm* between i and j.

The algorithm K@NetGA minimizes the *kcumd* measure to obtain a partitioning that considers both the attributes and the links between nodes. As genetic operators, K@NetGA exploits the locus-based adjacency representation [14], uniform crossover and neighbor-based mutation. With the locus-based, each individual of the population is represented with a vector of n genes. Each gene i can assume a value j in the range $\{1, \ldots, n\}$ meaning that the nodes i and j are connected. Then, a decoding phase is used to extact the connected components of the graph corresponding to single communities. The uniform crossover operator creates a binary mask with n elements which is used to generate an offspring. From the first parent, the genes in the positions where the mask is 0 are selected;

similarly, where the value of the mask is 1, the genes from the second parent are selected. Finally, the neighbor-based mutation swaps the value a gene with one of its neighbors at random.

The main steps of K@NetGA are the following. It receives in input the attributed graph G, the weighting factors W_{attr}, W_{links}, and W_{α}, the type of kernel k_{ID}, and:

1. runs the GA on G for a fixed number of iterations by using *kcumd* as fitness function to minimize, locus-based representation, uniform crossover and neighbor-based mutation as genetic operators;
2. obtains the partition $\mathcal{C} = \{C_1, \ldots, C_k\}$ that is the solution which minimizes the fitness value $kcumd(\mathcal{C})$;
3. merges two communities C_i and C_j if the number of connections between them is higher than the number of internal connections of C_i or C_j.

5 Experimental Setup

In this section, we discuss the experimental setup used, describing the datasets used to validate the proposal, the contestant algorithms and the performance indexes exploited for the evaluation of K@NetGA.

5.1 Datasets

We experimented K@NetGA on both synthetic and real-world attributed networks using MATLAB R2020a and the Global Optimization Toolbox.

For each simulation, in the *udm* formula, we assigned equal weight to attributes and links thus setting $W_{attr} = W_{links} = 0.5$. We did not investigate different settings of the weights because this problem is beyond the scope of the paper. The following subsections describe the experimental setup.

To generate the synthetic networks, we used the LFR-EA benchmark [9] which creates community networks by setting the *mixing parameter* μ and the *noise parameter* ν: the first manages the rate of intra- and inter- communities edges while the second determines the noise of attributes (i.e. the degree of common features between nodes). More specifically, low values of μ and ν generate synthetic networks with dense and well-structured communities where nodes of the same community have similar attributes. Viceversa, high values of the two parameters result in networks with a confused community organization with sparse communities where the internal nodes have different features.

For our experiments, we generated networks with 1000 nodes with the same parameters used in [15]. We selected the mixing parameter in the range [0.1; 0.9] and the attribute noise in the range [0; 0.9] and generated 10 network samples for a given couple (μ, ν).

Finally, we also analyzed two real-world citation networks, *Cora* and *Citeseer*[1] The first network has 2708 nodes, 5429 edges, and 1433 attributes, while the second has 3312 nodes, 4732 edges, and 3703 attributes.

[1] Both datasets are available at https://linqs.soe.ucsc.edu/.

5.2 Algorithms in Comparison

To validate our kernel-based algorithm, we first compared it to @NetGA to test the effectiveness of kernels in improving community detection. @NetGA and K@NetGA have been executed for 200 generations with population size 300, mutation rate 0.4 and crossover fraction 0.8. The parameters used for the 4 analyzed kernels have been $t = 0.5$, and $\alpha = 0.5$.

Then, we also made a comparison between K@NetGA and other benchmarks, Louvain [5], k-means [10] and other kernel-based methods. Specifically, we implemented a kernel-based version of the Louvain method. Classic Louvain is based on optimization process of the modularity [12] of a community division through a greedy technique, working only on the network structure and not considering the attributes. In this work, we used as input to Louvain also a kernel-based adjacency matrix.

Similarly, we implemented a kernel-based version of k-means, which is usually adopted as a clustering algorithm for attributes. Classically, k-means randomly assigns data points to a number k of clusters fixed in advance. Then, the centroid of each cluster is computed and every data point is assigned to its closest centroid. These steps are repeated until there are no assignments of data points to clusters, and a stopping criterion is reached.

Finally, we considered the kernel-based spectral community detection method by Aynulin and Chebotarev [4] specifically designed for attributed networks. For this algorithm, we equally weighted structure and attributes by setting $\beta = 0.5$ and considered Free Energy as suggested by their work.

5.3 Performance Indexes

For evaluating the quality of the detected communities, the Normalized Mutual Information (NMI) has been used and its cumulative version, the CNMI [9], which allows the integration of the well-known Normalized Mutual Information (NMI) over different settings of the structure mixing parameter μ and attribute noise ν as

$$CNMI = \frac{\sum^{\mu} \sum^{\nu} NMI}{S} \quad (5)$$

where S is the number of samples of the considered network graphs. NMI and CNMI take values from 0 to 1, and the larger the value, the better the matching of the found community with the ground truth.

6 Results

Experiment 1: kernel comparison. The aim of the first experiment has been (1) assessing the effectiveness of applying kernels in K@NetGA scheme and (2) evaluating the difference between the different kernels in terms of community detection performance.

We considered 3 synthetic networks, namely LFR-EA1, LFR-EA2 and LFR-EA3, with the (μ,ν) couple values (0.1, 0.5), (0.2, 0.5) and (0.3, 0.5), respectively.

Table 1. NMI results for @NetGA in LFR-EA1, LFR-EA2, and LFR-EA3 networks.

Algorithm	Dataset				
	LFR-EA1	LFR-EA2	LFR-EA3	Cora	Citeseer
@NetGA	0.9393	0.8813	0.8234	0.4088	0.3619

Table 2. K@NetGA NMI when kernels are applied to the unified distance measure.

Kernel	Dataset				
	LFR-EA1	LFR-EA2	LFR-EA3	Cora	Citeseer
Communicability	0.936	0.9015	0.8411	0.4075	0.3591
Heat	0.9439	0.9241	0.8444	0.4016	0.342
PageRank	0.9647	0.9327	0.8404	0.3922	0.3512
Free Energy	**1**	**1**	**1**	**0.4612**	**0.4011**

These structural and attribute settings resulted critical in @NetGA since the corresponding networks showed a lower NMI performance compared to other benchmarks [15].

Table 1 shows the average NMI obtained by @NetGA in the absence of kernels. In Table 2, we show what happens to K@NetGA which implements the kernels. Overall, all the kernels improve the NMI performance of @NetGA. The best-performing kernel is Free Energy over all the networks, followed by PageRank, Heat, and Communicability. Free Energy superiority is especially observable in the LFR-EA networks where the NMI is 1 independently from the community structure given by the (μ,ν) values. Therefore, we can conclude that the kernels applied to the unified distance measure matrix help in finding better community divisions.

Experiment 2: algorithms comparison. In a second experiment, we made a comparison between K@NetGA and other algorithms, by considering a wider set of LFR-EA networks.

We considered all the combinations of μ and ν values to produce different attributed graphs with a clear to ambiguous structure and/or attributes. Overall, we tested 900 networks generated by combining 9 different mixing parameters taken in the range [0.1, 0.9] with 10 different attribute noises taken in [0, 0.9], and producing 10 network samples for each (μ, ν) couple. In particular, we selected the best-performing kernel from the previous experiment, the Free Energy, we applied it to K@NetGA, and compared our algorithm to the following methods: @NetGA, classic Louvain and k-means, the ad-hoc versions of these two benchmarks here modified to be kernel-based methods, and the method by Aynulin and Chebotarev [4].

Table 3 shows that K@NetGA outperforms @NetGA achieving a CNMI value of 0.9957 vs 0.98 on the whole set of the synthetic networks, and also outperforms both Louvain and k-means. The classic versions of the two methods show a poor performance: on the synthetic networks, for example, Louvain executed on the structure achieves 0.6993, while the k-means on the attribute data achieves 0.3544. As expected, these results suggest that for detecting communities in

Table 3. Methods comparison in terms of CNMI (LFR-EA) and NMI (Cora, Citeseer).

Community Detection	Dataset		
	LFR-EA	Cora	Citeseer
K@NetGA	**0.9957**	**0.4612**	**0.4011**
@NetGA	0.98	0.4088	0.3916
Louvain	0.6993	0.4193	0.3434
k-means	0.3544	0.2892	0.317
Kernel-based Louvain	0.7142	0.4293	0.3504
Kernel-based k-means	0.3623	0.3123	0.332
Aynulin & Chebotarev	0.8253	0.4006	0.3619

attributed networks both the structure and the attribute data need to be taken into account. However, when giving in input to Louvain the kernel-based adjacency matrix and the kernel-based similarity matrix to k-means, their community performance improves both on the synthetic LFR-EA and the real-world Cora and Citeseer networks.

The algorithm by Aynulin & Chebotarev being kernel-based and an early-fusion method like K@NetGA performs better than Louvain and k-means, however, the *udm* metric we adopted in K@NetGA results more suitable for the community detection in the considered network.

7 Conclusion

In this study, we focused on kernel-based clustering on attributed networks by proposing K@NetGA, a genetic algorithm that optimizes a fitness function that embeds both the structure and the attribute information through a kernel-based unified distance measure. The fitness function combines both the effectiveness of the *unified distance measure* introduced by Papadopoulos et al. [13] and the power of kernels.

The results of the experiments on both real-world and synthetic networks show that kernels allow K@NetGA to outperform our previous work @NetGA in settings where @NetGA performance was not completely satisfying. In particular, when the Free Energy kernel is applied, the difference between K@NetGA and @NetGA is significant. In the synthetic networks considered, the Free Energy kernel is even able to improve the community detection by matching the ground-truth. Finally, by comparing K@NetGA with other kernel-based methods, it achieves the best performance demonstrating the suitability of the unified distance metric combined with kernels.

References

1. Avrachenkov, K., Chebotarev, P., Rubanov, D.: Kernels on graphs as proximity measures. In: Bonato, A., Chung Graham, F., Prałat, P. (eds.) WAW 2017. LNCS, vol. 10519, pp. 27–41. Springer, Cham (2017). https://doi.org/10.1007/978-3-319-67810-8_3

2. Aynulin, R.: Efficiency of transformations of proximity measures for graph clustering. In: Avrachenkov, K., Prałat, P., Ye, N. (eds.) WAW 2019. LNCS, vol. 11631, pp. 16–29. Springer, Cham (2019). https://doi.org/10.1007/978-3-030-25070-6_2
3. Aynulin, R.: Impact of network topology on efficiency of proximity measures for community detection. In: Cherifi, H., Gaito, S., Mendes, J.F., Moro, E., Rocha, L.M. (eds.) COMPLEX NETWORKS 2019. SCI, vol. 881, pp. 188–197. Springer, Cham (2020). https://doi.org/10.1007/978-3-030-36687-2_16
4. Aynulin, R., Chebotarev, P.: Measuring proximity in attributed networks for community detection. In: Benito, R.M., Cherifi, C., Cherifi, H., Moro, E., Rocha, L.M., Sales-Pardo, M. (eds.) Complex Networks & Their Applications IX. COMPLEX NETWORKS 2020 2020. Studies in Computational Intelligence, vol. 943, pp. 27–37. Springer, Cham (2021). https://doi.org/10.1007/978-3-030-65347-7_3
5. Blondel, V.D., Guillaume, J.L., Lambiotte, R., Lefevre, E.: Fast unfolding of communities in large networks. J. Stat. Mech. Theory Exp. **P10008** (2008)
6. Bothorel, C., Cruz, J.D., Magnani, M., Micenkova, B.: Clustering attributed graphs: models, measures and methods. Netw. Sci. **3**(3), 408–444 (2015)
7. Chunaev, P.: Community detection in node-attributed social networks: a survey. Comput. Sci. Rev. **37**, 100286 (2020)
8. Combe, D., Largeron, C., Egyed-Zsigmond, E., Géry, M.: Combining relations and text in scientific network clustering. In: 2012 IEEE/ACM International Conference on Advances in Social Networks Analysis and Mining, pp. 1248–1253. IEEE (2012)
9. Elhadi, H., Agam, G.: Structure and attributes community detection: comparative analysis of composite, ensemble and selection methods. In: Proceedings of the 7th Workshop on Social Network Mining and Analysis, pp. 1–7 (2013)
10. Hartigan, J.A., Wong, M.A.: Algorithm as 136: a k-means clustering algorithm. J. R. Stat. Soc. Ser. C (Appl. Stat.) **28**(1), 100–108 (1979)
11. Lancichinetti, A., Fortunato, S., Radicchi, F.: Benchmark graphs for testing community detection algorithms. Phys. Rev. E **78**(4), 046110 (2008)
12. Newman, M.E., Girvan, M.: Finding and evaluating community structure in networks. Phys. Rev. E **69**(2), 026113 (2004)
13. Papadopoulos, A., Pallis, G., Dikaiakos, M.D.: Weighted clustering of attributed multi-graphs. Computing **99**(9), 813–840 (2017)
14. Park, Y., Song, M.: A genetic algorithm for clustering problems. In: Proceedings of the Third Annual Conference on Genetic Programming, vol. 1998, pp. 568–575 (1998)
15. Pizzuti, C., Socievole, A.: A genetic algorithm for community detection in attributed graphs. In: Sim, K., Kaufmann, P. (eds.) EvoApplications 2018. LNCS, vol. 10784, pp. 159–170. Springer, Cham (2018). https://doi.org/10.1007/978-3-319-77538-8_12
16. Pizzuti, C., Socievole, A.: Multiobjective optimization and local merge for clustering attributed graphs. IEEE Trans. Cybern. **50**(12), 4997–5009 (2019)
17. Sommer, F., Fouss, F., Saerens, M.: Comparison of graph node distances on clustering tasks. In: Villa, A.E.P., Masulli, P., Pons Rivero, A.J. (eds.) ICANN 2016. LNCS, vol. 9886, pp. 192–201. Springer, Cham (2016). https://doi.org/10.1007/978-3-319-44778-0_23

Cultural Innovation Triggers Inequality in a Sharing Economy

Elpida Tzafestas(✉)

Laboratory of Cognitive Science, Department of History and Philosophy of Science, National and Kapodistrian University of Athens, University Campus, Ano Ilisia, 15771 Zografou, Greece
etzafestas@phs.uoa.gr

Abstract. In this work, we are studying the dynamics of wealth inequality in a simulated primitive economy of producer agents. These agents represent families that live, produce and move in an environment that can be unstable. The families have different technological abilities and can decide differently about sharing produced food with others – i.e., whether to share through public stores and how much – or keeping it all for themselves and become wealthier. We show through agent-based modelling that in a competitive environment with cultural evolution, where agents that survive pass their behavioral profiles to their descendants, wealth inequality between agents follows an initially upward and then downward trend before stabilizing around its final value. This trend is reminiscent of swings identified by economists. We study it in cases of "shocks", where after stabilization one of the parameters of the system is reinitialized (technology, environment, birth model and others). In all cases, the after-shock inequality movement shows the same overall trend but it is smaller in size, and this is true irrespective of the sharing outcome. We formulate the hypothesis that the size of the emerging maximum inequality is due to cultural reinitialization or innovation, where because of the shock the agents reinitialize their stances toward sharing, i.e. they innovate culturally. This hypothesis is supported by a series of experiments with varying degrees of cultural innovation as well as by an experiment with generalized cultural evolution. The global conclusion is that it is the cognitive/cultural lever of the economic relations, here the disposition toward sharing, that is responsible for the type and breadth of inequality that emerges.

Keywords: Sharing · Inequality · Shock · Culture · Innovation · Evolution

1 Introduction

Food transfer and sharing is very common among humans and especially pervasive in small-scale populations and societies, namely foragers or agricultural-foragers. Sharing is therefore one important field of study in anthropology, behavioral ecology, human evolution and other disciplines [1, 2]. The most prominent theories of evolution of sharing include: reciprocal altruism (either through aggressive sharing and assertive reciprocation [3] or through explicit reciprocity [4]), group cooperation (where groups of sharers have a selective advantage over groups of non-sharers [5–7]), direct reciprocity

C. De Stefano et al. (Eds.): WIVACE 2022, CCIS 1780, pp. 152–167, 2023.
https://doi.org/10.1007/978-3-031-31183-3_13

or risk reduction (where giving now allows to ask later if resources are highly volatile [8, 9]) and others. We have studied elsewhere the evolutionary origins of sharing [10] and we have presented a risk reduction model [8, 9, 11, 12] that entails a fundamental feature of food storage, public and/or private. We have shown there that, in general, high environmental instability and low technological ability promote the evolution of higher degrees of sharing. The relative capacities of the private and public stores are responsible for evolution or not of generalized sharing in the population. Although storage capacity is represented in the model as a technological parameter, it is not purely technological, but it is also the result of a social and cultural choice and it goes hand in hand with communal or individualist behavioral profiles and stances [10].

But inequality is ubiquitous in nature [13], and even if sharing is widespread, wealth differences and inequalities still arise and persist. We are interested in the trend of wealth inequality, bearing in mind that such a measure is an externally observed global measure, and that we possibly can see it now with our modern eyes but that is not and was not necessarily perceived explicitly as such by preindustrial or early industrial, non urbanized, populations. Our idea is that since some sort of inequality trends exist in primitive, preindustrial economies as well as in advanced, urbanized and industrial societies [14, 15], it would be interesting to draw comparisons and look deeper in order to better understand the nature of inequality. More specifically we want to identify ups and downs and cycles that we may relate to the swings and cycles in modern economies [16–18] despite the differences in the production model. Economists are debating on the forces that drive these movements and many options have been proposed that match the macroscopic observations: technological revolutions ([16] and successors), demographic pressure [14, 17], globalization [18] and spatial expansions [19] and others. At this stage, we are leaving aside the question of what general or other social conditions allow inequality to evolve [20, 21], or what the people feel about it [22]. We are only interested in the wealth inequality trends that emerge in the simple production system that allows sharing and what more generic repercussions these might have.

This paper is structured as follows. In Sect. 2, we present the agent and environment model and in Sect. 3 we present simulation results for the reference condition where wealth inequality shows an initially upward and then downward trend before stabilizing around its final value. Next, in Sect. 4 we report the results of a number of "shock" experiments, where after stabilization one of the parameters of the system is reinitialized (technology, environment, birth model and others). In all cases, the after-shock inequality movement shows the same overall trend but it is smaller in size, and this is true irrespective of the sharing outcome. In Sect. 5 we present and test our hypothesis that cultural innovation is responsible for emergence of higher inequality. In Sect. 6 we present an additional experiment on cultural evolution and discuss its implications and its potential for future studies. Finally, in Sect. 7 we conclude our research and give final thoughts.

2 The Model

We are studying a fairly simple agent-based situated production model that pertains directly to preindustrial or primitive agricultural populations, although it may be argued to apply to other forms of production economy as well, provided there is no trade. The model uses a spatial grid of 30 × 30 cells where a number of agents representing extended families/groups live, produce and occasionally move. In what follows, we use the terms agent and family interchangeably. Every cell has a level of fertility and each family has a level of technological ability that allows it to extract food. A cell therefore contains 0 or more agents depending on its overall capacity to sustain them, which is a function of cell fertility and agents' technological ability. The environment has a degree of instability, i.e., a probability with which a percentage of its production is lost (for climatic or natural resource reasons). Food that is not immediately consumed is stored for the future in private family stores or in public ones and the families decide about whether to store publicly and how much. All stores may be of limited or unlimited capacity.

Families grow or shrink with a constant birth and death rate, respectively. When the size of a family exceeds a size threshold, the family splits in two and the newborn family takes half the people and half the stored food. The newborn inherits all the parameters of its parent and there is a small probability of cultural mutation, i.e. of reinitialization of sharing gene and rate. If the fertility level of the current cell is insufficient, the newborn family migrates. If a family cannot sustain itself within the current cell either because fertility is insufficient or because its stored food has fallen below a security level, it migrates as well. If a migration cell with sufficient fertility cannot be found, a family may initiate a war against a neighbour. Such a survival war is initiated against the richest neighbour and the aggressor is supposed to be always successful. The aggressor and winner then steals all the privately stored food of the victim as well as its share in the public store. The victim migrates elsewhere or dies if this is not possible. The general algorithm is given below:

```
1.   Production locally according to size of family, technological
     means and place fertility
2.   Public sharing (if applicable): a proportion of the production
     (= sharing rate * production) is sent to the public store in the
     agent's position
3.   surplus = rest of production (after sharing) - current need
4.   If surplus > 0, then store it as food
5.   Else consume (-surplus) from publicly stored food in its
     position or in neighbouring positions
6.   If still in need,
             launch survival war against the richest neighbour
     If no such neighbour exists,
             try to migrate in a rich nearby place
     If nothing works, die (starve)
```

Table 1 presents the most important parameters of the model. Parameter diversity allows the emergence of population differences.

Table 1. Parameters of the basic model

Environment	
Environment size	30×30
Environmental instability	3 possible values: {0, 0.2, 0,5}
Environmental loss of production rate	3 possible values: {0, 0.5, 0.8}
Number of agents at t = 0	N = 100
Cell fertility	Uniform from 400 to 600
Maximum storage capacity per position	10 to 1000 or unlimited (−1)
Behavior of agents (extended families)	
Migration cost	1 or 2
Need for food	10 or 11
Vision	1 to 4 cells
War vision	1 to 4 cells
Food security level	1 or 2
Technology	Uniform from 1 to 5
Birth rate	0.3
Death rate	0.2
Maximum size	40 or 41
Private storage capacity	10 to 1000 or unlimited (−1)
Sharing gene	On/Off (with initial probability 50%/50%)
Sharing rate	Uniform from 0 to 0.5
Cultural mutation rate	0.05 (probability of sharing gene and rate reset at birth)
Cultural inheritance rate	1 by default (see Sect. 6)

All the agent parameters are inherited during family split. This basic model leads to Malthusian population evolution[1], where after a while agents fill the whole 2D-array of cells and the population size stabilizes around a value that may be regarded as the carrying capacity of the environment for certain instability characteristics and technological abilities (see Fig. 1). The environmental and behavioral parameters have been tuned to allow the population to stabilize fairly quickly (in 1000 to 2000 cycles) to the Malthusian state. The Malthusian state ensures that our results do not correspond to transient behaviors but represent the ultimate evolutionary trends of the system.

[1] Reverend Thomas Malthus (1766–1834) claimed that "the increase of population is necessarily limited by subsistence", i.e., when population grows faster than the increase of food production then the population will arrive at a plateau where famine, wars and diseases will prevail and where the population size is the maximum allowed by the actual production rates. His suggestion was to use higher production rates for increasing standards of living rather than letting the population grow unconstrained.

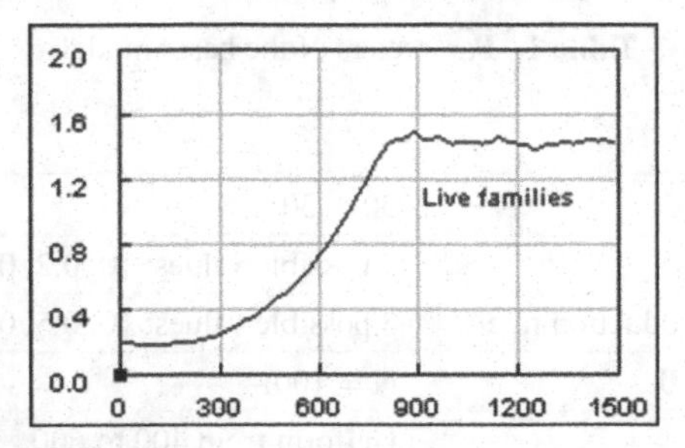

Fig. 1. (x: time, y: number of live families in 1000s) Typical outcome for initial N = 100, technology = 1 for all agents, highly unstable environment (instability = 0.5, loss rate = 0.8). The population stabilizes around 700 families that have filled the 30 × 30 cell grid (1 or 2 families per cell). The population will stabilize to the Malthusian[1] limit independently of the initial number of families. The speed of convergence to the limit may differ according to the various behavioral parameters.

The parameters have also been tuned to demonstrate these trends in terms of sharing. In the experiments that follow we initialize each family in the population with the sharing gene ON with 50% probability and with a sharing rate uniformly distributed between 0 and 0.5 (thus on average 0.25). The evolutionary dynamics of sharing are generally of one of two forms: either full sharing or no sharing emerges (see Fig. 2). Intermediate degrees of sharing are very rare and they are found almost exclusively in cases where the population has not stabilized within the experimental timeframe used (3000 cycles unless otherwise stated). The computational complexity of the simulation is fairly high with each run necessitating several minutes if thousands of agents are involved in the Malthusian condition. Therefore we have experimented with small size environments of 30 × 30 size, but we have performed some indicative but non exhaustive experiments with large worlds (100 × 100) to ensure that we get at first sight comparable results for sharing and inequality independently of size.

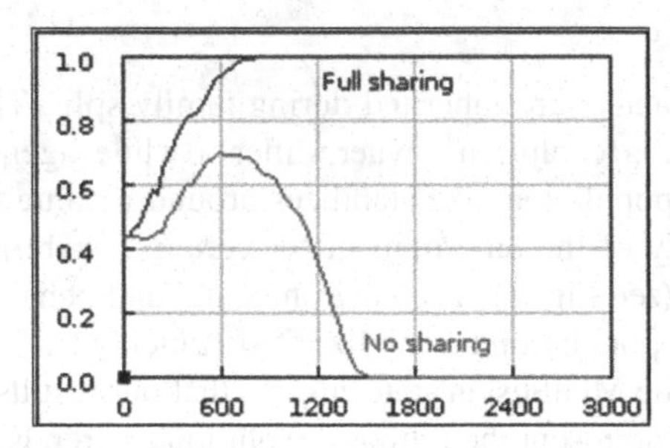

Fig. 2. (x: time, y: average sharing gene value in the population) Two typical sharing gene evolutionary movements. Lower (red) curve: A population where sharing disappears. Upper (blue) curve: A population where sharing is established. (Color figure online)

3 Reference Condition

We have run experiments to evaluate whether sharing evolves and to what degree in various environmental conditions and technological setups. In all cases, we start from 50% presence of the sharing gene in the initial population with an average rate of 25% (0.25, uniformly distributed between 0 and 0.5) and we measure and compare the average sharing gene and sharing rate in the final society after stabilization to a Malthusian state. We also measure wealth inequality in the population in a Lorenz-like manner, by computing the average deviation of agents' wealth from the average total value (details of computation are omitted for lack of space) and ranges from 0 (full equality) to 1 (full inequality). All results of experiments in this and the following sections are averages of 50 runs[2].

In the reference experiment shown in Table 2, we give the results of basic productive versus more advanced agents (technology = 1 uniformly in the population or technology uniformly distributed between 1 and 5) in stable (instability = 0) or extremely unstable environments (instability = 0.5, loss of production rate = 0.8). In our previous work [10], we have identified the conditions that define a strong tendency for emergence of sharing, namely environmental instability, low technological ability and especially much higher

Table 2. Reference results for two values of technology and two types of environment: Number of live families, Average sharing gene, Average sharing rate (these measures are taken in the end of the experiment), Min/Max/Average wealth inequality (these measures refer to the trend during the experiment). Averages of 50 runs.

	Live	**Sharing gene**	**Sharing rate**
	Ineq. (min)	**Ineq. (max)**	**Ineq. (avg)**
Tech 1 uniform			
No instability	1474.48	0.401	0.026
	0.353	0.758	0.52
High instability	727.38	0.89	0.069
	0.291	0.782	0.5
Tech 5 diverse			
No instability	4670.28	0.15	0.018
	0.331	0.736	0.423
High instability	3131.34	0.311	0.077
	0.291	0.745	0.42

[2] The number of runs was set empirically to ensure that the resulting standard deviations of all experiment series are very low compared to averages. This way, when comparing inequality measures, that are normalized by construction, we can judge significance in a quick, ad hoc manner. For example, for the second pair of lines of Table 3, the averages of 0.792 and 0.545 for Ineq(max) have a stdv of 0.041 and 0.061, respectively, therefore they are significantly different.

public storage capacity than private storage capacity, where these capacities express actual technological advances but also cultural values and processes. Here, however, public and private storage capacities are unlimited.

Firstly, we observe that sharing is far more common and with higher rates in unstable environments and for less productive agents. The numbers of live families depend on both these parameters (technological productivity and environmental instability) and are around the Malthusian limit in all cases. According to the previous section, the sharing gene results ought to be read as proportions of runs that have led to full sharing. We obtain similar results if we disable survival wars or if there is diversity in terms of a "survival war gene" and for other variations of the original model.

Secondly, we observe that the maximum inequality reached during society evolution with unlimited capacities is a little over 70% and this is consistent independently of technology and environmental characteristics. Figure 3 shows the movements of inequality in cases of emergence of full sharing or extinction of sharing, while Table 2 gives the minimum, maximum and average values of wealth inequality in all cases. This initially upward and then downward trend is reminiscent of the Kuznets curve [17]. Looking into the postwar history and the discussion on inequality trends and the changes that trigger them [14, 16–19], we can immediately think that a new socioeconomic environment may trigger such a movement. Candidate novelties that match postwar conditions are technological advancement, spatial expansion (and lately globalization), population collapse (such as immediately after the war), baby boom and family model change (from extended to nuclear family). All these changes have been rather abrupt compared to the more gradual changes before the eruption of the war and can thus be studied in the form of "shocks" applied to a system that has stabilized.

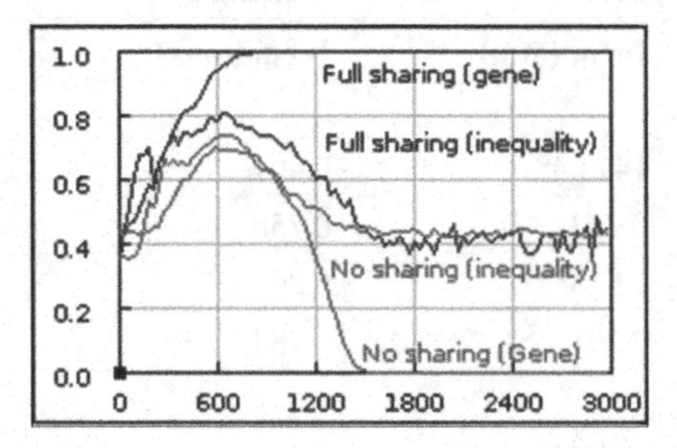

Fig. 3. (x: time, y: average sharing gene value in the population) For the experiments of Fig. 2, sharing gene and corresponding wealth inequality evolutionary movements. Lower (red) curves: A population where sharing disappears. Upper (blue) curves: A population where sharing is established. (Color figure online)

Before going on, one tricky observation has to be laid down, which concerns the relation between sharing and inequality. Looking into Fig. 3, we see that the rise of sharing parallels the rise of inequality, so that one may think that the rise of inequality causes sharing to rise. This is not right. If there is a causality, it is the opposite way, because sharing is the result of agent decisions, whereas inequality does not really exist in the system: it is an external measure we take and the agents have no idea such a thing exists, they do not perceive it, let alone do something about it. But causality is a philosophically-laden issue that we will not discuss here.

4 Shocks

As argued in the previous section, we experiment with a number of "shocks" to identify the type of inequality trend found in each case and which shock can be responsible for inequality rise. In all cases, after the system has stabilized, one of the parameters is changed and the system is re-run to eventual stabilization. The shocks defined and studied are the following:

```
Technology shock 2 (low to high):
Before: Technology=1, After: Technology=2
Technology shock 2 (high to low):
Before: Technology=2, After: Technology=1
These shocks can come with/without stored food loss (80% loss)
Population shock:
After: Loss of 25% of the population (number of families)
This shock can come with/without stored food loss and with/without
stored family members loss (70% loss)
Family model shock 1 (extended to nuclear family):
Before: Family size = 40, After: Family size = 10
Family model shock 2 (nuclear to extended family):
Before: Family size = 10, After: Family size = 40
Birth shock 1 (baby boom):
Before: Birth rate = 0.2, After: Birth rate = 0.3
Birth shock 2 (infertility):
Before: Birth rate = 0.3, After: Birth rate = 0.2
Environment shock 1 (becomes harsher):
Before: No instability, After: High instability
Environment shock 2 (becomes easier):
Before: High instability, After: No instability
Spatial shock 2 (world expansion):
Before: Environment size = 30 × 30. After: Environment size = 40 × 40
Spatial shock 2 (world shrinkage):
Before: Environment size = 40 × 40, After: Environment size = 30 × 30
Resource renewal:
After: All recource capacities reinitialized to corresponding
fertilities
```

Many of the shocks are actually indirect expressions of a technological shock: for example, an easier or harsher environment may be only the result of technological advance or collapse related to production, whereas world expansion or shrinkage may be only the result of technological advance or collapse related to transportation and mobility. Technological collapses are thinkable because of civilization collapse or conquest.

In Tables 3, 4, 5, 6, 7, 8 and 9, we report the inequality results for each one of the above shocks and we compare the inequality measures at the time of the shock (end of part 1) with those after the system has re-stabilized (end of part 2). In all cases the new inequality measures are consistently lower after the shock than before. This is surprising, because we were expecting at least the historically obvious candidates (world expansion, technological advance and easier environment) to allow at least the same levels of inequality. And actually the reason for experimenting with such pairs of symmetric shocks is to see whether the direction of change for a shock matters.

On the contrary, when an experiment is fully restarted after stabilization (which typically means that population is reinitialized from scratch to a low population size

Table 3. Technology shock results both for technology going from a low (=1) to a high value (=2) and going from a high (=2) value to a low (=1) one. No stored food loss.

Part 1	Ineq. (min)	Ineq. (max)	Ineq. (avg)
Part 2	**Ineq. (min)**	**Ineq. (max)**	**Ineq. (avg)**
Part 1: Tech = 1, Part 2: Tech = 2			
No instability	0.353	0.758	0.569
	0.239	0.573	0.43
High instability	0.303	0.792	0.556
	0.29	0.545	0.404
Part 1: Tech = 2, Part 2: Tech = 1			
No instability	0.352	0.751	0.511
	0.359	0.507	0.463
High instability	0.325	0.727	0.548
	0.23	0.479	0.376

Table 4. Population shock results. No stored food or family members loss. Technology = 1.

Part 1	Ineq. (min)	Ineq. (max)	Ineq. (avg)
Part 2	**Ineq. (min)**	**Ineq. (max)**	**Ineq. (avg)**
Part 1: Usual, Part 2: 25% population loss			
No instability	0.353	0.756	0.582
	0.278	0.583	0.481
High instability	0.322	0.784	0.584
	0.256	0.54	0.418

value compared to its Malthusian limit) the inequality levels obtained are as high as usual. This conflicts with the results from the population shock. The only other difference between full experiment restart and population shock is that in full restart all families randomly reinitialize their sharing values (gene ON with 50% probability and sharing rate uniformly distributed between 0 and 0.5, thus on average 0.25), like in the beginning of the experiment. On the contrary, shocks concern already established populations with more or less stable sharing profiles. Thus we form the hypothesis that the cultural reinitialization is responsible for the after-shock higher inequality levels reached. We dissect and study this hypothesis in the next section.

Table 5. Family model shock results both for extended family (max 40 members) becoming nuclear (max 10 members) and for nuclear family becoming extended. Technology = 1.

Part 1	**Ineq. (min)**	**Ineq. (max)**	**Ineq. (avg)**
Part 2	**Ineq. (min)**	**Ineq. (max)**	**Ineq. (avg)**
Part 1: Fam. Size 40, Part 2: Fam. Size 10			
No instability	0.355	0.756	0.573
	0.315	0.498	0.344
High instability	0.334	0.787	0.591
	0.275	0.505	0.329
Part 1: Fam. Size 10, Part 2: Fam. Size 40			
No instability	0.311	0.66	0.385
	0.288	0.486	0.446
High instability	0.271	0.656	0.376
	0.211	0.461	0.338

Table 6. Birth shock results both for baby boom (birth rate going from 0.2 to 0.3) and for fertility loss (birth rate going from 0.3 to 0.2). Technology = 1.

Part 1	**Ineq. (min)**	**Ineq. (max)**	**Ineq. (avg)**
Part 2	**Ineq. (min)**	**Ineq. (max)**	**Ineq. (avg)**
Part 1: Birth rate 0.2, Part 2: Birth rate 0.3 (Boom)			
No instability	0.349	0.757	0.583
	0.377	0.525	0.441
High instability	0.308	0.785	0.592
	0.242	0.494	0.368
Part 1: Birth rate 0.3, Part 2: Birth rate 0.2 (Infertility)			
No instability	0.352	0.749	0.512
	0.408	0.514	0.466
High instability	0.304	0.765	0.531
	0.269	0.511	0.408

5 Cultural Innovation

Cultural reinitialization is full reset of the sharing profile (sharing gene and sharing rate) of families after the shock. We experiment with degrees of cultural innovation in the population after the shock, where only a percentage of families reinitialize or innovate culturally, i.e. they reset their sharing profile. We try three such degrees: 20%, 60%, 100%.

Table 7. Environment shock results both for environment becoming harsher (from no to high instability) and for environment becoming easier (from high to no instability). Technology = 1.

Part 1	Ineq. (min)	Ineq. (max)	Ineq. (avg)
Part 2	Ineq. (min)	Ineq. (max)	Ineq. (avg)
Part 1: No instability, Part 2: High instability			
	0.35	0.762	0.589
	0.249	0.557	0.395
Part 1: High instability, Part 2: No instability			
	0.326	0.789	0.584
	0.327	0.658	0.505

Table 8. Spatial shock results both for world expansion (from 30 × 30 to 40 × 40 world) and for world shrinkage (from 40 × 40 to 30 × 30 world). Technology = 1.

Part 1	Ineq. (min)	Ineq. (max)	Ineq. (avg)
Part 2	Ineq. (min)	Ineq. (max)	Ineq. (avg)
Part 1: Small world, Part 2: Big world			
No instability	0.367	0.77	0.54
	0.416	0.536	0.465
High instability	0.287	0.791	0.523
	0.259	0.474	0.384
Part 1: Big world, Part 2: Small world			
No instability	0.349	0.757	0.553
	0.38	0.509	0.461
High instability	0.302	0.78	0.534
	0.215	0.478	0.37

Table 10 gives comparative results for inequality in the case of a population shock in an environment with high instability, where at shock time the corresponding percentage of families act as cultural innovators, i.e. they reinitialize their sharing profile. From the table it becomes obvious that higher degrees of innovation bring the after-shock system closer to the original levels of inequality and that only when all the population innovates are these levels the closest and actually comparable.

These results are similar for all other shocks: the original levels of inequality are only reached with full cultural innovation. This observation makes us suspect that since the type of shock is irrelevant, maybe there is no need to have any shock at all. Cultural innovation might do the job alone. We test this hypothesis by repeating the experiment with cultural innovation without any other change in the system, i.e., without shock. The

Table 9. Resource renewal shock results. Technology = 1.

Part 1	**Ineq. (min)**	**Ineq. (max)**	**Ineq. (avg)**
Part 2	**Ineq. (min)**	**Ineq. (max)**	**Ineq. (avg)**
Part 1: Usual, Part 2: Resource renewal			
No instability	0.331	0.748	0.575
	0.453	0.535	0.494
High instability	0.308	0.777	0.574
	0.32	0.503	0.427

Table 10. Population shock results in a high instability environment with various degrees of cultural innovation. Technology = 1.

Part 1	**Ineq. (min)**	**Ineq. (max)**	**Ineq. (avg)**
Part 2	**Ineq. (min)**	**Ineq. (max)**	**Ineq. (avg)**
Population shock, High instability			
20% innovators	0.328	0.79	0.591
	0.285	0.587	0.446
60% innovators	0.3	0.775	0.58
	0.279	0.653	0.456
100% innovators	0.328	0.795	0.594
	0.277	0.796	0.529

Table 11. No shock results in a high instability environment with various degrees of cultural innovation. Technology = 1.

Part 1	**Ineq. (min)**	**Ineq. (max)**	**Ineq. (avg)**
Part 2	**Ineq. (min)**	**Ineq. (max)**	**Ineq. (avg)**
No shock, High instability			
20% innovators	0.278	0.788	0.519
	0.252	0.521	0.383
60% innovators	0.303	0.778	0.535
	0.236	0.619	0.395
100% innovators	0.296	0.791	0.55
	0.238	0.782	0.452

results are shown in Table 11 and they confirm our idea that cultural reinitialization or innovation is the sole responsible for the movement and scale of inequality.

But, what is actually cultural innovation? And what are degrees of cultural innovation? We contend that as a response to a perceived new or very different environment, some people and some families may explore new behaviors, far more than is usually expected in times of no or little change. It makes therefore sense to think that the greater the change or the shock, the greater the number of people that will try new things. The world-changing events are precisely those that trigger novel behaviors from the majority of people. It is noteworthy that the direction of innovation is irrelevant because in the beginning there is no information about what is worthwhile and what not. For example, as a response to resource shrinkage, some families may think that sharing would increase their chances of survival whereas some others would become more selfish for the same reason.

6 General Cultural Evolution

We have performed a final experiment that concerns the cultural inheritance rate, i.e. the probability that a newborn will inherit the cultural values (sharing gene, sharing rate) of its parent. Up to now this was set to 1, i.e. there was always cultural inheritance and a small degree of cultural mutation at birth (set to 0.05 as Table 1 shows). The alternative to cultural inheritance would be for a newborn to be randomly reinitialized to a new such "personality". The cultural inheritance rate gives the probability that a newborn will inherit its parent, it is itself reinitialized uniformly between 0.2 and 0.8 and it may be inherited culturally together with the other sharing parameters. We have repeated the experiments of the previous sections with random, not 100%, cultural inheritance as just described and we have found that, (i) the results in terms of sharing and inequality do not differ, and (ii) the average cultural inheritance rate in the population consistently stabilizes to very high values (between 60% and 70% which is close to the theoretical maximum of 80%). The latter means that it is best for the population's survival and progress to build culture but also to maintain a degree of cultural individuality. These apply to the shock and innovation case as well, as is shown in the following Table 12 for the baby boom case and is verified for all other shocks as well. There, inheritance rate shows a slight upward trend after the shock, which is an indication for a Baldwin effect [23], i.e. it acts as an adaptation that serves to allow other adaptations. It should also be noted that there is no significant effect of the cultural mutation rate (initially 5%) and it could be safely set to 0.

Why is this interesting? One obvious reply is that it shows that the importance and effect of culture may be over-estimated, since while on the one hand culture represents an evolutionary strength, on the other hand people can massively and successfully deviate from it significantly.

Another thought is that if the people can discover a successful solution without culture and cultural evolution, then this might be evolutionarily dominant by Occam's razor. Some initial experiments that we performed show that less or no cultural inheritance leads to more ample sharing, to the detriment of non sharer families that go extinct rapidly. But nature is absolutely indifferent to "who" survives. Probably then, culture is

Table 12. Baby boom results in a high instability environment with cultural innovation and evolution of cultural inheritance rate.

Part 1	**Ineq. (min)**	**Ineq. (max)**	**Ineq. (avg)**	**Inheritance rate**
Part 2	**Ineq. (min)**	**Ineq. (max)**	**Ineq. (avg)**	
Baby boom shock, High instability				
20% innovators	0.352	0.778	0.585	0.659
	0.307	0.553	0.438	0.693
60% innovators	0.338	0.775	0.582	0.662
	0.296	0.632	0.436	0.706
100% innovators	0.344	0.781	0.582	0.653
	0.316	0.766	0.494	0.693

a sort of individual defense by imitation of the closest fittest (although proximate and ultimate fitness indicators can differ significantly). Or, culture has served some other evolutionary or psychological role and is recruited spontaneously but not necessarily profitably. All in all, why culture has evolved is still a mystery and experiments such as this help shed some light to the possible causes and mechanisms of evolution of culture.

7 Discussion

We have studied a simple situated production model that pertains directly to preindustrial agricultural populations and where sharing can evolve under certain conditions, starting from a random configuration. In this model, wealth inequality that emerges between agents follows a well-known, Kuznets-like, trend of initially upward and then downward movement before stabilizing. We prove that the idea of attributing this trend to the characteristics of the environment (natural, technological, social etc.) is false. This trend is the result of massive cultural innovation, i.e. of the agents' decisions to reset their sharing profiles as a response to a new or very different environment. Consequently, the resulting inequality emerges from the decision rather than from the triggering shock condition and it can theoretically emerge even in the absence of an external trigger. An assorted idea is to identify other phenomena that emerge spontaneously from a decision rather than from the apparent external or environmental condition. One obvious further direction of our study will be to test this hypothesis in more advanced economies, with intangible status, luxury production, trade and/or industry. Although we do not expect the economy-related decisions to act differently on inequality compared to our subsistence model, the emerging phenomena could also depend on individual or cultural assessments of goods and values, which are in turn responsible for behavioral decisions. Such a study would allow us to draw some inferences that could be closer to our modern complex economic environments. The issue of the world size also merits consideration. On the one hand, it is possible that scaling in very large environments (say 500 $\times$ 500) skews the results toward a different direction, such as has been observed for example in simple simulations of cultural assimilation and dissemination [24]. On the other hand, it is not realistic to assume that environments can be arbitrarily large, because people tend to quickly form "small worlds" and personal spaces, even if the current technological capabilities allow these small worlds to be bigger than before. Therefore a realistic scaling study should include some sort of partner selection in the social environment and development of a personal niche for each agent. The consideration of a family model (extended or nuclear) as in Sect. 4 may be crucial in this respect. Finally, as noted in the previous section, our model can be reworked to study the interaction between individual decisions and evolution of culture.

References

1. Gurven, M.: To give and to give not: the behavioral ecology of human food transfers. Behav. Brain Sci. **27**(4), 543–583 (2004)
2. Kaplan, H., Gurven, M.: The natural history of human food sharing and cooperation: a review and a new multi-individual approach to the negotiation of norms. In: Gintis, H., Bowles, S., Boyd, R., Fehr, E. (eds.) Moral Sentiments and Material Interests, The Foundations of Cooperation in Economic Life, Chapt. 3, pp. 75–113 (2005)
3. Moore, J.: The evolution of reciprocal sharing. Ethology Sociobiol. **5**(1), 5–14 (1984)
4. Jaeggi, A.V., Gurven, M.: Natural cooperators: food sharing in humans and other primates. Evol. Anthropol. **22**(4), 186–195 (2013)
5. Nettle, D., Panchanathan, K., Shakti Rai, T., Fiske, A.P.: The evolution of giving, sharing and lotteries. Curr. Anthropol. **52**(5), 747–756 (2011)
6. Boyd, R.: The evolution of reciprocity in sizable groups. J. Theor. Biol. **132**(3), 337–356 (1988)
7. Winterhalder, B.: Social foraging and the behavioral ecology of intragroup resource transfers. Evol. Anthropol. **5**(2), 46–57 (1996)
8. Kaplan, H.S., Schniter, E., Smith, V.L., Wilson, B.J.: Risk and the evolution of human exchange. Proc. R. Soc. Lond. B Biol. Sci. **279**(1740), 2930–2935 (2012)
9. Kaplan, H.S., Schniter, E., Smith, V.L., Wilson, B.J.: Experimental tests of the tolerated theft and risk-reduction theories of resource exchange. Nat. Hum. Behav. **2**(6), 383–388 (2018)
10. Tzafestas, E.: Relation between the public and the private and evolution of food sharing. In: Proceedings Social Simulation Conference, Milan (2022)
11. Hao, Y., Armbruster, D., Cronk, L., Aktipis, C.A.: Need-based transfers on a network: a model of risk-pooling in ecologically volatile environments. Evol. Hum. Behav. **36**(4), 265–273 (2015)
12. Aktipis, C.A., Cronk, L., de Aguiar, R.: Risk-pooling and herd survival: an agent-based model of a Maasai gift-giving system. Hum. Ecol. **39**(2), 131–140 (2011)
13. Scheffer, M., van Bavel, B., van de Leemput, I.A., van Nes, E.H.: Inequality in nature and society. Proc. Natl. Acad. Sci. **114**(50), 13154–13157 (2017)
14. Scheidel, W.: The Great Leveler. Violence and the History of Inequality from the Stone Age to the Twenty-First Century. Princeton University Press, Princeton (2017)
15. Kohler, T.A., Smith, M.E.: Ten Thousand Years of Inequality: The Archaeology of Wealth Differences. University of Arizona Press, Tucson (2018)
16. Kondratieff, N.: The Long Waves in Economic Life (1925). (in Russian)
17. Kuznets, S.: Secular movements in production and prices: their nature and their bearing about cyclical fluctuations. Houghton Mifflin (1930)
18. Milanovic, B.: Global Inequality: A New Approach for the Age of Globalization. Harvard University Press, Cambridge (2016)

19. Midlarsky, M.I.: The Evolution of Inequality: War, State Survival, and Democracy in Comparative Perspective. Stanford University Press, Redwood City (1999)
20. Ames, K.M.: On the evolution of the human capacity for inequality and/or egalitarianism. In: Price, T.D., Feinman, G.M. (eds.) Pathways to Power. Fundamental Issues in Archaeology, pp. 15–44. Springer, New York (2010). https://doi.org/10.1007/978-1-4419-6300-0_2
21. Mattison, S.M., Smith, E.A., Shenk, M.K., Cochrane, E.E.: The evolution of inequality. Evol. Anthropol. **25**(4), 184–199 (2016)
22. Starmans, C., Sheskin, M., Bloom, P.: Why people prefer unequal societies. Nat. Hum. Behav. **1**(4), 1–7 (2017)
23. Baldwin, J.M.: A new factor in evolution. Am. Nat. **30**, 441–451 (1896)
24. Axelrod, A.: The dissemination of culture: a model with local convergence and global polarization. J. Confl. Resolut. **41**(2), 203–226 (1997)

Metaheuristics, Robotics, and Machine Learning

Online Adaptation of Robots Controlled by Nanowire Networks: A Preliminary Study

Paolo Baldini[1(✉)], Michele Braccini[1], and Andrea Roli[1,2]

[1] Department of Computer Science and Engineering (DISI), Campus of Cesena, Alma Mater Studiorum Università di Bologna, Cesena, Italy
p.baldini@unibo.it

[2] European Centre for Living Technology, Venice, Italy

Abstract. The ability to adapt to changes and unexpected situations is a commonly acknowledged hallmark of autonomy and intelligence. In this work we take inspiration from biology for the definition of a robot able to continuously adapt to changes. Specifically, we define its control structure and the mechanism used to perform the adaptation. The former is based on the *Reservoir Computing* framework, on which the latter acts. The result is the design of an Online Adaptive Reservoir Computing system based on a novel memristive reservoir: the *Nanowire Network*. Finally, the robot is tested on three different tasks taking place in different arenas. The results are then discussed and compared with a baseline algorithm.

Keywords: Phenotypic plasticity · Online adaptation mechanism · Nanowire network robotics

1 Introduction

Recent technological advances have made it possible to build incredibly small robots, till the size of tens of nanometers. Nevertheless, the current smallest robots can perform only few predetermined actions, therefore they cannot attain the level of adaptivity required to accomplish complex missions. Conversely, AI robotic software has recently made tremendous advancements and has been proven capable of tackling difficult tasks with a high degree of reliability. This software, however, cannot be run onto small size robots. A viable way for filling this gap is provided by control programs based on unconventional computation, such as the ones based on artificial neural networks or models of genetic networks.

In this work we present a study on the viability of using Nanowire Networks to control robots subject to an online adaptive mechanism. This work is a first step towards the deployment of small robots capable of adapting their behavior to the environment in which they operate. This is indeed the main property required to actual autonomous embodied agents [20].

C. De Stefano et al. (Eds.): WIVACE 2022, CCIS 1780, pp. 171–182, 2023.
https://doi.org/10.1007/978-3-031-31183-3_14

2 Nanowire Networks

One of the main innovative points introduced in this work is the use of Nanowire Networks (NNs) for the control of a robot [3]. NNs are a novel kind of nanoscale electrical circuit, whose interest resides in their ability to produce a neuromorphic behavior. This is given by their self-organizing property and intrinsic structural and functional plasticity, mimicking biological networks [17,18]. The similarity resides in what is known as Hebbian Theory [12], a property that is commonly summarized as "neurons wire together if they fire together" [15]. The artificial equivalent of the synaptic strengthening is here represented by the behavior of the ion-silver-bridges that connect the wires of the network (see Fig. 1 B). When subject to a voltage difference, the ions aggregate in the junction increasing its conductance. When the stimulus is removed, the network slowly returns to its stable state. In this work, we consider the voltage stimulation to be the only way to modify the internal state of the NN. This dynamics rules the network plasticity and can be seen as a short-term memory, that can be used to process spatio-temporal data [7,9,11].

NNs are a promising technology in all the context that require complex computations and have strict consumption constraints, like edge computing and robotics.

3 Control System

NNs are powerful computing devices with promising properties for robotics, but their use requires some ingenuity. We decided to design a robotic architecture shaped on their specific characteristics and, in particular, based on the Reservoir Computing (RC) framework and including an adaptive mechanism to endow the robot with phenotypic plasticity.

3.1 Reservoir Computing

Reservoir Computing is currently accepted as the *de facto* framework for the use of unconventional dynamic systems for computation [19]. The working principle consists in perturbing a *reservoir* and analyze its resulting state through the use of a simple learning method, like regression or classification [13,16] (see Fig. 1 A). This exploits the ability of the reservoir to re-project spatio-temporal inputs to a higher dimensionality. The result is a faster and less power-consuming training. However, despite all its benefits the use of RC in robotics is still limited [4]. Additionally, online variants of this framework have been explored only to a limited extent [1].

In this work, we address some of these missing explorations. Specifically, we propose the use of NNs as reservoir. Their non-linear dynamics allow indeed sequences of data to be evaluated differently, exploiting the intrinsic memory arising by their structural plasticity. Additionally, we suggest in combination the use of an online adaptive mechanism. The goal is to create a robot able to adaptively perform advanced operations, exploiting very simple systems: the NNs.

3.2 Adaptive Mechanism

In order to perform an online adaptation, we decided to modify the classical RC architecture substituting the training of the readout with a stochastic optimization process operating on the inputs. The idea is that it is possible to learn how to perturb the reservoir in order to induce a desired internal state. The approach consists in performing a reconnection and weighting of the input signals to different nodes of the network (see Fig. 1 C). The affected connections are a stochastically chosen sub-set ranging from 10% to 40% of the total. Network nodes eligible for reconnection must differ from previous ones and be k-nodes distant from the outputs[1]. The weighting consists in attenuating or intensifying the input signal through the use of a multiplier. Its initial value is chosen with a normal distribution centered in 1, and is then adapted adding the result of a normal distribution centered in 0. This parameter allows to balance the influence of specific inputs in the computation. A possible application is when some sensors are more useful than others for a specific task, and we want to enhance their role (e.g., front sensors in collision avoidance). Alternatively, this feature may help to balance measures of different physical properties, or in different environmental conditions where the signals homogeneously change (e.g., average brightness during day or night).

The proposed strategy is a variant of a methodology used for the adaptation of robots controlled by Boolean Networks (BNs) [6]. The novelty consists in the weighting of the input signals. This modification takes advantage of the analogic working mode of the NNs, that adds both a complexity and a potential compared to BNs. As the inspiring vision of this work is that of a micro-robot, the adaptation mechanism used in the experiment is minimalistic, so as to facilitate the construction for real applications.

3.3 Robotic Architecture

The robotic architecture we designed is inspired by biological organisms. Specifically, our model is shaped on the Central Nervous System (CNS), which forwards and transforms the signals from the sensors to the cortex, and from the cortex to the muscles. The final version of this architecture (see Fig. 1 C) was obtained by an iterative refactoring, trying to match the technical with the biological part. The first step concerned the individuation of five macro-areas: *i.* sensing ($S_0, ..., S_n$), *ii.* input weighting and reconnection ($\alpha_0, ..., \alpha_n$), *iii.* reservoir, *iv.* output reconnection (β), *v.* actuation (LM, RM). Each of them is mapped into its biological correspondent (see Fig. 2). The sensing apparatus (*i.*) is equivalent to the biological sensors. The transmission and processing of the input signals (*ii.*) is represented by the thalamus; its role is indeed to forward all the incoming signals and to redirect feedbacks to the correct location [21]. Additionally, it controls the flow of information and transforms the signals acting like a filter [5,8].

[1] In our experiments k is 2.

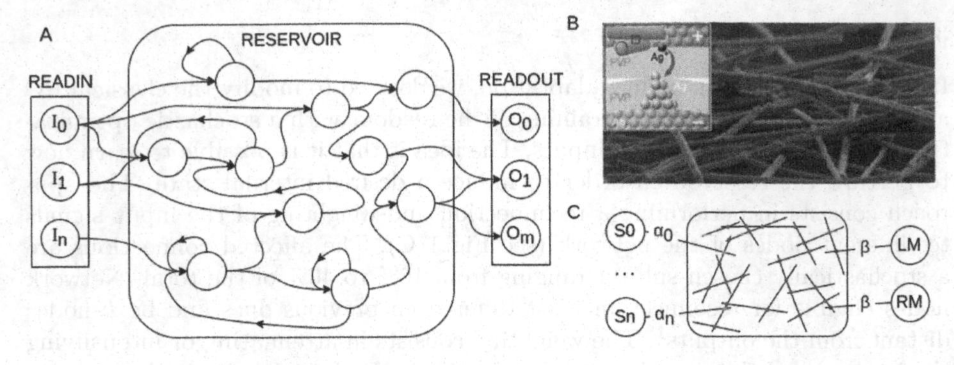

Fig. 1. A. The Echo State Network: one of the first instances of the Reservoir Computing architecture. **B.** Microscopy image of a Nanowire Network. Picture taken by [17] by courtesy of the authors. **C.** Schematic representation of the control architecture. $[S_0, ..., S_n]$ represent the robot sensors, while LM, RM represent the left and right motor. $[\alpha_0, ..., \alpha_n]$ represent the weighting factor, different for each sensor signal. β represents the influence of the motor resistance in the electrical equivalent system.

In fact, this is a simplification since in humans the transformation of the inputs does not happen in a single point, but along the entire afferent CNS[2].

The NN reservoir (*iii.*) performs most of the computations of the artificial control system. Because of that, we consider this component equivalent to the cortex. The transmission of the motor commands (*iv.*) from the cortex is represented by the pyramids. As for the afferent tract, this is also a simplification of the biological world. The efferent fibers[3] are indeed redirected to the muscles in many points of the descending tract. Finally, the motors (*v.*) are mapped on the biological muscles. This choice is straightforward, since both operate as an actuation apparatus.

In a nutshell, the basic idea of the control architecture is that the sensory inputs are weighted and forwarded to specific points of the network, influencing the "reasoning" of the robot. Accordingly, motion commands are taken from specific points and used to control the actuators.

3.4 Adaptation Cycle

The main goal of the adaptation mechanism is the emergence and optimization of a successful behavior, showing what is in biology known as phenotypic plasticity. This is the ability of a genotype to produce a visible response that depends on and adapts to the environmental conditions. To provide this ability the adaptation runs continuously during robot's life, and is therefore said to work *online* (see Fig. 3). The initial configuration is set at a random (i.e., a set of inward and outward connections and weights is sampled at random), and is evaluated during

[2] The tract of the CNS going from the sensors to the cortex.

[3] Fibers of the human body carrying signals from the cortex to the muscles.

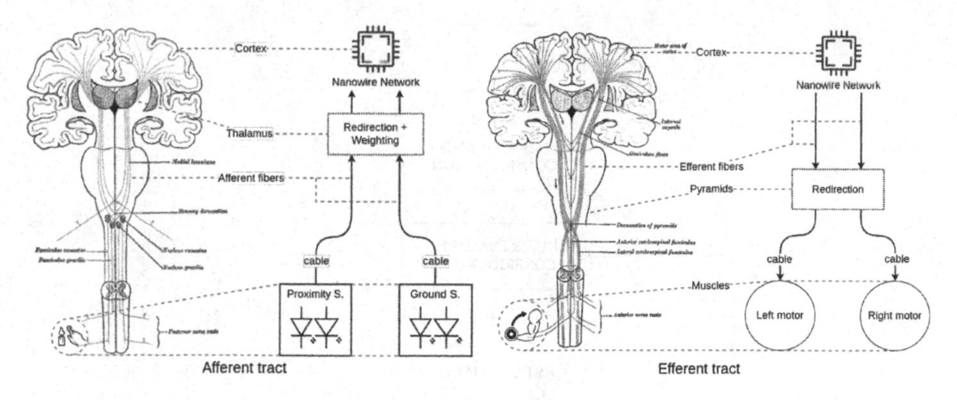

Fig. 2. Mapping between the control systems of the biological and artificial agent. On the left, the sensing path is represented. On the right, the control path is shown.

a fixed period of time and saved. This is then adapted by the adaptive mechanism previously described. The result is tested and, if better than the best one found, it is saved. Otherwise, the previously-found best-configuration is kept for future adaptations. The process repeats adapting the best found configuration and evaluating the quality of the adaptation. The quality is internally calculated by the robot itself in terms of a utility function, which is the agent internal driving force [2]. As the cycle continues perpetually, the robot has the opportunity to continuously improve its behavior and adapt it to possible changes in the working environment. As a consequence, each robot undergoes its own development, as typically happens in ontogenetic processes and historical processes in general [14].

During the experiments, we noticed that the optimization of some poor configurations slows down the adaptation. In order to speed up the process, we tested the use of a threshold value to early discard unsatisfactory solutions. Instead of them, new random ones are generated and evaluated.

4 Experiments

We evaluated the robotic system on three tasks taking place in three different simulated environments (see Fig. 4): *i.* Collision Avoidance (CA), *ii.* Area Avoidance (AA), *iii.* T-maze (TM). The experiments run within the Webots simulator and consist in the test of 250 unique NNs. The NN-based architecture is adapted 300 times, each generating a *configuration*. Each configuration is then tested during an *epoch*. The duration of each epoch depends on the task and on the size of the arena:

- 20s for the CA,
- 100s for the AA,
- 100s for the TM.

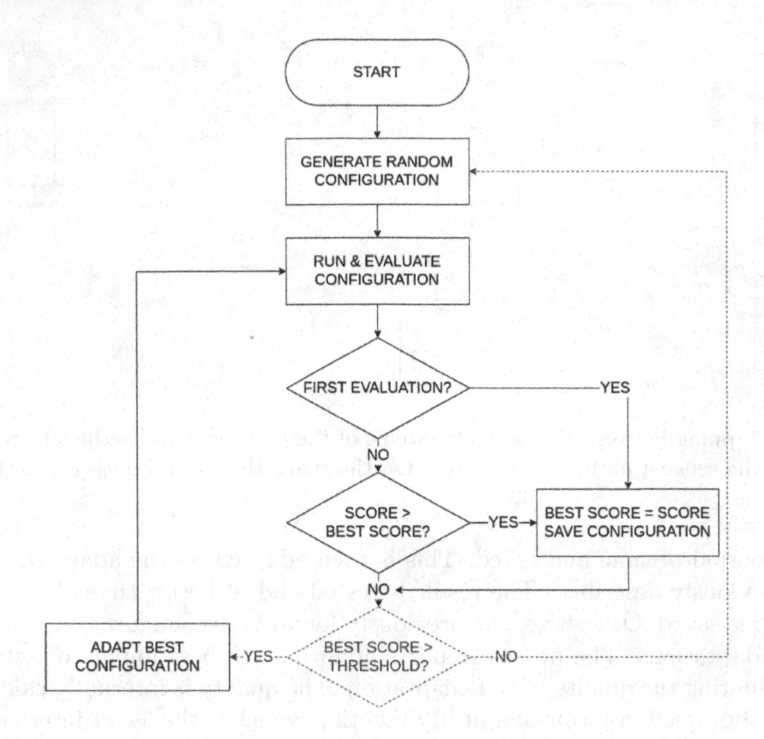

Fig. 3. The adaptation cycle that is continuously run by the robot. The gray dashed block and lines are an addition to the basic idea of adaptation.

Each assignment is defined by specific Objective Function (OF), that the robot self-evaluate and try to maximize.

The performance of this mechanism is compared with that of a stochastic mechanism, whereby the adaptation does not take place and the new configurations are generated randomly at each step. In other words, in the stochastic mechanism the best configuration is never adapted, but instead a new one is created from scratch at each epoch. This is used as a baseline to evaluate the quality of the adaptive mechanism.

4.1 Collision Avoidance

In the Collision Avoidance task the robot is required to avoid collisions while going as fast as possible on a straight line. Therefore, the corresponding OF penalizes excessive turns and time spent near to obstacles:

$$fitness = (1 - \sqrt{p_{max}}) \cdot (1 - |v_l - v_r|) \cdot \frac{v_l + v_r}{2} \tag{1}$$

where:

Fig. 4. The corresponding arena of each task: **A.** Collision Avoidance, **B.** Area Avoidance, **C.** T-Maze. **1** & **2** represent the T-maze end-points; **0** the starting-point).

$p_{max} \in [0, 1]$ is the normalized maximum proximity[4],
$v_l, v_r \in [0, 1]$ are respectively the normalized left and right motor speeds[5].

In Eq. 1, the proximity undergoes a square root in order to increase the sensibility of the OF to collisions[6]. The goal is to reward configurations that stay far from the obstacles, reducing the influence of the trajectory and speed.

The arena consists in a central obstacle and a set of external walls (see Fig. 4 A). The result is a circuit in which the robot has to run avoiding collisions. The passages on the left and right are stricter than the top and bottom ones, adding a minimal complexity to the task.

4.2 Area Avoidance

The Area Avoidance task is similar to the CA one, but requires the robot to avoid virtual areas identified only by the ground color. The obstacles stop to be physical and become instead intangible. This complicates the tasks because the robot has to react faster. A slow response would indeed push the robot deeper into the forbidden area, reducing the score of future actions. In this task the system is driven by a single sensor, allowing to test how much the weighting of the signal is useful in achieving a good result. Additionally, it shows that it is possible to obtain a wandering behavior and still avoiding obstacles with a computationally limited system. The performance of the robot at a specific step in the AA task is computed with the following OF:

$$fitness = (1 - |v_l - v_r|) \cdot \frac{v_l + v_r}{2} - 100 \cdot c \tag{2}$$

where:

[4] 0 represents a far object, 1 represents a near one.
[5] 0 represents an anticlockwise revolution, 0.5 a still state and 1 a clockwise revolution.
[6] Being the range of value in [0, 1], the square root increases the value of proximity.

$c \in \{0, 1\}$ is 1 if the robot is on the illegal area, 0 otherwise,
$v_l, v_r \in [0, 1]$ are respectively the normalized left and right motor speeds[7].

In Eq. 2, hovering an illegal area is associated with a strong penalty. This is independent of the direction of the robot or its speed. The result is that the adaptation leads to a behavior that mostly avoid the illegal areas, while the quality is tuned by the speed and direction of the movement.

The arena consists of few illegal areas that the robot cannot hover (white blocks in Fig. 4 B). Additionally, it is limited by the presence of virtual and physical borders, assigning a deserter robot negative scores while preventing it from going too distant. The goal is to allow every configuration to go back to a legal area, also if it starts in a disadvantageous position. Finally, every illegal area is surrounded by a neutral zone that informs the robot that is approaching a forbidden space.

4.3 T-Maze

In the T-Maze task the robot is required to reach the correct end-point of a T-shaped maze (see 1 & 2 in Fig. 4 C). The goal destination depends on the color of the floor at the start of the run (see 0 in Fig. 4 C): right for black, left for white. This requires the adaptive mechanism to exploit the NN plasticity to somehow memorize this initial input, and behave accordingly also when the signal stop to be perceived. In order to test the behavior, the robot is periodically kidnapped and placed back at the start of the maze. The OF for the TM task is the following:

$$fitness = 2 \cdot \overline{\alpha} \cdot \overline{\beta} - \alpha \tag{3}$$

where:

$\alpha \in \{0, 1\}$ is 1 if the ground color is the same as the starting one,
$\beta \in \{0, 1\}$ is 1 if the ground is gray.

The Eq. 3 highlights the presence of a gray color. This is the color of the ground outside the starting and ending points. In this region, the fitness of the robot does not change. Instead, the performance is penalized if the robot remains in the starting area or if it reaches the wrong end point. If the robot reaches the correct end point, its score is increased by 2 at each step of the permanence. This helps in slightly reducing the simulation time. One aspect not considered in Eq. 3 is the speed of the robot. The design of the task in combination with the epoch duration implicitly requires the robot to not be slow: in the opposite case, it would not have enough time to reach the end point.

5 Results

The results of the experiments are compared considering performance increment and distribution. The value at each time-step represents the average of the best

[7] 0 represents an anticlockwise revolution, 0.5 a still state and 1 a clockwise revolution.

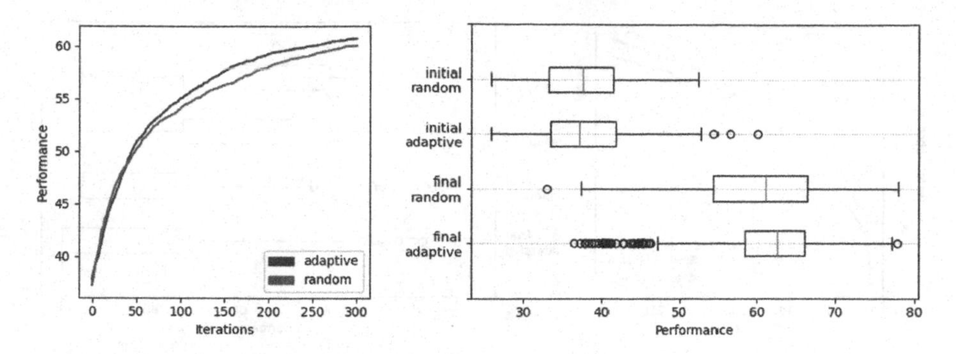

Fig. 5. CA results in terms of average fitness along the iterations (left) and distribution of the best results achieved in the 250 replicas of each experiment (right).

score obtained by each robot since the start of the simulation. This is therefore never decreasing, and is a good compromise to represent the ability to adapt [6].

The adaptive mechanism always shows higher performance than the random one. The difference between the results of the two approaches seems to be strictly related to the task that we are considering. In the CA, the results gap is not impressive (see Fig. 5). Indeed, although the adaptation leads to a slightly better performance, the random mechanism still allows to obtain good results. We can explain this behavior with the low complexity of the task and the high correlation of the signals from the sensors. Their disposition around the robot cause many of them to perceive the same obstacles. This allows the system to be intrinsically more resistant to faults, but also to perform well with just few sensors correctly connected. Additionally, in the given task the importance of the sensors is not homogeneous, with the front ones being more useful than the back ones.

When we consider the AA task the improvement is more evident, especially in the score distribution (see Fig. 6). This is due to the finer tuning of the weighting parameter, that helps to exploit all the information contained in the single input signal. Nevertheless, at the same time, the presence of a single sensor reduces the complexity of the connection, allowing also the random mechanism to perform well. This cause the effectiveness of the adaptation to be limited to the optimization of the input weight.

Finally, the TM task sees the highest improvement in performance (see Fig. 7). This is due to the increasing complexity, requiring a finer tuning. The ability to change direction is indeed strictly related to the way the system stores information in the NN. In order to make it influential, the mechanism has to accurately balance the stimulation from the ground sensor. This strongly suggests that the adaptation might become more useful as the complexity of the task increase, eventually becoming essential.

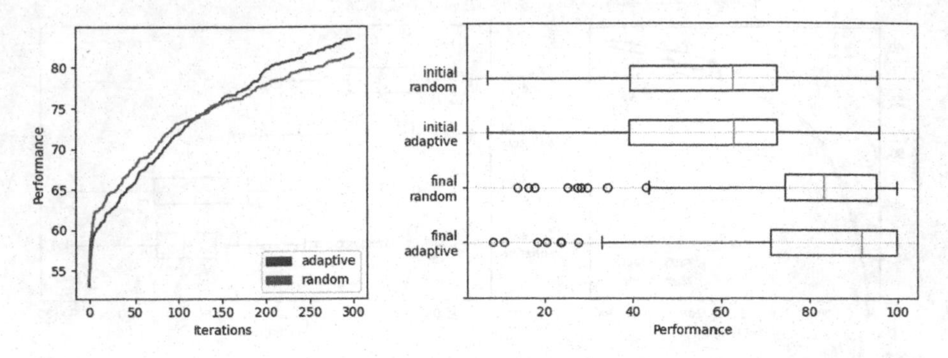

Fig. 6. AA results in terms of average fitness along the iterations (left) and distribution of the best results achieved in the 250 replicas of each experiment (right).

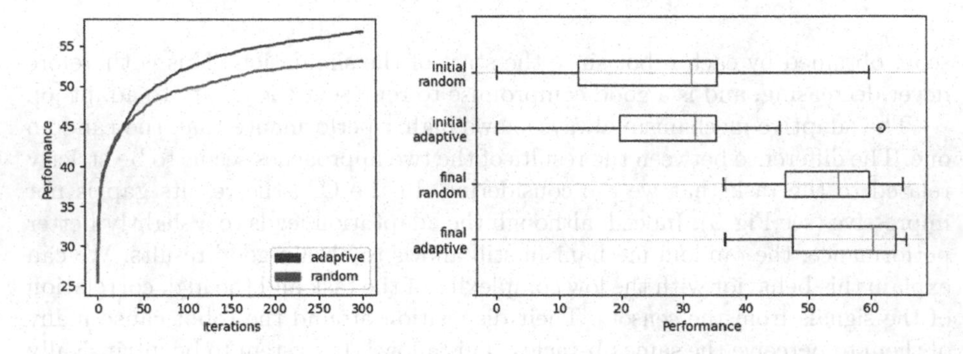

Fig. 7. TM results in terms of average fitness along the iterations (left) and distribution of the best results achieved in the 250 replicas of each experiment (right).

6 Discussion

The results presented in the previous section highlight a consistent gap between the scores obtained by the two compared mechanisms. The adaptation produces indeed overall better performances. Nevertheless, it is also evident that this difference is not impressive. One explanation is found in the amount of network nodes and epochs. Due to the complexity of the simulation, the tested NNs only contain a small amount of wires and junctions (see Table 1). Those are clearly much less than the number of epochs. This means that during the adaptation each sensor can possibly connect with any network node. Together with the intrinsic correlation of some sensor signals[8], and with their different influence[9], an effective solution may be found properly connecting just few sensors. In the full set of possible permutations of the input nodes, the amount of adequate

[8] Neighbor sensors often perceive same or similar information.

[9] Not all the sensors have the same utility in a task. For example, back sensors are usually less useful than front ones in the CA.

configurations is therefore fairly high. This means that the number of trials needed to obtain a good behavior is relatively small.

Table 1. Average number of wires and junctions in the used NNs. The difference between the AA task and the others is due to the creation density.

	Wires	Junctions
AA	68	120
CA	166	381
TM	166	381

The previous consideration is not completely satisfying, in that it would still suggest better performances of the adaptive method already from the initial epochs. The improvement is instead visible only after some tens of iterations. Therefore, another point to consider is the working mode of the NNs. Their relations with electrical circuits and the presence of the stimulation makes the choice of the connection node generally less important than in other network based systems (e.g., Boolean networks). Indeed, there is no risk for the input to disappear when reconnecting to neighbor nodes[10], but instead its influence on the actuators will only slightly change. Because of this working principle, the most important point in connecting to a NN seems to be related to the balance between the signals. This overall reduces the complexity of the wiring, but also limits the amount of complex behaviors that we can characterize from the network.

The result is that an adaptive reconnection, although useful, might not perform dramatically well compared to a random approach. It helps mainly when the complexity of the task is high or the size of the network grows, and when a more tuned balance between the input signals is needed. This idea is supported by the results obtained in the various tasks, seeing the performance gap increasing according to the complexity.

7 Conclusion

The goal of this work is to allow robots endowed with NNs to present the artificial correspondent of phenotypic plasticity. To achieve this goal we designed a control architecture based on NN and inspired by human CNS. The resulting system has been tested on three different tasks and environments. The results show that the robot is able to adapt its behavior to different tasks, attaining good performance. Additionally, they show that the adaptive mechanism allows to successfully exploit the intrinsic memory of the NNs. We conjecture that the limited advantage of an adaptive mechanism over a random one can be attributed

[10] Neighbor nodes in the graph representation of the NN. Two spatially near wires might indeed not be connected.

to the complexity of the task, the size of the NN, and its working mode. Therefore, we plan to test larger networks in harder tasks and environments. Other explorations may consider the use of local search techniques, exploiting a heuristic bias in the choice of the connections. As long term goal, we aim to design hardware adaptive mechanisms for the creation of microscopic robots, for example through the use of technologies like self-assembling wires [10].

References

1. Antonik, P., et al.: FPGA implementation of reservoir computing with online learning. In: 24th Belgian-Dutch Conference on Machine Learning (2015)
2. Ashby, W.: Design for a Brain: The Origin of Adaptive Behaviour, 2nd edn. Butler & Tanner Ltd. (1954)
3. Baldini, P.: Online adaptation of robots controlled by nanowire networks. Master's thesis, University of Bologna (2022)
4. Baldini, P.: Reservoir computing in robotics: a review (2022)
5. Basso, M., et al.: Cortical function: a view from the thalamus. Neuron (2005)
6. Braccini, M., Roli, A., Barbieri, E., Kauffman, S.: On the criticality of adaptive Boolean network robots. Entropy **24**(10), 1–21 (2022)
7. Christensen, D., et al.: 2022 roadmap on neuromorphic computing and engineering. Neuromorphic Comput. Eng. (2022)
8. Connelly, W., et al.: The thalamus as a low pass filter: filtering at the cellular level does not equate with filtering at the network level. Front. Neural Circuits (2016)
9. Demis, E., et al.: Nanoarchitectonic atomic switch networks for unconventional computing. Jpn. J. Appl. Phys. (2016)
10. Dueweke, M., et al.: Self-assembling electrical connections based on the principle of minimum resistance. Phys. Rev. E (1996)
11. Fu, K., et al.: Reservoir computing with neuromemristive nanowire networks. In: 2020 International Joint Conference on Neural Networks (IJCNN) (2020)
12. Hebb, D.: The Organization of Behavior: A Neuropsychological Theory. Psychology Press (2005)
13. Jaeger, H.: The "Echo State" Approach to Analysing and Training Recurrent Neural Networks. German National Research Center for Information Technology GMD Technical Report, Bonn, Germany (2001)
14. Longo, G.: How future depends on past and rare events in systems of life. Found. Sci. **23**(3), 443–474 (2018)
15. Löwel, S., Singer, W.: Selection of intrinsic horizontal connections in the visual cortex by correlated neuronal activity. Science (1992)
16. Maass, W., et al.: Real-time computing without stable states: a new framework for neural computation based on perturbations. Neural Comput. (2002)
17. Milano, G., et al.: Brain-inspired structural plasticity through reweighting and rewiring in multi-terminal self-organizing memristive nanowire networks. Adv. Intell. Syst. (2020)
18. Milano, G., et al.: Connectome of memristive nanowire networks through graph theory. Neural Netw. (2022)
19. Nakajima, K., Fischer, I. (eds.): Reservoir Computing: Theory, Physical Implementations, and Applications. NCS, Springer, Singapore (2021). https://doi.org/10.1007/978-981-13-1687-6
20. Pfeifer, R., Scheier, C.: Understanding Intelligence. MIT Press, Cambridge (2001)
21. Sommer, M.: The role of the thalamus in motor control. Curr. Opin. Neurobiol. (2003)

The Role of Dynamical Regimes of Online Adaptive BN-Robots in Noisy Environments

Michele Braccini[1(✉)], Edoardo Barbieri[1], and Andrea Roli[1,2]

[1] Department of Computer Science and Engineering (DISI), Campus of Cesena, Alma Mater Studiorum Università di Bologna, Cesena, Italy
m.braccini@unibo.it
[2] European Centre for Living Technology, Venice, Italy

Abstract. A novel online adaptation mechanism has been recently introduced, which is inspired by the phenotypic plasticity property present in biological organisms and which exploits the intrinsic computational capabilities of the Boolean network (BN) controlling the robot. In these robots, the BN is coupled with sensors and actuators and plays the role of the control system. The coupling is dynamically changed so as to increase a utility function. Recent results have shown that this mechanism can yield robots accomplishing tasks of a different nature and that, in general, critical networks attain the best performance compared to the other ones.

An analysis of this mechanism in noisy environments is needed to assess its performance in more realistic scenarios and to reduce the so-called reality gap. This work aims precisely to start investigating this question and to generalise the previously attained result.

Keywords: Phenotypic plasticity · Online adaptation mechanism · BNs robotics · Reality gap

1 Introduction

Experiments involving evolution and adaptation of artificial agents allow us to speculate about the prerequisites and general principles enabling the appearance (or the very existence) of living organisms with levels of capabilities comparable or superior to those attained in artificial contexts.

Boolean networks (BNs) [10] are promising models for the study of the properties of life in artificial contexts as they were introduced as models of gene regulation networks [4,8,9,14,20,21] and have proved capable of controlling robots, as in the Boolean network robotics (BN-robotics) approach [16–19].

A novel online adaptation mechanism that is inspired by the biological property of *phenotypic plasticity* has been recently proposed precisely with a twofold purpose [2]. In the short-term, it proposes itself as a mechanism capable of

C. De Stefano et al. (Eds.): WIVACE 2022, CCIS 1780, pp. 183–194, 2023.
https://doi.org/10.1007/978-3-031-31183-3_15

improving the adaptation capabilities of robots. In the long-term, it aims to find the general tenets behind the great adaptive capacity of living organisms.

Biologically, phenotypic plasticity is the ability of the same genotype to produce different phenotypes in response to the different environmental stimuli [7,11]. The artificial counterpart of phenotypic plasticity that we proposed tries to exploit the intrinsic computational capabilities of Boolean networks for controlling robots behaviours. In a recent work [3] we tested this mechanism by comparing it with different online adaptation techniques and in tasks of increasing difficulty. The main results can be summarised by saying that the ensembles of random Boolean network can be adapted without changing their internal structure and that criticality is an advantage condition for reaching higher performance.

In this paper, we build on previous work by introducing the concept of noise that perturbs the dynamics of the Boolean network that acts as robots' controller, aiming to generalise the results and considerations previously formulated. The new results show that under more realistic operating conditions the adaptation mechanism that exploits the inherent computational capabilities of Boolean networks produces, for the experimental conditions used, robots with higher maximum performance than the mechanism that modifies their structure. Furthermore, it is shown that the criticality property represents both an initial advantage over other dynamic regimes and a dynamic condition to which Boolean networks used as robot controllers and subject to online adaptation tend.

The paper is organised as follows. In Sect. 2 the methods used to answer the scientific question raised are described. In particular, the adaptive mechanisms and the tasks employed are described, together with the experimental details and the specific noise introduced. Section 3 shows the results and the related discussion. Lastly, Sect. 4 includes the conclusion and the possible future works.

2 Method

In this section we present the mechanisms used by the BN-robots during their adaptation phase and the specific tasks on which the robots will be tested. In addition, we provide the details of the experimental settings along with the particular definition of noise used in the experiments.

2.1 Adaptation Mechanisms

The online adaptation mechanisms implemented in our tests are as follows:

In-Out mapping—The initially random generated BN is not changed throughout the adaptation period, which is subdivided into evaluation epochs; at the end of each epoch, the current coupling between BN and actuators and sensors is perturbed. In this way, the inherent bouquet of dynamics the BN can express is exploited. Note that by doing so, the initial dynamical regime of the Boolean network is not changed.

Mutation—This one modifies the BN as follows: while the couplings sensors-BN and BN-actuators stay the same throughout the adaptation, both functions and connections of the BN are randomly perturbed. With the aim of finding a more suitable Boolean network configuration, at the end of each epoch the mutation mechanism logically negates $\frac{1}{100}$ of the Boolean function entries, and redistributes $\frac{1}{100}$ of the arcs, both in random fashion. Unlike the previous mechanism, here we will observe as a side effect a change in the dynamical regime of the networks.

2.2 Tasks Description

The adaptive pressure is represented for each task by an objective function. In detail, at the end of each epoch the objective function returns a score representing the quality of the current controller. The score is accumulated for each simulation step and reset at the beginning of each new epoch. A robot attempts to maximise its score along the epochs, through the use of the adaptive mechanisms presented above. At the end of each epoch, only those changes that do not decrease robot's performance are retained. So, a controller that obtain a score equal to F_{t+1} at the epoch $t+1$ is discarded only if $F_{t+1} < F_t$: sideways moves in the search space of Boolean networks are thus permitted.

The details of the chosen tasks for the realisation of the experiments are presented, including the arena and objective function used.

Task I—Simple navigation with collision avoidance. The arena used for this task is that present in Fig. 1. The objective function used for evaluating the performance at this task is as follows:

$$\frac{100}{E}\sum_{n=1}^{E}(1-\theta(n))\cdot(1-\sqrt{|\,l(n)-r(n)\,|})\cdot\frac{l(n)+r(n)}{2} \tag{1}$$

where $1-\theta(n)$ is the normalised distance between the robot and the closest obstacle, computed by means of the highest value returned by the proximity sensors; $l(n)$ and $r(n)$ denote the linear speed of the wheels (left and right, respectively). As it can be seen from the formula, robot's performance is evaluated by rewarding robots proportionally to their distance to the obstacles and their speed.

To allow robots to perceive obstacles, we used the 24 proximity sensors of the *foot-bot* [1] model as robots' sensors, aggregated into groups of 3 (resulting in 8 effective Boolean signals) and binarised after the application of a threshold on their maximum values. Then, we used a threshold equal to 0.1 for binarising the resulting values, which will be equal to 1 if they are greater than the threshold and 0 vice versa.

Task II—This task is a complication of Task I, as it adds two virtual regions that partition the arena. Each robot has one region assigned for navigating with obstacle avoidance, and consequently the other is forbidden. Its objective function will therefore take into account a penalty equal to 1 in the event of

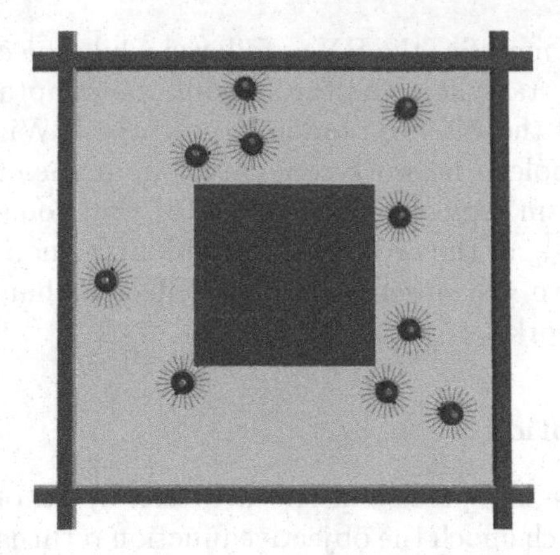

Fig. 1. Top view of the arena employed for the Task I. The obstacles are the gray objects (the walls and the square block at the center), while robots are the blue circles. (Color figure online)

navigation in the prohibited region and a reward of value 1 when the robot is in the correct region. The following is the resulting objective function, largely similar to Eq. 1:

$$\frac{100}{E}\sum_{n=1}^{E}\sigma(n)\cdot(1-\theta(n))\cdot(1-\sqrt{|\,l(n)-r(n)\,|})\cdot\frac{l(n)+r(n)}{2} \quad (2)$$

The penalty/reward factor has been introduced by means of the σ parameter. The correct region is the one where the robot is initially placed during the deployment phase. To provide the robot with feedback on its current position during adaptation, a virtual sensor was used to perturb the dynamics of the Boolean network. The arena used for this task with the two virtual regions is shown in Fig. 2.

Task III—The third and last task is a foraging task. The robots must "grab" a virtual object in the *blue zone* and release it to the *green zone* while performing obstacle avoidance. The robot positions are encoded with 2 Boolean signals as follows: **00** for the neutral zone, **01** for the green zone and **10** for the blue zone. To represent these action we have provided the robot with a virtual hook controlled by an output binary signal, which encodes the *pick up* or *deposit* actions. A light source provides a gradient and helps robots find their way around the arena. This information is passed to the robots by 8 additional signals. To sum up, for this task $8+8+2=18$ (proximity, light, region) input signals that perturb the dynamics of the Boolean network are used and $1+2$ (motors and hook) output signals useful for controlling the actuators.

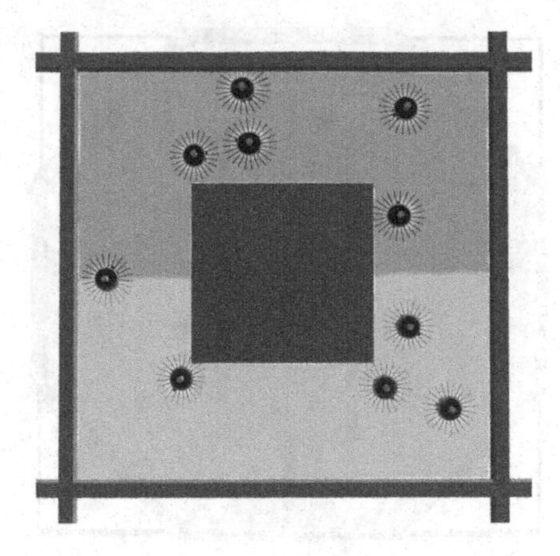

Fig. 2. Top view of the arena employed for the Task II. The two virtual areas are represented in red and blue. (Color figure online)

A reward system r for the correct *pick up* action and correct and wrong *deposit* actions has been added to the objective function and here reported:

$$r = \begin{cases} +50, & \text{if } \textit{pick up} \text{ while the robot is in the blue region} \\ +50, & \text{if } \textit{deposit} \text{ while the robot is in the green region} \\ -100, & \text{if } \textit{deposit} \text{ in other cases} \end{cases} \quad (3)$$

So the resulting objective function for this task is as follows:

$$\frac{100}{E}\sum_{n=1}^{E} r(n) + (1-2\theta(n)) \cdot (1-\sqrt{|\, l(n)-r(n)\,|}) \cdot \frac{l(n)+r(n)}{2} \quad (4)$$

Note that *Task III* is the most complex, also because it requires to maintain a memory in the dynamics; therefore, simple reactive controllers, such as Braitenberg-like robot controllers [5], cannot accomplish the task.

2.3 Experimental Setting

In this study, BNs are used as robot control program of a *foot-bot* robot model [1]. In particular, we make use of BNs with 1000 nodes and $k = 3$ inputs per node. According to the theory [13], initially ordered ensembles of BNs can be generated using bias values in $p = \{0.1, 0.9\}$, critical with $p = \{0.21, 0.79\}$ and, finally, chaotic with $p = \{0.5\}$. As the *mutation* mechanism changes the structure of the network, the actual dynamical regime of the specific network could change significantly during the adaptation phase. Due to the nature of the changes that the mutation mechanism introduces, the initial dynamical regimes of the Boolean networks will progressively shift towards chaotic regions as the adaptation process proceeds. In fact, in particular, the random flips that act on truth tables

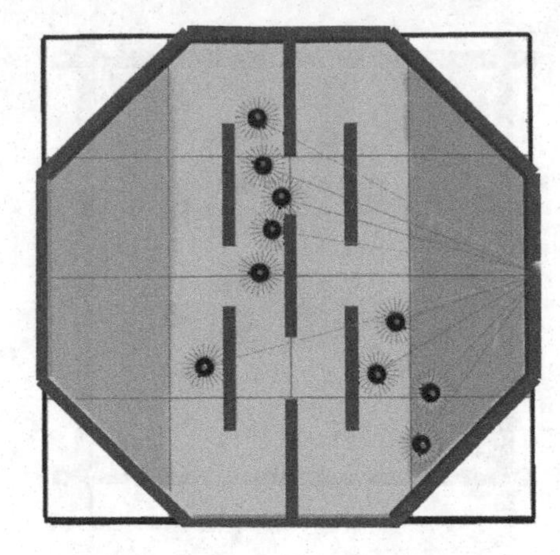

Fig. 3. Top view of the arena employed for the Task III. The blue area represents the collecting area, while the green ones indicates the nest, i.e. where collected virtual objects are to be released. To allow the orientation of the robots, a light source was introduced on the outer border of the green area. (Color figure online)

tend to bring the overall bias of the networks to a value close to $p = 0.5$; this, in combination with the value used for the number of input nodes per node $k = 3$, produces BNs in the chaotic regime. Since our previous work [3] confirmed that adaptive pressure wins over the shift towards chaos imposed by the mutation mechanism in shaping the robots' controller, in the following we will study the **actual operating dynamical regime** of the BNs and relate it to the performance achieved by them.

To produce significant statistics about the performances of the different types of controllers subject of this work, we executed 50 replicas of each task in the ARGoS simulator [15]. In each task, 10 robots, each one controlled by a different Boolean network, try to adapt autonomously—without shared memory—using the presented mechanisms.

A replica is composed of 500 epochs, each one lasts 80 seconds with 10 synchronous updates of the BN per second. The specific kind of noise introduced is described in the next section.

2.4 Noise

The tasks described above are performed while the Boolean network controlling a robot is perturbed by noise in the form of logical negation of the state of a BN node during the robot adaptation phase. The occurrence of these events follows an exponential distribution with parameter ν referring to the average number of perturbations per second. Thus given the current instant t_0 to calculate the instant of the next perturbation t_e the following law is used:

$$t_e = \frac{-ln(1-x)}{\nu} + t_0 \tag{5}$$

where x is a value drawn at random between $]0,1[$ with uniform distribution. When the time exceeds t_e, a node of the network is randomly chosen and its state is negated. In light of the results of preliminary exploratory investigations, we decided to use $\nu = 10$: this proved to be a good compromise between lower values, which did not cause significant changes to the robot dynamics obtained in the absence of noise in all the different dynamic regimes tested, and, conversely, higher values, resulted detrimental for the performances of all network ensembles.

Therefore, the results regarding the performance of the robots in the three tasks will take into account the influence of the specific definition of noise that we have introduced.

3 Results

To compare the performance of the adaptation mechanisms along with the different dynamical regimes of BNs we collect the following statistics:

Maximum scores. We collect the maximum scores achieved by each robot in each replica;

Derrida values. Taking into account the best epoch of every single robot, we gather the states that make up their trajectories during the adaptation process, and we calculate the Derrida value [13] indicating the **actual operating dynamical regime** expressed by the BNs. For this purpose we consider *ordered networks* those with Derrida value $D < 1 - \epsilon$, *critical* if $D \geq 1 - \epsilon \wedge D \leq 1 + \epsilon$, and, lastly, *chaotic* if $D > 1 + \epsilon$, with $\epsilon \in \{0.2, 0.3\}$.

Figure 4 summarises the results concerning the dynamical regime shift imposed by the combined effect of adaptive pressure and mechanisms. In particular, we note that, for the adaptation-by-mutation and for initially ordered and critical BNs cases, all the tasks present distributions of Derrida values characterised by higher values with respect to those adapted through the *in-out mapping* mechanism. Indeed, as previously mentioned, the *in-out mapping* mechanism does not change the Boolean network and therefore does not modify the dynamical regime and can be used as a baseline for the comparison. Noteworthy is the great leap that initially ordered networks undergo due to the action of the *mutation* mechanism, which makes us to consider them as full-fledged dynamically critical networks.

In Figs. 5 and 6 we find the comparison of the best performances, in the case of $\epsilon = 0.2$ and $\epsilon = 0.3$, respectively, for the definition of the *actual operating dynamical regime*. First of all, we can observe that with both $\epsilon = 0.2$ and $\epsilon = 0.3$, except for the chaotic case, the *in-out mapping* outperforms the *mutation* mechanism in terms of average performance. We remark also that this trend is more pronounced in the case of $\epsilon = 0.3$. Moreover, as the complexity of the task increases (remembering that Task I $\prec$ Task II $\prec$ Task III), critical networks show a mean performance advantage over other regimes.

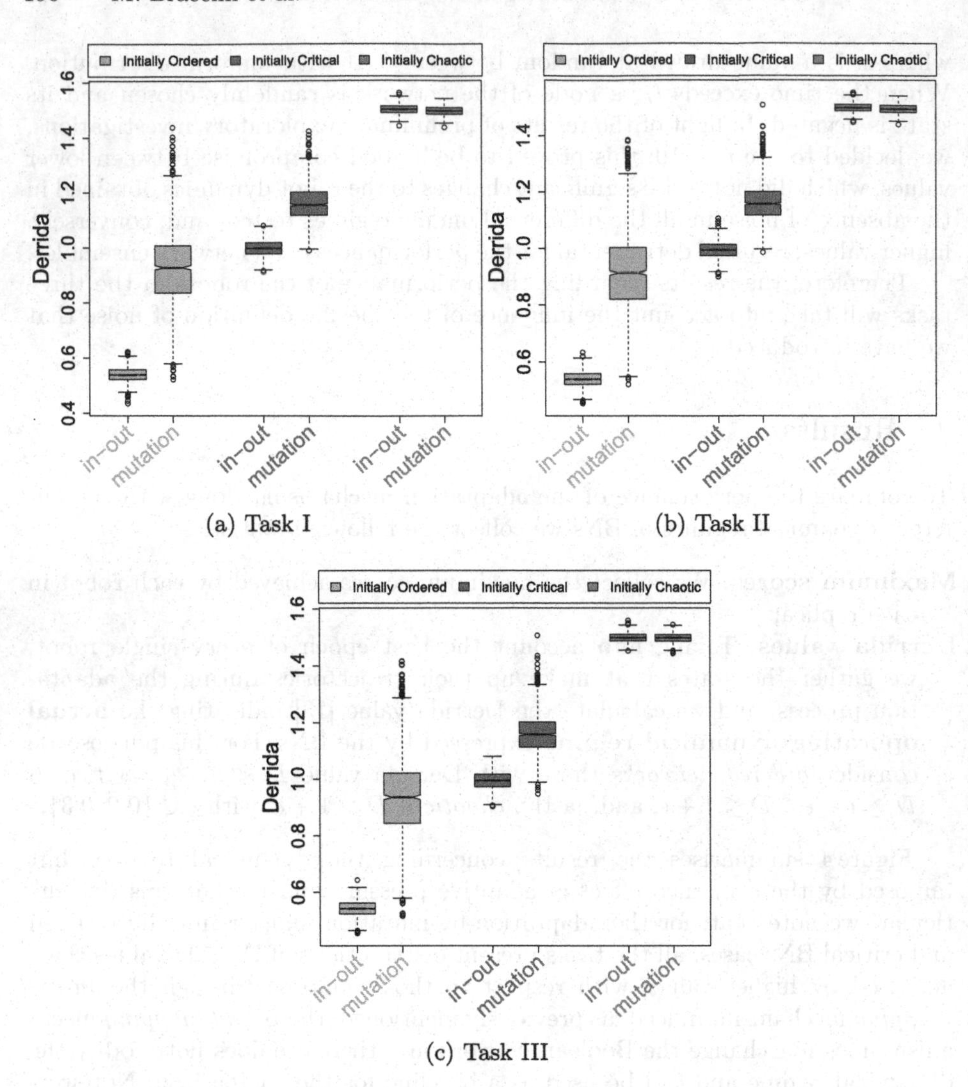

Fig. 4. (4a), (4b) and (4c) report the statistical change of the dynamical regimes, estimated by computing Derrida values, after the adaptation processes.

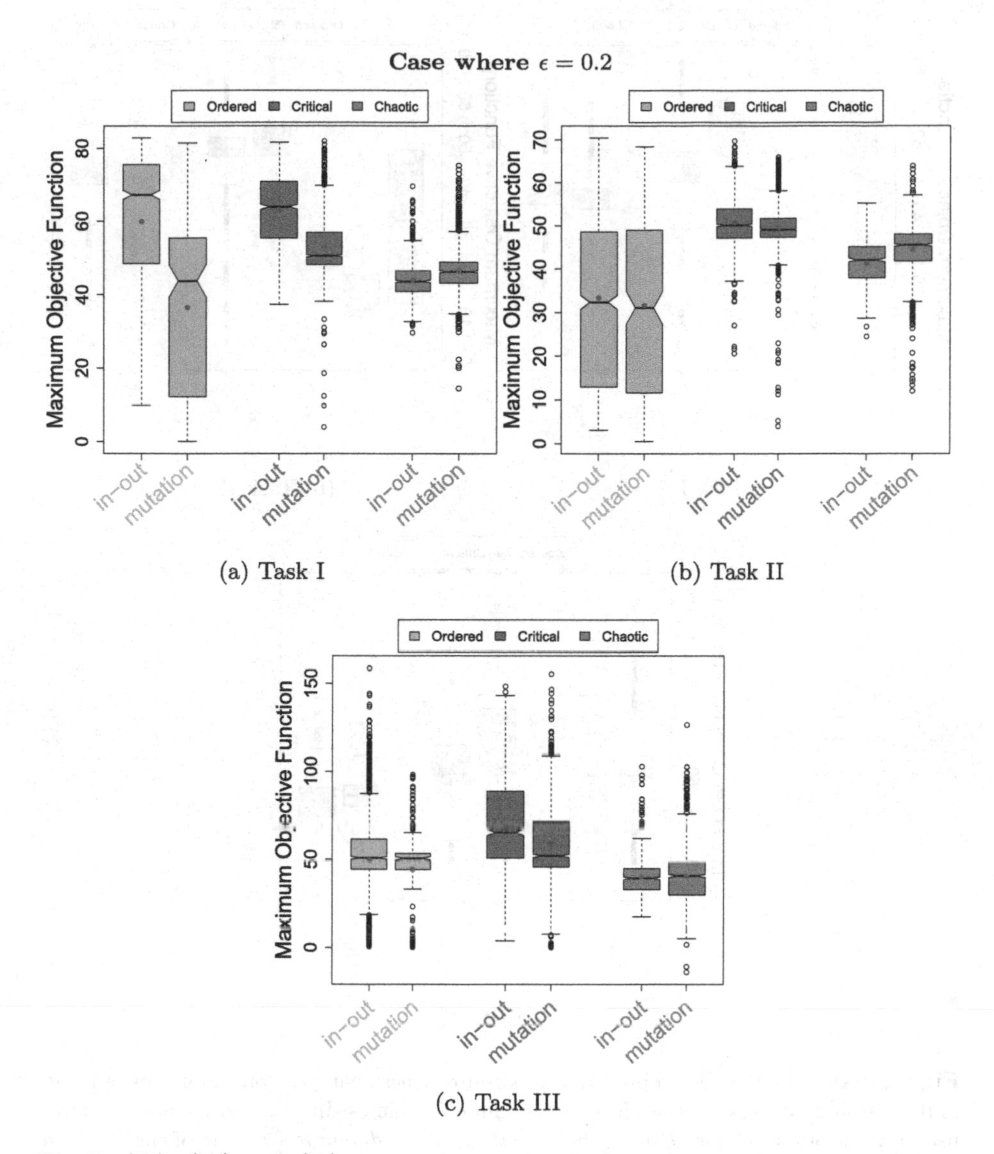

(a) Task I

(b) Task II

(c) Task III

Fig. 5. (5a), (5b) and (5c) report the maximum fitness comparison among in-out (abbreviated to "in-out") and mutation mechanisms for the different dynamical regimes. For these graphs, the higher the value reached, the better, in general, the performance of the robot that has achieved it. The purple dot represents the mean of the distribution. For this last analysis BNs have been considered: ordered if $D < 1 - \epsilon$, critical if $D \geq 1 - \epsilon$ and $D \leq 1 + \epsilon$, and, lastly, chaotic if $D > 1 + \epsilon$. In this comparison a value equal to 0.2 has been used for the ϵ parameter.

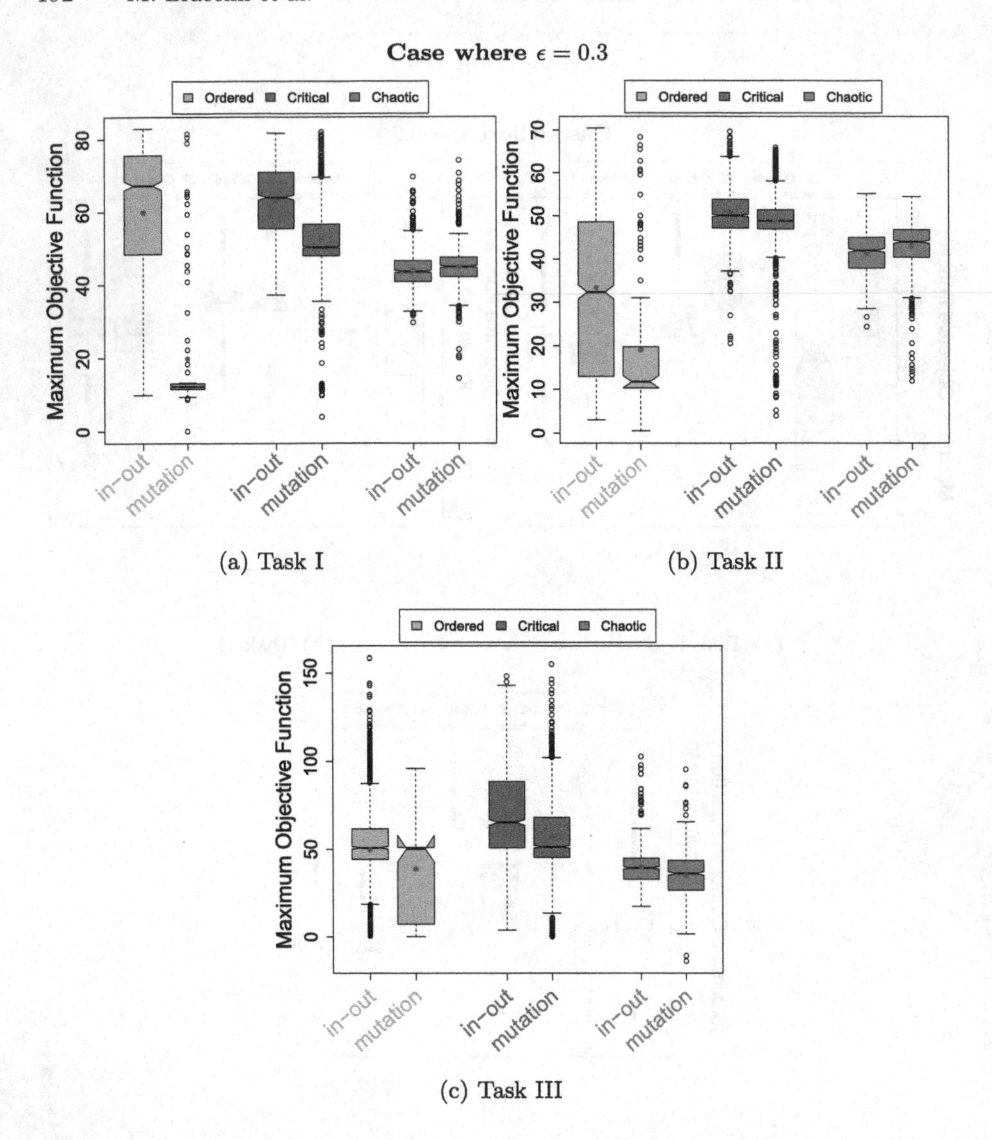

(a) Task I

(b) Task II

(c) Task III

Fig. 6. (6a), (6b) and (6c) report the maximum fitness comparison among in-out and mutation mechanisms for the different dynamical regimes. In this comparison we have used a value of $\epsilon = 0.3$ for defining the *actual operating dynamical regime* of the Boolean networks.

4 Conclusion

In this work we performed a comparison of two online adaptation mechanisms in scenarios involving BN-robots performing tasks of different complexity in noisy environments. The introduction of noise allowed us to verify, in more general experimental scenarios, whether there was an advantage of one mechanism over the other and under what conditions this could occur. Given its universality, we were particularly interested to verify whether criticality could prove to be an advantageous condition in terms of performance even with noisy environments.

Overall, the experiments carried out in this work reinforce and at the same time generalise the results obtained in previous works [2,3]. In summary, we can argue that these results support the hypothesis that the criticality in BN-robotics online adaptation scenarios does not only represent, on average, an advantageous starting condition, but also that criticality is a property to which adapted phenotypes (observable manifestations of the interaction between BN, robot physicality and environment; behaviours in short) tend. Lastly, limited to the experimental conditions tested, it can also be asserted that random Boolean networks, in the role of robot controllers, possess all the characteristics to adapt themselves to perform the tasks they have been subjected to: this was made evident by the tendency of the *in-out mapping* mechanism to perform better than the *mutation* mechanism. As further evidence in support of these hypotheses, we report that these conclusions were found to be robust to several definitions of *actual operating dynamical regime*, i.e. with different values of ϵ.

Since Pask's first pioneering experiments on the artificial ear [6], the online adaptation of artificial entities has implications concerning the cybernetic concepts of the creation of novelty, meaning-creation and sensor evolution. Therefore, it is our intention to introduce other bio inspired mechanism for online adaptation and to integrate them with offline ones to better frame, respectively, the phylogenetic versus ontogenetic forces in the creation of novelty.

Furthermore, we plan to quantitatively assess the meaning-creation process taking place in robots undergoing adaptation by means of measures derived from Information Theory, such as Semantic Information [12].

References

1. Bonani, M., et al.: The marXbot, a miniature mobile robot opening new perspectives for the collective-robotic research. In: 2010 IEEE/RSJ International Conference on Intelligent Robots and Systems, 18–22 October 2010, pp. 4187–4193. Taipei, Taiwan. IEEE (2010)
2. Braccini, M., Roli, A., Kauffman, S.: Online adaptation in robots as biological development provides phenotypic plasticity. arXiv preprint arXiv:2006.02367 (2020). To be published in Proceedings of WIVACE 2021, LNCS Series, Springer
3. Braccini, M., Roli, A., Barbieri, E., Kauffman, S.A.: On the criticality of adaptive Boolean network robots. Entropy **24**(10), 1368 (2022). https://doi.org/10.3390/e24101368

4. Braccini, M., Roli, A., Villani, M., Serra, R.: Dynamical properties and path dependence in a gene-network model of cell differentiation. Soft. Comput. **25**(9), 6775–6787 (2020). https://doi.org/10.1007/s00500-020-05354-0
5. Braitenberg, V.: Vehicles: Experiments in Synthetic Psychology. MIT Press, Cambridge (1986)
6. Cariani, P.: To evolve an ear. Epistemological implications of Gordon Pask's electrochemical devices. Syst. Res. **10**(3), 19–33 (1993)
7. Fusco, G., Minelli, A.: Phenotypic plasticity in development and evolution: facts and concepts. Philos. Trans. R. Soc. B **365**, 547–556 (2010)
8. Huang, S., Eichler, G., Bar-Yam, Y., Ingber, D.: Cell fates as high-dimensional attractor states of a complex gene regulatory network. Phys. Rev. Lett. **94**(128701), 1–4 (2005)
9. Huang, S., Ernberg, I., Kauffman, S.: Cancer attractors: a systems view of tumors from a gene network dynamics and developmental perspective. In: Seminars in Cell and Developmental Biology, vol. 20, pp. 869–876. Elsevier (2009)
10. Kauffman, S.: Metabolic stability and epigenesis in randomly constructed genetic nets. J. Theor. Biol. **22**(3), 437–467 (1969)
11. Kelly, S., Panhuis, T., Stoehr, A.: Phenotypic plasticity: molecular mechanisms and adaptive significance. Compr. Physiol. **2**(2), 1417–1439 (2011)
12. Kolchinsky, A., Wolpert, D.: Semantic information, autonomous agency and non-equilibrium statistical physics. Interface Focus **8**(6), 20180041 (2018)
13. Luque, B., Solé, R.: Phase transitions in random networks: simple analytic determination of critical points. Phys. Rev. E **55**(1), 257 (1997)
14. Montagna, S., Braccini, M., Roli, A.: The impact of self-loops on Boolean networks attractor landscape and implications for cell differentiation modelling. IEEE ACM Trans. Comput. Biol. Bioinform. **18**(6), 2702–2713 (2021). https://doi.org/10.1109/TCBB.2020.2968310
15. Pinciroli, C., et al.: ARGoS: a modular, multi-engine simulator for heterogeneous swarm robotics. Swarm Intell. **6**(4), 271–295 (2012)
16. Roli, A., Villani, M., Serra, R., Benedettini, S., Pinciroli, C., Birattari, M.: Dynamical properties of artificially evolved Boolean network robots. In: Gavanelli, M., Lamma, E., Riguzzi, F. (eds.) AI*IA 2015. LNCS (LNAI), vol. 9336, pp. 45–57. Springer, Cham (2015). https://doi.org/10.1007/978-3-319-24309-2_4
17. Roli, A., Benedettini, S., Birattari, M., Pinciroli, C., Serra, R., Villani, M.: A preliminary study on BN-robots' dynamics. In: Proceedings of the Italian Workshop on Artificial Life and Evolutionary Computation (WIVACE 2012), pp. 1–4 (2012)
18. Roli, A., Braccini, M.: Attractor landscape: a bridge between robotics and synthetic biology. Complex Syst. **27**, 229–248 (2018). https://doi.org/10.25088/ComplexSystems
19. Roli, A., Manfroni, M., Pinciroli, C., Birattari, M.: On the design of Boolean network robots. In: Di Chio, C., et al. (eds.) EvoApplications 2011. LNCS, vol. 6624, pp. 43–52. Springer, Heidelberg (2011). https://doi.org/10.1007/978-3-642-20525-5_5
20. Serra, R., Villani, M., Graudenzi, A., Kauffman, S.: Why a simple model of genetic regulatory networks describes the distribution of avalanches in gene expression data. J. Theor. Biol. **246**(3), 449–460 (2007). https://doi.org/10.1016/j.jtbi.2007.01.012
21. Villani, M., Barbieri, A., Serra, R.: A dynamical model of genetic networks for cell differentiation. PLoS ONE **6**(3), e17703 (2011)

A Novel Evolutionary Approach for Neural Architecture Search

Alessandro Bria, Paolo De Ciccio, Tiziana D'Alessandro, and Francesco Fontanella(✉)

Department of Electrical and Information Engineering (DIEI), University of Cassino and Southern Lazio, Cassino, Italy
fontanella@unicas.it

Abstract. Convolutional Neural Networks (CNNs) have proven to be an effective tool in many real-world applications. The main problem of CNNs is the lack of a well-defined and largely shared set of criteria for the choice of architecture for a given problem. This lack represents a drawback for this approach since the choice of architecture plays a crucial role in CNNs'performance. Usually, these architectures are manually designed by experts. However, such a design process is computationally intensive because of the trial-and-error process and also not easy to realize due to the high level of expertise required. Recently, to try to overcome those drawbacks, many techniques that automize the task of designing the architecture neural networks have been proposed. To denote these techniques has been defined the term "Neural Architecture Search" (NAS). Among the many methods available for NAS, Evolutionary Computation (EC) methods have recently gained much attention and success. In this paper, we present a novel approach based on evolutionary computation to optimize CNNs. The proposed approach is based on a newly devised structure which encodes both hyperparameters and the architecture of a CNN. The experimental results show that the proposed approach allows us to achieve better performance than that achieved by state-of-the-art CNNs on a real-world problem. Furthermore, the proposed approach can generate smaller networks than the state-of-the-art CNNs used for the comparison.

1 Introduction

In the last decade, Deep Neural Networks (DNNs) have demonstrated their great success in many real-world applications, including image classification, natural language processing, and speech recognition, among others. Convolutional Neural Networks (CNNs) represent a subcategory of DNNs, particularly exploited in computer vision. CNNs allow the extraction of meaningful features directly from the raw data without using any feature engineering approach. There are several types of CNNs, differing from each other in the way layer blocks are arranged and linked, and in the choice of hyperparameters. The performance of a CNN mainly depends on two aspects: the choice of the architecture and

C. De Stefano et al. (Eds.): WIVACE 2022, CCIS 1780, pp. 195–204, 2023.
https://doi.org/10.1007/978-3-031-31183-3_16

the related hyperparameters, and the values of the neuron weights. Weights are usually tuned through a training procedure based on gradient-based algorithms that iteratively minimize a loss function which measures the difference between the actual output and the ground truth provided for the training data. On the other hand, finding the optimal architecture is a much harder task since this kind of problem cannot be formulated by a continuous function. Consequently, effective CNN architectures are manually designed by people with very rich expertise. Indeed, many state-of-the-art and popular networks, e.g. VGG, ResNet and DenseNet, have been manually designed by researchers with rich knowledge in both neural networks and image processing. However, in practice, most end users are not with such kinds of knowledge, and they tend to use already developed and validated models without changing the structure and the related default values for the hyperparameters. Furthermore, in most cases, CNN architectures are problem-dependent. Then even a small change in the data distribution may require redesigning the architecture accordingly.

A promising way to address the above-mentioned challenges is to automate the optimization of the network architecture and the related hyperparameters by using a Neural Architecture Search (NAS) algorithm. This task can be modelled as an optimization problem in the search space consisting of all the possible neural architectures. The key components of a NAS algorithm are the encoding strategy to represent the architecture and the hyperparameter to be optimized, a search strategy, and an evaluation function. In [14], the authors provided a comprehensive and systematic survey showing several ways to apply a NAS. This survey mentions NAS-RL [18], based on Reinforcement Learning, and MetaQNN [2], based on the Markov decision process, as the pioneers in this field, which reached an interesting performance on image classification tasks. Another set of NAS approaches relies on intuition to design the global search space into a modular search space; Some of the most popular are NASNet [17] and ENAS [12].

Evolutionary computation (EC) techniques have recently received much attention from the NAS community since they are well-known for their global search ability in complex and huge search spaces. This attention has paved the way to Evolutionary Neural Architecture Search (ENAS) [11]. In this context, the optimization of CNNs architecture becomes a problem that focuses on the following aspects: the hyperparameters of each layer, the depth of the architecture and the connections between layers. A first example of the application of Evolutionary Algorithms (EA) to the neural architecture search is Large-scale Evolution [13], whose aim is to automatically learn an optimal architecture, trying to reduce human intervention. This approach starts initializing a large population with simple CNNs and obtains the best individual by reproducing, mutating, and selecting the population. In [16], instead, the authors propose a new neural architecture encoding scheme, GeNet, where the neural architecture is represented as a fixed-length binary string. It randomly initializes a group of individuals and applies a predefined set of genetic operations to modify the binary string and generate new individuals. Finally, after an evaluation step, it selects the most competitive individual as the final neural architecture. One of the problems of the mentioned approaches is the huge search space that implies the necessity of more time and resources to find the best individual. To shrink the search space, new techniques were introduced that explore block [15] or cell [4] based encoding spaces.

In this paper, we present a novel approach for optimizing both hyperparameters and the architecture of CNNs for image classification. The proposed approach is based on a specifically devised encoding scheme as well as the related crossover and mutation operators. The proposed encoding scheme allows us to evolve linear and non-linear architectures (i.e. those containing multiple paths) whose size does not need to be a priori set. The experimental results confirmed the effectiveness of our approach on real-world data containing images of the handwriting of people affected by Alzheimer's disease.

2 The Proposed Approach

Recent technological advances have allowed the development of more and more complex and deep CNN architectures, implying an increasing number of hyperparameters to be set. Typically CNNs are built using three types of layers:

- Convolutional layers,
- Pooling layers,
- Fully Connected layers.

where each type may have some variants and several hyperparameters to be set.

Variants and hyperparameters of the different layers are related to each other, so, as a consequence, finding the optimal set is a hard task.

The proposed ENAS-based approach has been developed in such a way as to optimize both the network architecture and the related hyperparameters and works as follows:

1. Randomly generate the initial population of individuals;
   - Decode the chromosomes and synthesis of neural networks;
   - Evaluate the fitness (validation accuracy) of each individual (net) in the population. Every population's net is trained, and the validation accuracy is evaluated.
2. Repeat the following steps until the termination condition is satisfied:
   - Select the individuals for reproduction (Parents); we used the tournament selection strategy;
   - Breed new individuals through crossover and mutation operations to give birth to offspring;
   - Evaluate the fitness of the offspring;
   - Replace the least-fit individuals of the population with new individuals (next generation).

2.1 Solution Encoding

As mentioned above, in the proposed approach, each chromosome encodes a network architecture and the related group of hyperparameters, so its chromosome consists of two parts, named L_H and L_A. L_H is a fixed-length list of genes, each encoding one of the following hyperparameters:

- Learning rate (real): governs the pace at which an algorithm updates or learns the values of a parameter estimate;
- Max Learning rate (real);
- Learning rate gamma (real): implements a learning rate decay mechanism during the training phase;
- Gradient clipping (real): limits the gradient range of variation;
- Weight decay (real): implements a weights decay mechanism during the training phase;
- Batch size (integer): number of samples processed before the model is updated;
- Dropout (boolean): useful to counteract the undesired phenomenon of overfitting.
- Fully connected size (integer): number of neurons making up the fully connected layer.

On the other hand, L_A is a variable-length list of genes, each encoding a network layer. A gene contains the following information:

- type (categorical): the type of layer (convolutional or residual);
- param (integer): the number of convolutional layers or the type of residual block (basic or bottleneck); it depends on the value of the first element;
- double channel (boolean): whether to double or not the number of output features.

Every individual is initialized with values chosen among a discrete group of possible values.

Furthermore, the size of L_A is limited to the interval $[l_{min}, l_{max}]$, by discarding the individuals whose L_A has a size out of this interval. This choice allows us to avoid evolving too small or too large networks.

2.2 Evolutionary Operators

We implemented the following evolutionary operators:

- **Mutation** operates differently on L_H (i.e. the part of the chromosome encoding the hyperparameter values) and L_A (architecture part). Hyperparameters are mutated by choosing a new value among a discrete set of possible values. The architecture is mutated by choosing one of the following actions: elimination of a block, change of the type of layer, and addition of a block.
- Also the **Crossover** operator [8], operates differently on L_H and L_A. For L_H we implemented a single cut-off point crossover, whereas for L_A we implemented a two-point crossover.

Table 1. Values of the evolutionary parameters used in the experiments.

Parameter	Value
Population size	40
Number of generations	20
Elitism	2
Tournament size	3
Crossover probability	0.8
Mutation probability	1.0
Minimum length of $L_A(l_{min})$	2
Maximum length of $L_A(l_{max})$	5

2.3 Fitness Evaluation

Once the offspring are generated, the evaluation of every individual goes through the following steps:

1. Chromosome Decoding;
2. Synthesis of the neural network, setting of architecture and hyperparameters;
3. Training of the model;
4. Fitness evaluation, where the fitness function is the validation accuracy.

3 Experimental Results

To assess the effectiveness of our system, we performed two sets of experiments. The aim was to assess the effectiveness of the proposed approach on images of different sizes and natures with a different number of samples and classes. During a preliminary phase, we tested our system on two well-known benchmark datasets, namely MNIST [10] and CIFAR-10 [9], widely used to test DNNs (see Table 2). Then, we evaluated our system on a dataset consisting of images containing the handwriting of people affected by Alzheimer's and healthy people. The evolutionary parameters have been set after some preliminary trials (see Table 1).

The experiments were performed by using an Intel Xeon i7-7700 CPU @ 3.60 Intel Xeon Silver 4110 @ 2.10 GHz 377 GB of RAM with a GPU Tesla V100. As software frameworks, we used Pytorch 3.6.9 and Ubuntu 18.04.3 LTS.

3.1 Preliminary Results

To preliminary assess the effectiveness of the proposed approach, we analysed two well-known datasets, widely-used to test CNNs, namely CIFAR-10 and MNIST. The main characteristics of those datasets and the comparison results are shown in Table 2. From the table, we can see that on CIFAR-10 we achieved an accuracy equal to 86%, which is far from that of the state-of-the-art (99.61% [3]), whereas on MNIST we reached an accuracy of 99.7%, a which is quite comparable with the state-of-the-art (99.91% [1]).

The above results can be deemed satisfactory since the state-of-the-art on those datasets is represented by the best results among hundreds of network architectures and hyperparameter tunings, which have been proposed and tested in the last decade by many researchers.

3.2 Real-World Data

To assess the effectiveness of our approach on real-world data, we analysed a third dataset containing handwriting data. The dataset contains data from the execution of multiple handwriting tasks by two groups of participants: 85 Healthy controls (HC) and 89 Patients (PT) suffering from Alzheimer's, for a total of 174 participants. We used six out of twenty-five tasks of the original dataset [5] to evaluate the performance of our system. The tasks are described in Table 3. Further details about the data acquisition can be found in [5].

For each task, a sample contained the spatial coordinates and the pressure of the handwriting movements done by the participant to perform that task, sampled at a frequency 200 Hz. Starting from this information we generated synthetic images by approximating the original curve (i.e. elementary handwritten trait) by interpolating the sampled points. We also used kinematic information (pressure, velocity and jerk) encoded in the RGB channels. More details can be found in [6,7].

We used the 5-fold Cross-validation strategy to evaluate the performance of our approach and partitioned the dataset as follows: training set, with the 70% of samples; validation set, with the 10% of samples and test set with the 20% of samples. Figure 1 shows the trend of the training and the validation accuracy of the best individual selected by our approach for each task. Looking at this figure it seems that the individuals selected by the proposed system didn't suffer from the undesired phenomenon of overfitting, except for tasks #2 and #4.

In order to assess the effectiveness of our approach, we compared its performance with some well-known and widely-used CNNs: VGG19 (V19 in the following), ResNet50 (R50), InceptionV3 (IV3), InceptionResNetV2 (IR2). Table 4

Table 2. Benchmark datasets and comparison results. Accuracy is expressed in percentage.

Dataset	#samples	Size	type	#classes
CIFAR-10	60,000	32×32	RGB	10
MNIST	60,000	28×28	BW	10
Comparison results				
	Our approach		State-of-the-art	
CIFAR-10	86.0		99.6	
MNIST	99.7		99.91	

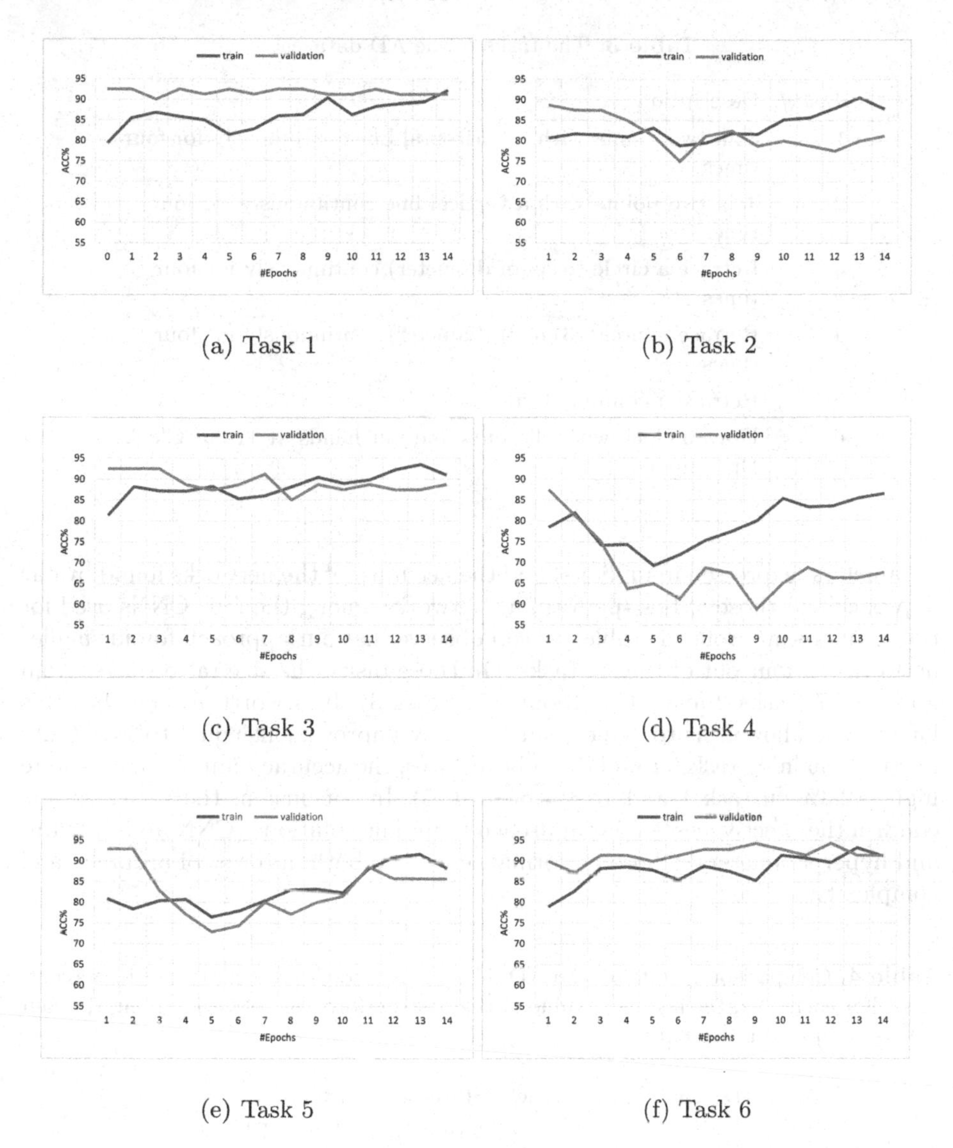

(a) Task 1 (b) Task 2

(c) Task 3 (d) Task 4

(e) Task 5 (f) Task 6

Fig. 1. Training and validation accuracy of the best individuals for each task

shows the accuracies achieved by those networks and by our approach on the five test folds. From the table, we can observe that our approach outperformed the four CNNs on all tasks. With respect to the best-performing CNNs (see the starred values in the table), the average accuracy improvement is equal to about 9%. The largest improvement was achieved on task 6 (15.4%), whereas the smallest ones were achieved on tasks 4 and 5.

To further assess the effectiveness of our approach we also analyzed the complexity of the networks generated. For each task, Table 5 shows the number of

Table 3. The tasks of the AD dataset.

Task#	Description
1	Join two points with a horizontal line continuously for four times
2	Join two points with a vertical line continuously for four times
3	Retrace a circle (6 cm of diameter) continuously for four times
4	Retrace a circle (3 cm of diameter) continuously for four times
5	Retrace a complex form
6	Draw a clock with all hours and put hands at 11:05 (Clock Drawing Test)

parameters (expressed in millions) and the accuracy of the networks found by our approach and those of best-performing networks among the four CNNs used for the comparison. From the table, we can observe that our approach found smaller networks on four out of the six tasks. On those tasks, the size ratio ranges from about 0.37 (tasks 2 and 5) to about 0.73 (task 3). It is worth noting that this latter ratio allowed us to achieve an accuracy improvement equal to 11.7%. As for the remaining tasks (1 and 6), in both cases, the accuracy improvements were high (12.3% on task 1 and 15.4% on task 6). In conclusion, the above results confirm the effectiveness of our approach in finding solutions (CNN architectures and hyperparameters) whose performance is good both in terms of accuracy and complexity.

Table 4. Comparison results on the AD dataset. Accuracies are expressed in percentages. For each task, the accuracy value of the best-performing network among the four CNNs compared is starred.

Task #	Our approach	State-of-the-art			
		V19	R50	IV3	IR2
1	79.9	67.6*	59.3	61	64.4
2	69.7	57.6	60.5	54.7	61.5*
3	84.6	69.4	70.4	73.0*	70.0
4	71.6	67.2	64.4	67.7*	62.4
5	72.0	64.6	65.7	68.1*	67.7
6	83.6	67.5	68.2*	61.5	64.5

Table 5. Complexity comparison (in terms of the number of parameters) between our approach and the best-performing network among the state-of-the-art CNNs used for the comparison. Accuracy (Acc) and complexity (Cmp) are expressed in percentages and millions of parameters, respectively.

Task #	Our approach		State-of-the-art		
	Acc	Cmp	Acc	Cmp	Net
1	79.9	34	67.6	25	V19
2	69.7	23	61.5	62	IR2
3	84.6	22	72.9	30	IV3
4	71.6	17	67.7	30	IV3
5	72.0	11	68.1	30	IV3
6	83.6	34	68.2	32	R50

4 Conclusions and Future Work

In this paper, we have introduced a novel approach based on Evolutionary Network Architecture Search (ENAS) for optimising the hyperparameters and architectures of CNNs. The proposed approach allows us to evolve linear or non-linear architectures (i.e. those containing multiple paths). We have tested the proposed approach on two widely used benchmark datasets (MNIST and CIFAR-10) and a dataset containing handwriting data for diagnosing AD. In the first case, our system achieved a performance that can be compared to state-of-the-art only on the MNIST dataset. It is worth noting that this result can be deemed satisfactory since, on those datasets, hundreds of network architectures and hyperparameters have been tested by many researchers. On the other hand, on the second dataset, our approach outperformed four state-of-the-art networks both in terms of accuracy and the number of parameters. It was also worth pointing out that the results presented in this work are very promising since they come from exploratory work. Therefore, we expect for them a large room for improvement.

In order to improve the performance of the proposed approach, future work will focus on different fitness functions as well as new encoding strategies. The first will evaluate individuals taking into account network complexity and variance of the validation accuracy, whereas the second will be used to optimize the number of channels of each layer of the encoded CNNs.

References

1. An, S., Lee, M., Park, S., Yang, H., So, J.: An ensemble of simple convolutional neural network models for mnist digit recognition. arXiv preprint arXiv:2008.10400 (2020)
2. Baker, B., Gupta, O., Naik, N., Raskar, R.: Designing neural network architectures using reinforcement learning. In: International Conference on Learning Representations (2017)

3. Bruno, A., Moroni, D., Martinelli, M.: Efficient adaptive ensembling for image classification. arXiv preprint arXiv:2206.07394 (2022)
4. Chu, X., Zhang, B., Ma, H., Xu, R., Li, Q.: Fast, accurate and lightweight super-resolution with neural architecture search. In: 2020 25th International Conference on Pattern Recognition (ICPR), pp. 59–64. IEEE Computer Society, Los Alamitos (2021)
5. Cilia, N., De Stefano, C., Fontanella, F., Scotto Di Freca, A.: An experimental protocol to support cognitive impairment diagnosis by using handwriting analysis. In: Procedia Computer Science, Proceeding of the 8th International Conference on Current and Future Trends of Information and Communication Technologies in Healthcare (ICTH), pp. 1–9. Elsevier (2019)
6. Cilia, N.D., D'Alessandro, T., De Stefano, C., Fontanella, F., Molinara, M.: From online handwriting to synthetic images for alzheimer's disease detection using a deep transfer learning approach. IEEE J. Biomed. Health Inform. **25**(12), 4243–4254 (2021)
7. Cilia, N.D., D'Alessandro, T., Stefano, C.D., Fontanella, F.: Deep transfer learning algorithms applied to synthetic drawing images as a tool for supporting alzheimer's disease prediction. Mach. Vis. Appl. **33**(3), 49 (2022)
8. Hussain, A., Muhammad, Y.S., Nawaz, A.: Optimization through genetic algorithm with a new and efficient crossover operator. Int. J. Adv. Math. **2018**, 14 (2018)
9. Krizhevsky, A.: Learning multiple layers of features from tiny images (2009). https://www.cs.toronto.edu/~kriz/learning-features-2009-TR.pdf
10. Lecun, Y., Bottou, L., Bengio, Y., Haffner, P.: Gradient-based learning applied to document recognition. Proc. IEEE **86**(11), 2278–2324 (1998)
11. Liu, Y., Sun, Y., Xue, B., Zhang, M., Yen, G.G., Tan, K.C.: A survey on evolutionary neural architecture search. IEEE Trans. Neural Netw. Learn. Syst. **34**(2), 550–570 (2023)
12. Pham, H., Guan, M., Zoph, B., Le, Q., Dean, J.: Efficient neural architecture search via parameters sharing. In: Dy, J., Krause, A. (eds.) Proceedings of the 35th International Conference on Machine Learning. Proceedings of Machine Learning Research, vol. 80, pp. 4095–4104. PMLR (2018). https://proceedings.mlr.press/v80/pham18a.html
13. Real, E., et al.: Large-scale evolution of image classifiers. In: Proceedings of the 34th International Conference on Machine Learning, vol. 70, pp. 2902–2911. ICML'17, JMLR.org (2017)
14. Ren, P., et al.: A comprehensive survey of neural architecture search: challenges and solutions. ACM Comput. Surv. (CSUR) **54**(4), 1–34 (2021)
15. Sun, Y., Xue, B., Zhang, M., Yen, G.G.: Completely automated CNN architecture design based on blocks. IEEE Trans. Neural Netw. Learn. Syst. **31**(4), 1242–1254 (2020)
16. Xie, L., Yuille, A.: Genetic CNN. In: Proceedings of the IEEE International Conference on Computer Vision (ICCV) (2017)
17. Zoph, B., Vasudevan, V., Shlens, J., Le, Q.V.: Learning transferable architectures for scalable image recognition. In: 2018 IEEE/CVF Conference on Computer Vision and Pattern Recognition (CVPR), pp. 8697–8710. IEEE Computer Society, Los Alamitos (2018)
18. Zoph, B., Le, Q.: Neural architecture search with reinforcement learning. In: International Conference on Learning Representations (2017)

Single and Multi-objective Genetic Programming Methods for Prediction Intervals

Karina Brotto Rebuli[1(✉)], Mario Giacobini[1], Niccolò Tallone[1], and Leonardo Vanneschi[2]

[1] University of Turin, Largo Paolo Braccini 2, 10095 Grugliasco, TO, Italy
karina.brottorebuli@unito.it

[2] NOVA University of Lisbon, Campus de Campolide, 1070-312 Lisbon, Portugal

Abstract. A PI is the range of values in which the real target value of a supervised learning task is expected to fall into, and it should combine two contrasting properties: to be as narrow as possible, and to include as many data observations as possible. This article presents an study on modelling Prediction Intervals (PI) with two Genetic Programming (GP) methods. The first proposed GP method is called CWC-GP, and it evolves simultaneously the lower and upper boundaries of the PI using a single fitness measure. This measure is the Coverage Width-based Criterion (CWC), which combines the width and the probability coverage of the PI. The second proposed GP method is called LUBE-GP, and it evolves independently the lower and upper boundaries of the PI. This method applies a multi-objective approach, in which one fitness aims to minimise the width and the other aims to maximise the probability coverage of the PI. Both methods were applied with the Direct and the Sequential approaches. In the former, the PI is assessed without the crisp prediction of the model. In the latter, the method makes use of the crisp prediction to find the PI boundaries. The proposed methods showed to have good potential on assessing PIs and the results pave the way to further investigations.

Keywords: Prediction Interval · Crisp prediction · Modelling uncertainty · Machine Learning

1 Introduction

The prediction of Machine Learning (ML) methods is typically presented as a single value, also called crisp prediction. Generally, we assume that exists a mathematical relationship among the input variables able to predict the output and that the ML model can accurately approximate it [1]. However, assessing the uncertainty of the model's prediction can be considered as important as the prediction itself, since it informs how trustworthy that prediction is. In the past decades some studies have been proposed to incorporate the uncertainty into

C. De Stefano et al. (Eds.): WIVACE 2022, CCIS 1780, pp. 205–218, 2023.
https://doi.org/10.1007/978-3-031-31183-3_17

ML models (see Sect. 2). This article presents an exploratory study on modelling uncertainty through Prediction Intervals (PI) with Genetic Programming (GP). A PI is simply the range of values in which the real value is expected to be. Inside the PI the predicted value can assume any value with the same probability, which is not the case for statistical intervals. Despite this limitation, working with PIs has the advantage that they do not assume any distribution of the residuals of a crisp prediction and, therefore, they can account for a wider variety of data. Most of the ML studies on the uncertainty of the model's prediction optimise the PI instead of the confidence or credibility intervals used in statistical models [6–12]. The main difficulty in the optimisation of a PI is that the best PI has two contrasting properties, it is as narrow as possible but at the same time it includes as most data observations as possible.

In GP it is possible to allow the assessment of PIs with a single or a multi-objective approach. The first proposed method uses a single fitness measure that combines the width and the probability coverage of the PI to evolve simultaneously the lower and the upper boundary models of the PI. The second proposed method uses the multi-objective approach in which one fitness measure aims to minimise the width and the other aims to maximise the probability coverage of the PI. The proposed multi-objective GP for PI prediction can use Nondominated Sorting Genetic Algorithm II (NSGA-II) [2] and Nested Tournament (NT) [3] selection methods.

The manuscript is organised as follows: Sect. 2 presents the framework on the assessment of uncertainty in ML methods; Sect. 3 contains the details of the proposed methods; Sect. 4 presents the experimental study conducted with these methods; and Sect. 5 concludes the manuscript with the main remarks and future work considerations.

2 Assessing Uncertainty

The uncertainty in the results of a model can be aleatoric (also called ontic or intrinsic) and/or epistemiologic. The aleatoric uncertainty is the intrinsic randomness of a phenomenon and the epistemic uncertainty is caused by lack of knowledge in the model [4]. The latter can be due to few data, to limitations of the model framework or to a poor parametrisation of the model.

Results of statistical models are usually presented as the mean value, or more strictly speaking the expected value, *and* its confidence or credibility interval, depending if the approach is frequentist or Bayesian, respectively. These intervals around the expected value intend to give to the user the degree of uncertainty of the prediction. PIs can be considered an indirect approach to assess the real uncertainty of a prediction.

2.1 ML Approaches in Assessing Uncertainty

Mainly three approaches have been used to assess PIs with ML methods. The first, called *direct*, models the entire mass of data directly, without first having determined the crisp prediction. The second, called *sequential*, consists in modelling the residuals of the crisp prediction. In the sequential approach, the PI can be assessed either by fitting another ML model on the residuals or by a computationally expensive method, like Monte Carlo simulations or Bootstrap resampling. The third approach, much less used, consists in including the statistical paradigms into the ML frameworks, which is the case of Bayesian Neural Networks (BNN). The prediction of this model is not a single value, it is a probability distribution, the so called *posteriori*. As in the traditional Bayesian statistics, the *posteriori* gives the credibility interval in terms of the probability of the predicted value inside the range of an interval of values. Cartagena et al. [5] presented a revision of the studies on uncertainty for ML and Fuzzy models. The next paragraphs mention some of the references cited in this article and others that were not reviewed by these authors.

One of the most important examples using the direct approach is presented by Khosravi et al. [6]. They proposed a Neural Network (NN) model, called Lower Upper Bound Estimation (here referred as LUBE-NN), that outputs a PI. The output layer of the NN is composed of two output nodes, each one to predict one bound of the PI. The cost function of the NN is modified to account for the two objectives of a PI, *i.e.*, to maximise its coverage probability and to minimise its width. The coverage probability of the PI is assessed by the PI Coverage Probability (PICP) (Eq. 1) and the PI width is assessed by the Normalised Mean Prediction Interval Width (NMPIW) (Eq. 2).

$$PICP = \frac{1}{N}\sum_{1}^{N} c_i, \quad \text{where } c_i = \begin{cases} 1 & \text{if } t_i \in [\ L(x_i), U(x_i)\ ] \\ 0 & \text{otherwise} \end{cases} \tag{1}$$

$$NMPIW = \frac{1}{NR}\sum_{1}^{N}(L(x_i) - U(x_i)) \tag{2}$$

where N is the number of observations, t_i is the true value of the ith observation, $L(x_i)$ and $U(x_i)$ are the lower bound and upper bound of the PI for the ith observation and R is the range of the underlying target values. The new cost function combines the PICP and the NMPIW into a single measure and it is called Coverage Width-based Criterion (CWC)(Eq. 3).

$$CWC = NMPIW(\ 1 + \gamma(PICP)\ )e^{(-\eta(PICP-\mu))} \tag{3}$$

where μ is the desired probability coverage, η is a penalisation for the difference between PICP and μ, $\gamma(PICP) = 1$ if $PICP < \mu$ or 0 otherwise. The hyperparameter η needs to be optimised and if the PICP is smaller than μ, the CWC becomes the NMPIW. Since this cost function is not differentiable, the updating of the NN weights is done with the Simulated Annealing algorithm. However, this can be done with a different optimisation algorithm. For instance, Taormina and Chau [7] proposed a version of LUBE-NN in which they use a Particle Swarm optimisation to update the weights of the NN.

Another important example of the direct approach was presented by Cruz et al. [8]. It is a variation of the LUBE-NN method called Joint Supervision (JS). In

JS, the NN has three nodes in the output layer, the Lower Bound (LB) node, the Upper Bound (UB) node and a the crisp prediction node. More significantly, the cost function of the LB and UB nodes is entirely different. It is called Total Loss (TL) (Eq. 4) and it is a linear combination of the Mean Squared Error (MSE) and a second metric, called Interval Loss (IL).

$$TL^l = MSE^l + \lambda\, IL^l \text{ , where } l = \{Lower, Upper\} \tag{4}$$

where MSE^l is the MSE of each output neuron, IL^l is the IL of each output neuron and λ is a hyperparameter that weights these two loss functions. The MSE (Eq. 5) minimises the distance between the node outputs and the targets, while the IL (Eq. 6) minimises the distance between the predicted boundary of the PI and the targets.

$$MSE^l = \frac{1}{N}\sum_{k=1}^{N} e_l^2(k) \text{ , where } l = \{Lower, Crisp, Upper\} \tag{5}$$

$$IL^l = \frac{1}{N}\sum_{k=1}^{N} E^l[\, e_l(k)\,] \text{ , where } l = \{Lower, Upper\} \tag{6}$$

where N is the number of observations, e_i is the error of the prediction for the kth observation (Eq. 7) and $E^l[\, e_l(k)\,]$ is the interval error for the kth observation (Eq. 8).

$$e_l(k) = y(k) - \hat{y} \tag{7}$$

$$E_{Lower} = \begin{cases} [e_{Lower}(k)]^2 & \text{if } e_{Lower} > 0 \\ 0 & \text{otherwise} \end{cases} \quad \text{and} \quad E_{Upper} = \begin{cases} [e_{Upper}(k)]^2 & \text{if } e_{Upper} < 0 \\ 0 & \text{otherwise} \end{cases} \tag{8}$$

The major contribution of these new cost functions is that they are differentiable and, thus, the NN can learn with gradient descent. The hyperparameter λ also needs to be optimised.

Some contributions using the sequential approach that are worth mentioning are described next. Shrestha and Solomatine [9] presented a method in which the dataset is grouped by a Fuzzy clustering and the mean of the prediction errors of each cluster is evaluated. The PI is then calculated by the weighted average of the mean errors of the clusters to which the new observation belongs. Shrestha et al. [10] presented a method in which they modeled the errors of a NN model as a dependent variable of the input variables of another NN model. Zhanga et al. [11] proposed a NN model in which the residuals are calculated and classified using the Chebyshev distance-based agglomerative hierarchical clustering algorithm. The UB and the LB of the PI are then estimated with the quantiles of the prediction residuals of the corresponding cluster. Khosravi et al. [12] used Boostrap samples to generate NN model predictions and their residuals. Afterwards, they used these residuals to fit the PI with a NN similar to the LUBE-NN.

The most important methodological contributions to assess the uncertainty of ML algorithms using the statistical paradigm are based on Bayesian Neural Networks (BNN). In these networks, the weights are probability distributions whose parameters are updated over the NN iterations. The main difficulty with these methods is to handle with the mathematical components of the Bayesian approach. Blundell et al. [13] presented a suitable version of the BNN in which the update of the parameters of the probability distributions in done by approximation using the Kullback-Leibler divergence between the *posteriori* and a known distribution. Some proposals on how to improve it have been made. For example, Lakshminarayanan et al. [14] proposed an alternative to Bayes by Backprop that is simpler to implement and that uses fewer hyperparameters. As an alternative, Maddox et al. [15] presented a variation of the Stochastic Weight Averaging (SWA) method, by fitting a Gaussian distribution using the SWA solution as the first moment of stochastic gradient descent (SGD) and a low rank plus diagonal covariance derived from the SGD iterations, thus approximating the *posteriori* distribution over the NN weights.

Besides these approaches, it is worth to mention that conformal prediction has been used with ML methods. It regulates the interval prediction using a data split called calibration set. An example with GP is found in [16].

It is interesting to notice that the majority of the approaches aggregate the two objectives in a scalar measure, which is the case of LUBE-NN and the JS methods. Fewer examples in the literature work directly with the two objectives, as in Taormina and Chau [7].

3 Prediction Interval GP Proposed Models

This work presents two GP models, CWC-GP and LUBE-GP, for PI assessment. For each of them, the direct and the sequential approaches were applied.

3.1 CWC-GP

The CWC-GP is implemented with a single-objective GP, using the CWC measure as fitness function. It uses the main components of the classical GP, with modified solution and crossover operator. The tree solution of this method has two other trees in the branches of its root node, one to model the LB and the other to model the UB of the PI. The CWC measure of the PI formed by the LB and the UB trees is the fitness of the solution. The genetic operator crossover is applied only for parents of the same bound of the PI. Therefore, the LB and UB optimisations are independent. Furthermore, given that the trees are created and assigned at random to the LB and UB branches, three critical situations can happen. (i) The LB and UB solutions can coincide. If the LB and the UB solutions have the same structure, both are mutated with the same mutation settings of the main algorithm. (ii) The LB and UB solutions can be inverted. It can happen that the solution assigned to the LB would be better for the UB of the PI and vice-versa. In this case, the LB and the UB solutions are exchanged between them. (iii) The PI generated with the CWC Tree can be outside the data region. It can happen that both solutions do not generate an interval that contains the data. In this case, if the PI is above the data region, the LB solution of the CWC Tree is exchanged by the LB solution of the CWC Tree with the best PICP in the population of the current generation. The opposite is applied if the PI is below the data region.

3.2 LUBE-GP

LUBE-GP stands for Lower Upper Bound Estimation with GP and it estimates the lower and upper bounds of the PI preserving the multi-objective nature of the problem, *i.e.*, through the optimisation of the two PIs properties, its probability coverage and its width. The selection method plays an important role in multi-objective GP-based algorithms and NSGA-II and NT selection methods were implemented. Due tothe possibility of working with the optminisation of more than one objective at time, this method also permits that the solution for each boundary of the PI to have its own evolution process, and the proposed LUBE-GP applies this approach. The GP solution for the LB uses the Lower Interval Loss (Eq. 6) and the MSE (Eq. 5) as fitnesses. Correspondingly, the GP solution for the UB uses the Upper Interval Loss and the MSE as fitnesses. The other components of the LUBE-GP are the same as those used in the classical GP. In NT selection method, the first tournament selects the individuals based on the PI boundary loss fitness, *i.e.* maximises the probability coverage of the PI, and afterwards by the MSE fitness, which minimises the width of the PI.

3.3 Direct and Sequential Versions

The direct versions of the proposed models are applied as described above. For the sequential versions, first the crisp prediction is generated with a classical GP learning. Afterwards, the observations whose crisp predicted outcomes are smaller than the target values are used to fit the LB solution and, correspondingly, the observations whose crisp predicted outcomes are greater than the target values are used to fit the UB solution.

4 Experimental Study

The main goal of the experiments was to explore the behaviour of the algorithms' hyperparameters related to the PIs, namely the PI desired coverage (μ), the parameter η of the CWC measure and the multi-objective selection methods. Each experiment was run 30 times with random data partitioning with 70% of the data for training.

The primitive functions used to build the GP trees were $+$, $-$, $\times$ and protected $\div$ (denominator replaced by the constant 1.0×10^{-6} when it was zero). The terminal set was composed by ephemeral constants from $]5-, 5[$, in addition to the dataset feature. Trees were initialised with the ramped-half-and-half method [17]. To compare the results of the proposed methods, the LUBE-NN [6] also was implemented. The values of the hyperparameters were chosen empirically in preliminary studies and are presented in Table 1.

The main results of the experiments are the probability coverage (PICP measure) and the width (NMPIW measure) of the generated PI. The ideal PICP is equals to the desired coverage, so the results of PICP are presented and discussed in terms of the difference between PICP and μ. In the sake of the clarity of the text, this difference is called δ.

Table 1. Default settings used in the experiments. SA refers to the Simulated Annealing hyperparameters used in the LUBE-NN method.

	CWC-GP	LUBE-GP		LUBE-NN
Tree max initial depth	4	3	N. of layers	2
Tree max depth	5	5	N. of neurons	32
Population size	40	40	Learning rate	0.01
Generations	200	200	Epochs	1000
XO propability	0.2	0.2	SA - Initial temperature	20
Mutation probability	0.8	0.8	SA - Temp. decay rate	0.001
Selection method	tournament size 3	NSGA-II	SA - Max iterations	1000
Elitism	True	True	SA - Max mutation step	0.3
			SA - κ	0.01

4.1 Datasets

The experiments were run with artificial datasets (N = 200 observations) built based on the dataset used by the authors of the LUBE-NN method [6]. The general form of the function was kept, $f(x) = x^2 + sin(x) + 2 + \epsilon$, but the residuals and the domain of the function were modified. The motivation for this was to study the limitation of the models to predict the boundaries of the PI depending on the pattern of the residuals. Three different residuals were created: homocedastic (*hom*) with $\epsilon_{hom} \sim N(0,5)$, heterocedastic (*het*) with $\epsilon_{het} \sim N(0, \frac{x}{5})$ and asymmetrical (*asym*) with $\epsilon_{asym} \sim N(0, \tau)$, where $\tau = 2\epsilon_{het}$ if $\epsilon_{het} < 0$ or $\tau = 0.2\epsilon_{het}$ otherwise. Besides that, another variation of the dataset was generated with non-uniform (*nu*) distribution of the values in the domain of the function, *i.e.* in the features space, with $x \sim N(0,5)$. For the other datasets, $x \sim U(-10, 10)$.

4.2 Desired Coverage Experiment

This experiment was carried out with the aim of exploring the behaviour of the models in correctly predicting the PI for different values of the desired coverage (μ). The following values of μ were tested: $\mu \in \{0.10, 0.50, 0.90, 1.00\}$. The Direct

CWC-GP, Sequential CWC-GP, Direct LUBE-GP, Sequential-GP and LUBE-NN were included in this experiment.

Figure 1 shows the δ results for all tested methods. As can be seen in these plots, the methods performed similarly with the *hom*, *het* and *asym* datasets, indicating that for these datasets the pattern of the residuals was not important for the performance of the models. On the other hand, for CWC-GP and LUBE-GP some reuslts with the *nu* dataset were slightly worse. This shows that the performance of the proposed methods was worse when the distribution of the data in the features space is not uniform.

For Direct CWC-GP with $\mu = 0.10$, the difference between PICP and μ was almost always equal to -0.10. That is, this method was not able to find the data region with $\mu = 0.10$. With the other values of μ this method presented better results. The Sequential CWC-GP also presented an inferior performance for $\mu = 0.10$. However, it was better than the Direct CWC-GP, as the model found the data region (δ values ranging from -0.10 to 0.08).

For Sequential LUBE-GP and LUBE-NN, the bigger the μ, the smaller the δ. However, their results are indeed distinct. For Sequential LUBE-GP, the δ values were mostly positive. For LUBE-NN, they were mostly negative and the magnitude of the δ values were usually close to the value of the desired probability coverage (μ). This indicates that the PIs of Sequential LUBE-GP included more data than the proportion set by the desired probability coverage. On the other hand, the prevalence of negative values of δ with the magnitude close to the desired coverage for LUBE-NN shows that this method did not include most of the data observations in the PI. Complementarily, as can be seen in Supplementary Material Table SM-1[1], the widths of the PIs generated by the LUBE-GP were always very big, but they increased with the increase of μ. While the widths of the PIs generated by the LUBE-NN were always very similar, independently of the value of μ. The bigger the μ, the wider the PI is expected to be. Thus, this reinforces the δ results, showing that the PIs generated by LUBE-GP were excessively large, therefore including more data than they should, and that LUBE-NN failed in fitting the data.

The δ values of the PIs for Direct CWC-GP and Direct LUBE-GP in train and test partition were very different (see Supplementary Material, Figure SM-1). This difference was not observed for Sequential CWC-GP, Sequential LUBE-GP and LUBE-NN methods. This indicates that the direct versions of CWC-GP and LUBE-GP were more prone to overfit.

4.3 Balance Between PICP and NMPIW in CWC Measure (η) Experiment

The CWC measure balances PICP and NMPIW through the parameter η. As it can be seen in Eq. 3, when the PICP is bigger than the desired coverage (μ), the CWC becomes the NMPIW. Otherwise, when PICP is smaller than μ, the difference between PICP and μ is included in the CWC evaluation. In this case,

[1] Supplementary material available at https://bit.ly/3zsRfGP.

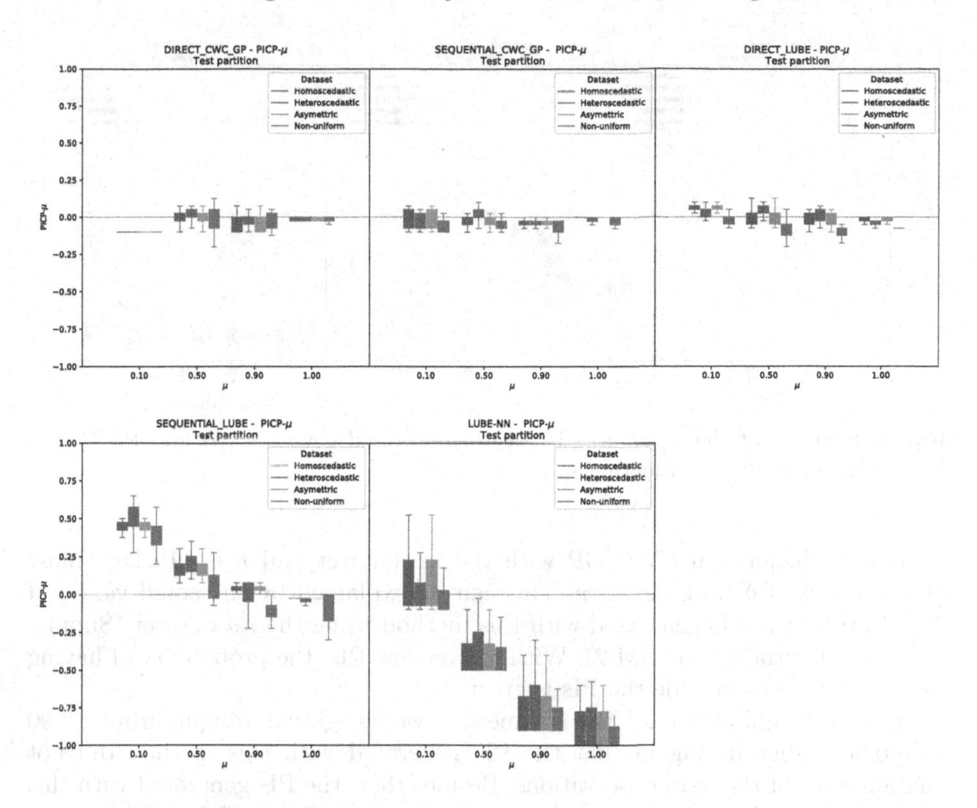

Fig. 1. Boxplots of the differences between the desired coverage (μ) and PICP in Desired Coverage Experiments for all tested methods.

the PICP-μ difference is weighted by η: the larger the η, the bigger the weight of the PICP-μ difference in the value of CWC. The choice of η is empirical and this experiment aimed to study how this hyperparameter affects the performance of the methods for the simulated datasets. The Direct CWC-GP, Sequential CWC-GP and LUBE-NN were included in this experiment and the following values of η were tested: $\eta \in \{5, 10, 20\}$.

As can be see in Fig. 2, the values of δ for the Direct and Sequential CWC-GP were similar to those observed in the Desired Coverage experiment, except for $\eta = 5$. With this value of η, the δ values for Direct CWC-GP were all close to -0.90. It corresponds to the magnitude of the desired coverage used in the experiments, showing that with this value of η the method could not find the data region. Otherwise, for this value of η, the δ values of the Sequential CWC-GP were closer to zero (ranging from -0.90 to 0.05), indicating that this method performed better with a smaller weight on the difference between the $PICP$ and μ. This is related to the fact that the Sequential CWC-GP tended to produce narrower PIs (see Supplementary Material Figure SM-2). With a smaller NMPIW, the weight of the $e^{-\eta(\ PICP - \mu\ )}$ term in the CWC (Eq. 3) could be smaller.

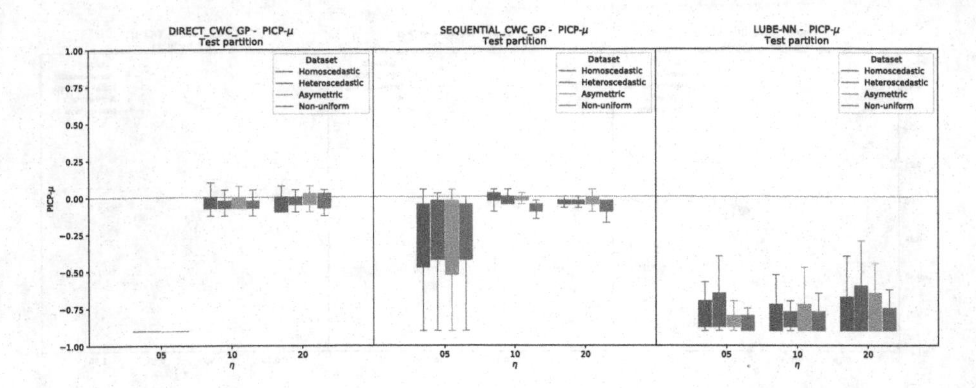

Fig. 2. Boxplots of the differences between the desired coverage (μ) and PICP in η Experiments for all tested methods.

For the Sequential CWC-GP with the *nu* dataset and $\eta \in \{10, 20\}$, most of the values of δ were negative. This can be explained by the small values of $NMPIW$ of the PIs genetared with this method using the *nu* dataset (Supplementary Material Figure SM-2). With too narrow PIs, the probability of having data observations outside the PIs is greater.

All values of δ for the LUBE-NN method were negative, ranging from -0.90 to -0.30, indicating again that the PIs generated with this method did not contain most of the data observations. Besides that, the PIs generated with this method were all very large (Supplementary Material Figure SM-2). Thus, even producing large PIs, LUBE-NN did not properly model the data.

4.4 Selection Methods Experiment

This experiment aimed to study how the multi-objective selection methods of LUBE-GP affect its performance in predicting PIs. The size of the tournament changes the selection pressure for the objective corresponding to that tournament, thus, different sizes may be needed for different objectives. Therefore, in addition to the comparison between NT and NSGA-II, different tournament sizes were tested for the first and the second tournaments of the NT selection method. Only Direct LUBE-GP and Sequential LUBE-GP were included in this experiment and the following tournament sizes were tested: $[3, 2], [4, 3], [5, 2], [5, 3]$, respectively for the first and second tournament.

Figure 3 shows the boxplots of the δ values for Selection Method experiments for Direct and Sequential LUBE-GP. The Direct LUBE-GP with all selection methods produced smaller values of δ in train than in test partition, indicating that, with regard to the $PICP$ measure, this method overfitted.

Except for the *nu* dataset, the δ values of the Direct LUBE-GP using NSGA-II selection method in the test partition were centered in zero, while when using the NT selection method most of the δ values were negative. This can be con-

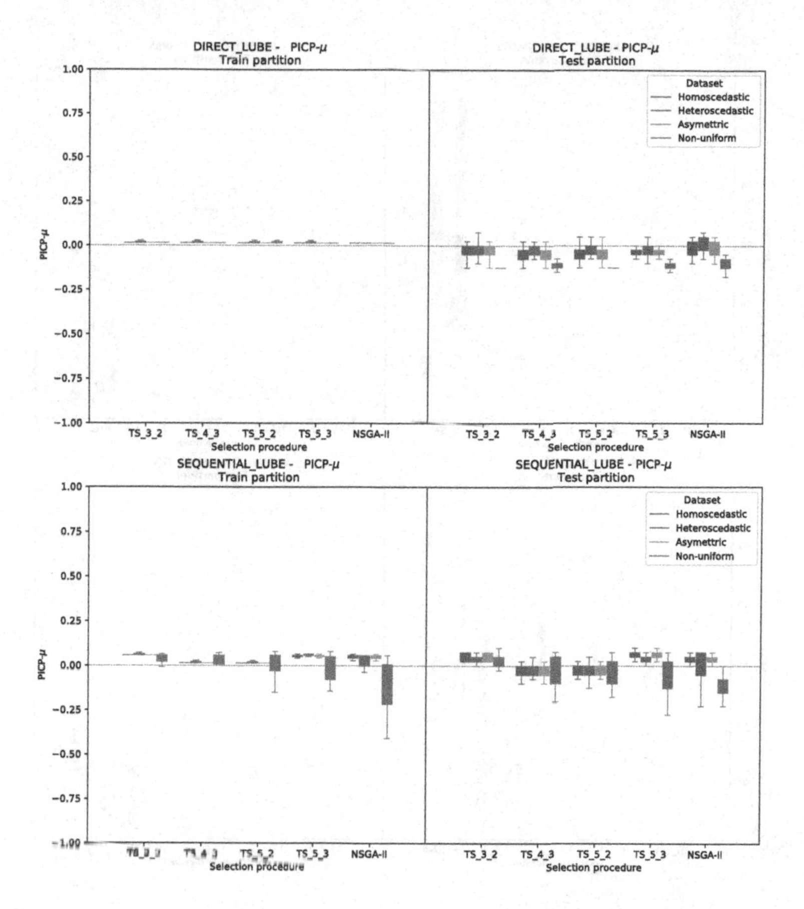

Fig. 3. Boxplots of the differences between the desired coverage (μ) and PICP in Selection Methods Experiments for all tested methods in train and test partitions.

sidered an advantage of NSGA-II over NT, as there is no bias between in the errors in train and test partitions when using NSGA-II selection method.

For Sequential LUBE-GP, the δ values varied more with *nu* dataset, using all selection methods. The highest variability was observed with the NSGA-II selection method in the train partition. In the test partition, the use of NSGA-II resulted in the most negative δ values, showing that this selection method failed in modelling the PI for this dataset.

Figure 4 shows the boxplots of the $NMPIW$ values for selection method experiments with Direct LUBE-GP and Sequential LUBE-GP. In these boxplots it is possible to see that LUBE-GP using NSGA-II produced excessively large PIs, but this did not happen when using the NT selection method. This can explain why the δ results when using NSGA-II selection method were better than when using the NT selection method: they contained more data just because they were too large, not because they were properly modelling the PI. Interestingly, for

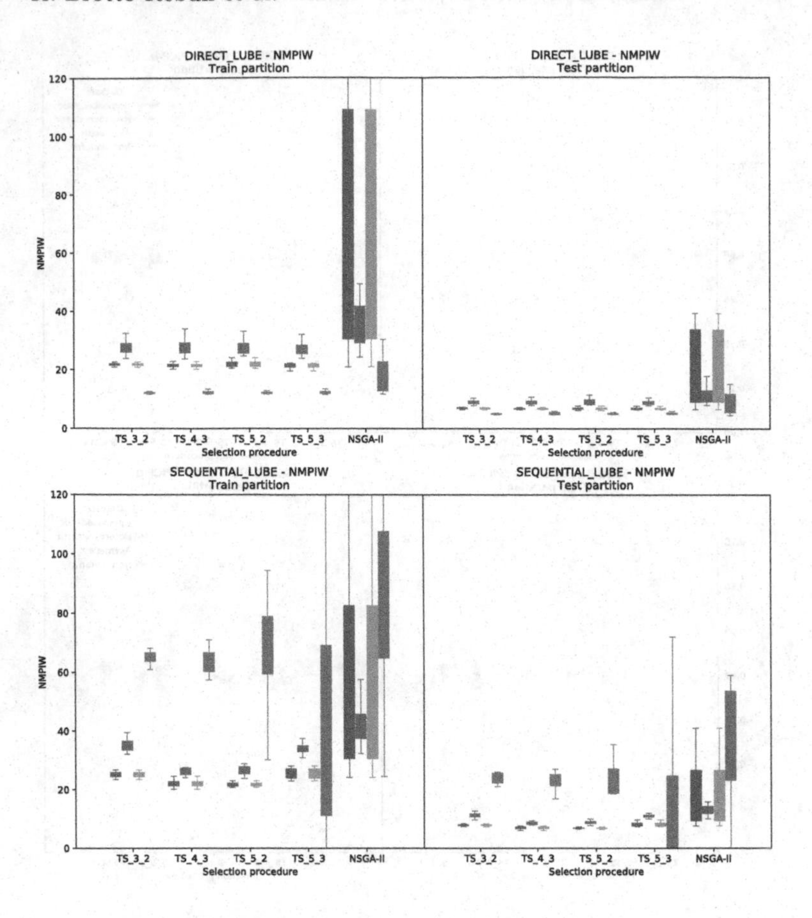

Fig. 4. Boxplots of the NMPIW in Selection Methods Experiments for all tested methods in train and test partitions.

the Direct LUBE-GP all selection methods produced PIs with smaller $NMPIW$ for the nu dataset. On the contrary, for the Sequential LUBE-GP all selection methods produced PIs with larger $NMPIW$ for this dataset, considering both train and test partitions. Regardless this difference in the performance of the Direct and Sequential methods related to the nu dataset, the use of NSGA-II produced much wider PIs than if using the NT selection method. Therefore, even with worse results in terms of δ, the use of NT selection method produced better results for both Direct and Sequential LUBE-GP.

5 Conclusions and Future Work

This work presented two GP methods for constructing Prediction Intervals (PIs), which is an indirect measure of the uncertainty of the predictions of a model. The CWC-GP method employed a single-objective GP, using as fitness the CWC measure, which combines the coverage and the width of the PI. The LUBE-GP used a multi-objective GP in which the coverage and the width of the PI are optimised separately. Both methods were implemented in two versions: Direct and Sequential. In the Direct version, the PI is estimated without the single-value prediction of the algorithm. In the Sequential version, first the single-value prediction is generated and afterwards it is used to find the lower and upper boundaries of the PI. These methods were compared with the LUBE-NN [6], in experiments to study the behaviour of the methods with regard to the Desired Coverage (μ), the Balance between PICP and NMPIW in CWC measure (η) and the multi-objective Selection Methods.

The experiments showed a trend of improving the performance of the methods with the increase of the desired coverage. This trend happens because the models are trained with the proportion of the data set as the desired probability coverage. Therefore, the smaller the value of this hyperparameter, the less data is used to train the model. In practice, this hyperparameter is chosen according to the use of the PI by the domain expert or the decision maker.

In general, the methods performed similarly with the *hom*, *het* and *asym* datasets. On the contrary, many results were worse when using the *nu* dataset. This shows that the proposed methods are robust with regard to the distribution of the residuals of the model, but not to the distribution of the data over the features space. When the distribution of the data on the features space is not uniform, the uncertainty of the prediction is greater in the regions in which there is less data, as these are less informative regions. Therefore, the worst results for *nu* dataset showed that, even though the methods modelled the variability of the predicted values, they did not modelled the uncertainty of the predictions.

Some results of the Desired Probability Coverage and Selection Methods experiments showed that the Direct version of the proposed methods were more prone to overfit, compared to the Sequential versions of these methods. The LUBE-GP method tended to produce excessively large PIs when using the NSGA-II selection method, but not when using the NT selection method. Thus, the NT selection method can be considered more suitable for modelling PIs. The LUBE-NN method failed in modelling the PIs, probably because the proper set of values of the hyperparameters of this method could not be find. This was due the large number of hyperparameters of this method. The ideal value of the η hyperparameter for the CWC-GP method was shown to be dependent of the combination of problem and method to be used to model the PI.

These preliminary results highlight the need of further investigations, like finding new strategies to use the dataset in order to train the model for small values of μ, inverting the order of the fitnesses and using different values of tournament sizes in NT selection method, adopting different strategies to chose the winner solution in the 1*st* Pareto Front of the NSGA-II selection method and testing other values of the SA hyperparameters in the LUBE-NN method. Moreover, in the future we will apply the model to more complex and real dataset and to implement a LUBE-GP version that uses straightly the PICP and the NMPIW as fitnesses measures.

References

1. Kendall, A., Gal, Y.: What uncertainties do we need in bayesian deep learning for computer vision?. In: 31st Conference on Neural Information Processing Systems (NIPS 2017), Long Beach, CA, USA (2017)
2. Deb, K., Pratap, A., Agarwal, S., Meyarivan, T.: A fast and elitist multiobjective genetic algorithm/ NSGA-II. IEEE Trans. Evol. Comput. **6**(2), 182–197 (2002)
3. Vanneschi, L., Castelli, M., Scott, K., Popovic, A.: Accurate high performance concrete prediction with an alignment-based genetic programming system. Int. J. Concr. Struct. Mater. **12**, 72 (2018)
4. Kiureghian, A.D., Ditlevsen, O.D.: Aleatoric or epistemic? Does it matter? Struct. Saf. **31**(1), 105–112 (2009)
5. Cartagena, O., Parra, S., Muñoz-Carpintero, D., Marín, L.G., Sáez, D.: Review on fuzzy and neural prediction interval modelling for nonlinear dynamical systems. IEEE Access **9**, 23357–2338 (2021)
6. Khosravi, A., Nahavandi, S., Creighton, D., Atiya, A.F.: Lower upper bound estimation method for construction of neural network-based prediction intervals. IEEE Trans. Neural Netw. **22**(3), 337–346 (2011)
7. Taormina, R., Chau, K.: ANN-based interval forecasting of streamflow discharges using the LUBE method and MOFIPS. Eng. Appl. Artif. Intell. **45**, 429–440 (2015)
8. Cruz, N., Marín, L.G., Sáez, D.: Neural network prediction interval based on joint supervision. In: 2018 International Joint Conference on Neural Networks (IJCNN) (2018)
9. Shrestha, D.L., Solomatine, D.P.: Machine learning approaches for estimation of prediction interval for the model output. Neural Netw. **19**(2), 225–235 (2006)
10. Shrestha, D.L., Kayastha, N., Solomatine, D.P.: ANNs and other machine learning techniques in modelling models' uncertainty. In: 19th International Conference, Limassol, Cyprus (2009)
11. Zhanga, C., Zhaoa, Y., Fanb, C., Lia, T., Zhanga, X., Lia, J.: A generic prediction interval estimation method for quantifying the T uncertainties in ultra-short-term building cooling load prediction. Appl. Therm. Eng. **173**, 115261 (2020)
12. Khosravi, A., Nahavandi, S., Srinivasan, D., Khosravi, R.: Constructing optimal prediction intervals by using neural networks and bootstrap method. IEEE Trans. Neural Netw. Learn. Syst. **26**(8), 1810–1815 (2015)
13. Blundell, C., Cornevise, J., Kavukcuoglu, K., Wierstra, D.: Weight uncertainty in neural networks. In: Proceedings of the 32nd International Conference on Machine Learning (ICML 2015), Lille (2015)
14. Lakshminarayanan, B., Pritzel, A., Blundell, C.: Simple and scalable predictive uncertainty estimation using deep ensembles. In: 31st Conference on Neural Information Processing Systems (NIPS 2017), Long Beach, CA (2017)
15. Maddox, W.J., Garipov, T., Izmailov, P., Vetrov, D., Wilson, A.G.: A simple baseline for bayesian uncertainty in deep learning. In: 33rd Conference on Neural Information Processing Systems (NIPS 2019), Vancouver, CA (2019)
16. Thuong, P.T., Hoai, N.X., Yao, X.: Combining conformal prediction and genetic programming for symbolic interval regression. In: Proceedings of the Genetic and Evolutionary Computation Conference (Berlin, Germany) (GECCO '17), Berlin, DE (2017)
17. Koza, J.R.: Genetic programming: on the programming of computers by means of natural selection. Stat. Comput. **4**, 87–112 (1992)

Green Design of Single Frequency Networks by Multiband Robustness and a Hybrid Metaheuristic

Fabio D'Andreagiovanni[1,2](✉), Hicham Lakhlef[2], and Antonella Nardin[3]

[1] French National Centre for Scientific Research (CNRS), Paris, France
[2] Heudiasyc UMR 7253, Sorbonne Universités, Université de Technologie de Compiègne, CNRS, CS 60319, 60203 Compiègne, France
{d.andreagiovanni,hicham.lakhlef}@hds.utc.fr
[3] Department of Engineering, Università degli Studi Roma Tre, via Vito Volterra 62, 00146 Roma, Italy
antonella.nardin@uniroma3.it

Abstract. The passage to a second generation of broadcasting Single Frequency Networks has generated the need for reconfiguring and redesigning existing networks. In this work, we present a robust optimization model for the green design of such networks based on the digital television DVB-T2 standard. Our robust model pursues protection against uncertainty of signal propagation in a complex real-world environment. As reference model, we adopt Multiband Robustness and we propose to solve the resulting model by a hybrid metaheuristic that combines mathematically strong formulations of the optimization model, with an exact large neighborhood search. We report computational tests based on realistic instances, showing that the multiband model grants highly protected solutions without reducing service coverage and without leading to a high price of robustness.

Keywords: Green Wireless Networks · Single Frequency Networks · Integer Programming · Robust Optimization · Hybrid Metaheuristics

1 Introduction

All around the world, in the last century, cities have experienced a huge expansion, becoming critical socio-economic hubs, and intergovernmental organizations like the United Nations estimates that, by 2050, two out of every three people in the world will live in cities [44]. In this evolution process of cities, Information and Communication Technologies (ICT) have played a crucial role, supporting a better digital interaction of the inhabitants with the urban environment, government and services, leading to the introduction of the concept of "Smart City" [31,42]. Among the fundamental elements of a smart city, telecommunication infrastructures and services have gained a major role and provide a

C. De Stefano et al. (Eds.): WIVACE 2022, CCIS 1780, pp. 219–231, 2023.
https://doi.org/10.1007/978-3-031-31183-3_18

fundamental basis for the six pillars of a smart city (see e.g., [24,32]). The introduction of the 5th Generation of wireless networks (5G) has in particular been recognized as an accelerator of the realization of actual smart cities, allowing to support the launch of new services with unprecedentedly high performance [12,26,41,43,45]. However, besides such last generation services, elderly broadcasting wireless services like radio and television continue to be regarded as fundamental mass media that must be guaranteed to people for an immediate and plain access to news. A milestone in the evolution of broadcasting radio and television communications was constituted by the passage from analogue to digital transmissions, which has enabled an enhanced exploitation of the frequency spectrum and allowed to furnish services of higher quality [3,23,28]). Within this new digital context, it has been possible to introduce the concept of Single Frequency Networks (SFNs), namely a network that uses only one single frequency channel for broadcasting and in which all the broadcasting stations transmit the same bits on this single channel (see e.g., [25,34,36,39]). The vast majority of SFNs providing digital television broadcasting services are based on the standard DVB-T (Digital Video Broadcasting - Terrestrial): first published in 1997, this standard has quite recently been improved through its second generation (DVB-T2), in order to increase its data carrying capacity, support major operation flexibility and signal reception robustness and allow a major number of broadcasters to co-exist in the same band. Due to the change of the standard, there was the possibility for new broadcasting enterprises to enter the market and old companies have had to update the configuration of their networks. Such reset of the networks has led to revive the attention towards approaches that could automatically design SFN networks on the besis of Mathematical Programming techniques.

In this work, we address the question of developing a Robust Optimization model for designing SFN networks based on the DVB-T2 standard, while taking into account the uncertainty that naturally affects wireless signal propagation. Specifically, our original contributions are:

1. We present a binary linear programming model for representing the robust counterpart based of an SFN design problem including signal-to-interference constraints. The objective of the problem is to pursue a green optimization of the network, finding the minimum total power emission that allows to serve a target fraction of the population of a region. The counterpart is defined according to the principles of Multiband Robust Optimization [8], a refined version of the classical Γ-Robust Optimization model [6].
2. Since the resulting robust optimization model may result very challenging even for a state-of-the art optimization software like IBM ILOG CPLEX, we define a hybrid metaheuristic for its solution, proposing to combine a probabilistic variable fixing procedure with an exact large variable neighborhood search. The probabilistic fixing exploits the precious information that can be derived from a tight linear relaxation of the model adopted to represent the SFN design problem, whereas the exact search consists of exploring a solu-

tion neighborhood formulating the search as an optimization problem that is solved at the optimum.

3. We highlight the performance of our new modelling and algorithmic approach by means of tests conducted on realistic SFN/DVB-T instances, showing the superior performance of the multiband approach with respect to both a benchmark robust and deterministic model.

We remark that, while the deterministic (i.e., not considering data uncertainty) optimal design of wireless networks based on signal-to-interference ratios has received wide attention, the use of optimization under uncertainty techniques, such as Robust Optimization and Stochastic Programming, has received less attention and has especially considered the effects of traffic uncertainty. This is also true for the case of DVB-T, in which optimization approaches have tended to not address data uncertainty by means of optimization under uncertainty techniques (e.g., [2,17,30,33,34,37,38]). To the best of our knowledge, just the work [18] has tried to adopt an optimization under data uncertainty method, in the form of a heuristic min-max regret approach, for tackling signal propagation uncertainty and this paper is the first work that discusses the adaption of a robust optimization approach to DVB-T and presents a hybrid metaheuristic for the solution of the resulting complex problem.

2 Optimal SFN Design

To derive an optimization model, we refer to an SFN network based on the DVB-T standard: in such network, all broadcasting stations synchronously transmit identical data on the same frequency channel according to the OFDM modulation scheme [23]. Broadcasting services are spread over a target territory to reach the receiving devices of a population. Following the recommendations of telecommunications regulatory bodies (e.g., [1,27]), we discretize the target territory into a raster of pixels: each pixel represents a fragment of territory so small that the signal strength measured at its center (testpoint) can be considered representative for the strength of signals in any other point of the pixel. If we denote by S the set of broadcasting stations and by T the set of testpoints, the network design problem can be essentially described as that of i) setting the power emission of every station and b) selecting the serving station of each pixel/testpoint, with the objective of minimizing the total emitted power (green perspective) under the condition of granting service to a given fraction of the people located in the target territory.

The previous problem belongs to the family of *Wireless Network Design Problems* (e.g., [14,37]) and, in particular, it constitutes a variant of the Scheduling and Power Assignment Problem, known to be NP-Hard [11,37]. We can model the two decisions taken in the considered problem by means of two sets of binary variables, namely:

- to represent the power emission of a station $s \in S$, we introduce a set of equally spaced discrete power values $P = \{p_1, p_2, \ldots, p_n\}$ on which each station may emit. We also introduce the set $L = \{1, 2, \ldots, n\}$ to represent the index

of the discrete power values, which we call power levels. Given the set P, we introduce a set of binary variables $y_{s\ell} \in \{0,1\}$ to represent whether a station $s \in S$ emits on power value $p_\ell \in P$, such that $y_{s\ell} = 1$ when s emits on p_ℓ and $y_{s\ell} = 0$ otherwise (in what follows, we also alternatively write that a station $s \in S$ emits on power level $\ell \in L$).
- to represent whether a testpoint $t \in T$ is served by a station $s \in S$, we introduce a binary variable $x_{ts} \in \{0,1\}$ such that $x_{ts} = 1$ when t is served by s and $x_{ts} = 0$ otherwise.

In order to evaluate whether a testpoint $t \in T$ is covered with service with reference signal $s \in S$, the signal-to-interference ratio must be above a threshold $\delta > 0$ ([40]):

$$SIR_{ts} = \frac{\sum_{\sigma \in U(t,s)} a_{\sigma t} \cdot \left(\sum_{\ell=1,\ldots,n} p_\ell \cdot y_{\sigma\ell}\right)}{N + \sum_{\tau \in I(t,s)} a_{\sigma t} \cdot \left(\sum_{\ell=1,\ldots,n} p_\ell \cdot y_{\sigma\ell}\right)} \geq \delta$$

in which i) $N > 0$ is the noise of the system, ii) $\delta > 0$ is the minimum SIR value requested for considering the tesptpoint served, iii) the power emitted by a station $\sigma \in S$ is expressed by the combination of the power values p_ℓ by the corresponding binary power activation variable $y_{\sigma\ell}$ (i.e., $\sum_{\ell=1,\ldots,n} p_\ell \cdot y_{\sigma\ell}$). The summation distinguish between useful and interfering signal according to a DVB-T time-window detection explained in detail in [23,35,37].

The SIR lies at the core of every wireless network design problem (see e.g., [11,14,16,22,29,37,40]), and for the specific case that we consider, the model, which we denote by DVB-MILP, is:

$$\min \sum_{s \in S} \left(\sum_{\ell \in L} p_\ell \cdot y_{s\ell}\right) \tag{1}$$

$$\sum_{\sigma \in U(t,s)} a_{\sigma t} \cdot \left(\sum_{\ell \in L} p_\ell \cdot y_{\sigma\ell}\right) - \delta \sum_{\tau \in I(t,s)} a_{\sigma t} \cdot \left(\sum_{\ell \in L} p_\ell \cdot y_{\sigma\ell}\right) + M(1 - x_{ts}) \geq \delta \cdot N \qquad t \in T, s \in S \tag{2}$$

$$\sum_{s \in S} x_{ts} \leq 1 \qquad t \in T \tag{3}$$

$$\sum_{t \in T} \sum_{s \in S} \pi_t \cdot x_{ts} \geq \alpha \cdot \sum_{t \in T} \pi_t \tag{4}$$

$$\sum_{\ell \in L} y_{s\ell} = 1 \qquad s \in S \tag{5}$$

$$y_{s\ell} \in \{0,1\} \qquad s \in S, \ell \in L \tag{6}$$

$$x_{ts} \in \{0,1\} \qquad t \in T, s \in S \tag{7}$$

in which, a) the objective function (1) pursues the minimization of the total power emission; b) the quality-of-service conditions are expressed by the SIR formulas (easily reorganized by simple algebra operations) and including the big-M term $M(1 - x_{ts})$ to activate or deactivate the constraint depending on whether testpoint t is served by station s (see [14,17,19,29] for details); c) constraints (5) imposes that each station emits by exactly one power value whereas (3) imposes that each testpoint is served by at most one station. The constraints (4) impose that at least a fraction $\alpha \in [0, 1]$ of the total population of the region must be covered with service (π_r is the population of one pixel) . Finally, (6) and (7) are the decision variables previously defined.

3 Protecting Against Propagation Uncertainty

The fading coefficients a_{ts} that are part of the SIR constraints (2) are naturally subject to uncertainty because of the wide range of factors that influence signal propagation in a real environment (e.g., landscape, obstacles, weather, etc.) and that are hard to precisely assess [40]. These coefficients are commonly computed by (empirical) propagation models that, using extensive field propagation measurements, provide a formula for computing the coefficient values on the basis of factors like the distance between the communicating points, the portion of the spectrum adopted for transmissions, and the characteristics of the propagation environment (e.g., with many obstacles like tall buildings or in line of sight). As well-known by telecommunication professionals, the actual propagation values may be sensibly different from the values returned by the propagation models and it is thus very important to protect design solutions from possible fluctuations in these values.

Since the fading coefficients constitute *uncertain data*, i.e. data whose value is not exactly known when the problem is solved, we protect against fluctuations in their value that could cause infeasibility or sub-optimality of produced solutions by Robust Optimization (RO). RO is possibly the most successful optimization under uncertainty methodology and is essentially based on defining a robust counterpart of the original problem that identifies the best solution under worst data deviations, allowing to control the price of robustness (i.e., the deterioration in the value of optimal solutions due to excluding non-robust solutions from the feasible set) typically by setting one parameter as in the Γ-RO model (see [5,6].

As Robust Optimization model, we propose here to adopt *Multiband Robust Optimization* (MB) introduced in [8–10] to generalize and refine classical Γ-Robustness [6]: MB uses multiple deviation bands for better modeling arbitrary discrete distributions, under the form of histograms, which are commonly considered by professionals to analyze deviations in the input data in real-world optimization problems (as also illustrated in [4]). Also, MB allows to take into account "good" deviations, typically neglected in canonical RO approaches. As basis, we assume that the actual value of a generic uncertain fading coefficient a_{ts} belongs to the symmetric interval $[\bar{a}_{ts} - d_{ts},\ \bar{a}_{ts} + d_{ts}]$ (here, $\bar{a}_{ts}$ is the nominal value of the uncertain coefficient, while d_{ts} is its maximum allowed deviation). Practically, $\bar{a}_{ts}$ could be the value provided by a propagation model, while d_{ts} could be set as the maximum deviation that the network planner wants to consider according to its risk aversion. Following the principles of MB, for the uncertain fading coefficient we define the following MB Uncertainty Set:

1. we partition the overall deviation range $[-d_{ts}, d_{ts}]$ into K bands, defined on the basis of K deviation values:
$-d_{ts} = d_{ts}^{K^-} < \cdots < d_{ts}^{-1} < d_{ts}^0 = 0 < d_{ts}^1 < \cdots < d_{ts}^{K^+} = d_{ts}$;
2. through these deviation values, K deviation bands are defined, namely: a set of positive deviation bands $k \in \{1, \ldots, K^+\}$ and a set of negative deviation bands $k \in \{K^- + 1, \ldots, -1, 0\}$, such that a band $k \in \{K^- + 1, \ldots, K^+\}$ corresponds to the range $(d_t^{k-1}, d_t^k]$, and band $k = K^-$ corresponds to the single value $d_t^{K^-}$. Note that $K = K^+ \cup K^-$;
3. we define a lower and upper bound on the number of values that may experience a deviation of value in each band: for each band $k \in K$, two bounds $l_k, u_k \in \mathbb{Z}_+$: $0 \leq l_k \leq u_k \leq |T| \cdot |S|$ are introduced.

The linear robust counterpart of an uncertain SIR constraint defined for a couple (s, t) is obtained according to the theoretical results of Multiband Robust Optimization substituting each SIR constraint (2) of (s, t) with constraints:

$$\sum_{\sigma \in U(s,t)} a_{t\sigma} \cdot \left(\sum_{\ell=1,\ldots,n} p_\ell \cdot y_{s\ell} \right) - \delta \sum_{\sigma \in I(s,t)} a_{t\sigma} \cdot \left(\sum_{\ell=1,\ldots,n} p_\ell \cdot y_{s\ell} \right)$$

$$- \left(\sum_{k \in K} \theta_{ts}^k \cdot w_{ts}^k + \sum_{s \in S} z_{ts} \right) + M(1 - x_{ts}) \geq \delta \cdot N \qquad (8)$$

$$w_{ts}^k + z_{ts} \cdot \left(\sum_{\ell=1,\ldots,n} p_\ell \cdot y_{s\ell} \right) \geq d_{ts}^k \left(\sum_{\ell=1,\ldots,n} p_\ell \cdot y_{s\ell} \right) \qquad k \in K \qquad (9)$$

$$w_{ts}^k \geq 0 \qquad k \in K \qquad (10)$$

$$z_{ts} \geq 0 \qquad (11)$$

which includes the additional dual constraints (9) and variables (10), (11) for linearly reformulating the original (non-linear) robust multiband SIR constraints.

The robust optimization problem that we consider and that we denote by Robu-DVB-MILP is obtained by DVB-MILP substituting each SIR constraints with (8) and the auxiliary dual constraints and variables (9), (10), (11).

4 A Hybrid Solution Algorithm

The previous robust problem results challenging to be solved even for a state-of-the-art optimization solver like CPLEX [13], especially because of the presence of the complicating (robust) SIR constraints. To solve it, we thus propose a hybrid metaheuristic that combines heuristic exploration of the feasible set with the adoption of exact optimization methods (i.e., guaranteeing convergence to an optimal solution) for suitable subproblems of the complete problem. Specifically, we propose a metaheuristic that follows the algorithmic principles presented in [15,20], to which we refer the reader for more details. It is mainly based on a *probabilistic variable fixing* procedure integrated with an *exact large neighborhood search*. The probabilistic fixing procedure combines an a-priori and an a-posteriori fixing measures. In our case, the a-priori measure is provided by a linear relaxation of the robust model (model (1)–(7) with SIR constraints (2) replaced by (8)–(11)), denoted by Robu-DVB-MILP, while the a-posteriori measure is given by a (tighter) linear relaxation of the model (1)–(7), denoted by DVB-MILP (where a subset of variables has been fixed in value). At the end of each cycle of variable fixing, the a-priori fixing measure is updated, evaluating how good were the applied fixing. Once reached a time limit, the fixing cycle stops and an exact search runs for trying to improve the best solution found.

In the probabilistic fixing procedure, a number of solutions are built iteratively: at every iteration, a partial solution (i.e., a solution where only a subset of variables has its value fixed) is available and we can fix the value of an additional variable. Once the value of all the variables has been fixed, we obtain a complete solution whose quality is evaluated by means of its objective value. The fixing procedure is based on the observation that once the power emission variables have been fixed in value, it is possible to easily check which testpoints are covered with service by some station and compute the value of the objective function. We thus base the fixing procedure on deciding the values assumed by the power variables. At a generic iteration of the construction cycle of a feasible solution, we have at disposal a partial solution to the problem (obtained by having chosen the power emissions of a subset of stations $S^{\text{FIX}} \subseteq S$ by fixing their variables y_{sl} while respecting (5)). We probabilistically choose the next station whose power is fixed by the formula: whose power emission is fixed by means of the following formula, defined $\forall s \in S \backslash S^{\text{FIX}}, l \in L$:

$$p_{sl} = \frac{\alpha\ \tau_{sl} + (1-\alpha)\eta_{sl}}{\sum_{s \in S \backslash S^{\text{FIX}}} \sum_{\lambda \in L} [\alpha\ \tau_{\sigma\lambda} + (1-\alpha)\ \eta_{\sigma\lambda}]}\,, \tag{12}$$

which expresses the probability of fixing the power emission of station $s \in S \backslash S^{\text{FIX}}$ to power level P_l by considering all the couples $\sigma \in S \backslash S^{\text{FIX}}$, $\lambda \in L$ of stations whose emission is not yet fixed. In the formula, τ_{sl} is the a-priori attractiveness measure obtained from the optimal value of Robu-DVB-MILP including power-indexed variables, while η_{sl} is given by the value of a tight linear relaxation of DVB-MILP including fixing of variables done in previous iterations. The two measures are combined by a coefficient $\alpha \in [0, 1]$. If we set $y_{sl} = 1$ for some (s, l), due to (5) we can set $y_{s\lambda} = 0$ for all $\lambda \in L : \lambda \neq l$.

After having defined the power emissions of all stations (assume this is denoted by a binary power vector $\bar{y}$), all the SIR ratios can be easily computed. On the basis of the value of these ratios, we can also easily check which testpoints are covered with service and thus derive a valorization of the server assignment variables $\bar{x}$. The resulting solution $(\bar{y}, \bar{x})$ which is feasible for DVB-MILP is accepted as robust when it maintains its feasibility also when the fading coefficients are deviating to their worst value. Once a round of construction of feasible solutions has been operated, the a-priori measures are updated through formula:

$$\tau_{sl}(h) = \tau_{sl}(h-1) + \sum_{SOL=1}^{\gamma} \Delta\tau_{sl}^{SOL}$$

$$\Delta\tau_{sl}^{SOL} = \tau_{sl}(0) \cdot \left(\frac{OG(v^{AVG}, u) - OG(v^{SOL}, u)}{OG(v^{AVG}, u)} \right) \tag{13}$$

where $\tau_{sl}(h)$ is the a-priori measure of fixing station s at power level P_l at the h-th execution of the cycle and $\Delta\tau_{sl}^{SOL}$ is the modification to the value of the a-priori measures, computed over a summation that considers the last γ solutions that have been constructed. Moreover, u is an upper bound on the optimal value of the problem, v^{SOL} is the value of the SOL-th feasible solution built in the last construction cycle, v^{AVG} is the average of the values of the last γ solutions that have been constructed. The optimality gap OG(v,u) measures how far is the value v of a solution from the upper bound u and is defined as $OG(v, u) = (u - v)/v$. The role of formula (13) is to update the a-priori measure rewarding (penalizing) those fixing that have lead to a solution with lower (higher) optimality gap in comparison to the moving average value v^{AVG}.

At the end of the construction cycle, with the aim of improving the best robust solution found, an exact neighborhood search is conducted, i.e. we explore a (very large) neighborhood of the best solution, formulating the search as an optimization problem which is optimally solved by a state-of-the-art solver (see e.g., [7,21]). The adoption of exact searches is motivated by the fact that, while it can be difficult and long for a solver to solve the complete problem, it is instead possible to efficiently solve to optimality some subproblems. The large neighborhood that we define is built from a robust solution $(\bar{y}, \bar{x})$ allowing to change the power emission of all stations by either 1) turning off a station s (i.e., setting $y_{s0} = 1$ or 2) allowing a modification of the power emission to the adjacent power level set by $\bar{y}$ (i.e., if $y_{sl} = 1$ then it is allowed to set $y_{sl-1} = 1$ or $y_{sl+1} = 1$. The exact search is then conducted by expressing the previous conditions as linear constraints that are added to Robu-DVB-MILP and the resulting problem is solved by an exact solver.

The pseudocode of the matheuristic for solving Robu-DVB-MILP is presented in Algorithm 1. The first step consists of solving the linear relaxation of Robu-DVB-MILP including the power fixing of each couple (s, l) with $s \in S$ and $l \in L$. The obtained optimal values are employed to initialize the a-priori measures $\tau_{sl}(0)$. Then a solution construction cycle is executed until reaching a time limit. In each execution of the cycle, a number of feasible solutions are built first by fixing the power emission binary variables through formula (12), then deriving the corresponding valorization of variables x and finally checking their robustness. At the end of each execution of the cycle, the a-priori measures τ are updated on the basis formula (13). As last step, once the construction time limit is reached, the exact large neighborhood search is conducted, using as basis the best robust feasible solution defined during the construction cycle.

Algorithm 1

1: compute the linear relaxation of the power-indexed version of Robu-DVB-MILP for all $y_{sl} = 1$ and initialize the values $\tau_{sl}(0)$ with the corresponding optimal values
2: let (x^*, y^*) be the best robust feasible solution found
3: **while** a global time limit is not reached **do**
4: **for** $SOL := 1$ to γ **do**
5: construct a feasible power vector $\bar{y}$ using the probabilistic fixing formula (12)
6: check which SIR inequalities are satisfied and derive the corresponding $\bar{x}$ vector
7: check the robustness of the feasible solution $(\bar{x}, \bar{y})$
8: **if** the coverage granted by $(\bar{x}, \bar{y})$ is better than that of (x^*, y^*) **then**
9: update (x^*, y^*) with $(\bar{x}, \bar{y})$
10: **end if**
11: **end for**
12: update τ according to (13)
13: **end while**
14: execute the exact large neighborhood search using (x^*, y^*) and the modified power-indexed version of Robu-DVB-MILP as basis
15: return (x^*, y^*)

5 Preliminary Computational Results

The robust optimization approach was tested on 15 instances including realistic data defined from regional DVB-T networks deployed in Italy, including up to about 300 stations and 4000 testpoints. The revenue associated with covering a testpoint is represented by the population of the testpoint, so, in what follows, the value of the best solution found by an algorithm is expressed as the percentage of the population covered with service. As optimization software, we used IBM ILOG CPLEX [13] and the algorithms were tested on a Windows machine with 2.70 GHz Intel i7 and 8 GB of RAM. The hybrid metaheuristic of Algorithm 1 ran with a time limit of 1 h (50 min are devoted to the solution construction and 10 min are reserved to the execution of the exact neighborhood search). The parameters α and γ are set equal to 0.5 and 5, respectively. The robust model takes into account a deviation range that allows deviation up to 20% of the value of the fading coefficients and that is partitioned into 5 deviation bands. In order to evaluate the performance of the multiband robustness model, we considered the coverage of the population that is able to guarantee and the corresponding Price of Robustness (PoR), which we recall to be the reduction in solution optimality that we must pay in order to guarantee protection against uncertain coefficients. We also generated 1000 scenarios of realizations of the uncertain fading coefficients for evaluating the protection that the best found robust solution is able to guarantee. The preliminary results of the computational tests are presented in Table 1, where: i) ID identifies the instance; ii) *COV* is the percentage coverage of the population associated with the best solution found within the time limit and is reported for three models of the design problem, namely *Det*, which is the model not considering the presence of uncertain fading coefficient, *Full*, which considers the model including all the fading coefficients set to their worst value, and *Multi*, which is the Multiband Robust Optimization model; ; iii) *PROT* is the percentage of scenarios in which the best solution found results feasible (specified for the three considered models); iv) *PoR%* is the price of robustness, expressed as percentage increase in the total power value emitted by all stations.

Table 1. Computational results

ID	COV%			PROT%			PoR%		
	Det	Full	Multi	Det	Full	Multi	Det	Full	Multi
1	96.5	81.6	95.7	83.4	100	100	0	32.5	15.5
2	96.7	83.0	96.2	86.0	100	100	0	24.8	13.7
3	95.6	84.2	95.3	88.1	100	100	0	21.3	12.8
4	95.4	83.4	95.9	86.7	100	100	0	30.2	14.3
5	96.3	85.5	96.4	87.5	100	100	0	24.8	12.9
6	95.8	82.9	95.4	87.6	100	100	0	25.7	13.2
7	96.9	83.3	96.3	86.7	100	100	0	25.2	12.4
8	95.6	84.7	95.7	88.8	100	100	0	28.7	14.1
9	95.7	81.5	95.4	87.4	100	100	0	29.3	16.0
10	96.3	85.6	96.8	86.3	100	100	0	27.1	13.7
11	96.2	83.3	96.2	92.4	100	100	0	25.5	14.6
12	95.8	81.0	96.1	90.3	100	100	0	24.2	12.7
13	96.4	84.8	95.7	87.9	100	100	0	27.9	11.6
14	95.5	82.9	95.6	89.6	100	100	0	28.8	13.6
15	95.8	84.6	96.3	88.9	100	100	0	25.7	16.8

Looking at the table, a first observation that can be made is that the percentage coverage granted by the solution associated with full robustness is much lower than those by the deterministic and multiband models (on average only an unsatisfying 83% of the population is covered). This is not so surprising, since imposing full robustness forces the model to take into account all worst data deviations occurring simultaneously and this leads to a substantial shrinkage of the feasible set and to the identification of robust solutions that are unnecessarily conservative (it is indeed highly unlikely that all data jointly deviate to their worst value). In contrast, multiband robustness allows to guarantee a percentage coverage of the population that is very close to that of the deterministic model (on average 96.0% granted by the deterministic model versus 95.9% of the multiband model). This (superior) performance of the multiband model must be observed also taking into account the protection that is offered: the multiband model is able to offer the same full 100% protection of the full robustness model, which is much higher than that associated with the deterministic model, whose solutions turn out to be infeasible for about 12% of the cases. Finally, if we look at the price of robustness, the multiband model is able to entail a percentage increase in total power which is about halved on average with respect to full robustness (naturally, the deterministic model is associated with null price of robustness since it does not provide any protection). Looking jointly at the three performance indicators, multiband robustness is thus able to guarantee a full protectiona against deviations in propagation while maintaining the same level of coverage of the deterministic model and granting a substantial reduction in the price of robustness with respect with the full robustness model.

As future work, we intend to widen the computational experience to a larger set of instances, also conducting a study about the impact of parameter tuning. Moreover, we intend to also better study the impact of different characterization of the uncertainty set on the robustness of solutions.

References

1. AGCOM. Specifications for a DVB-T planning software tool (2009). http://www.agcom.it (in Italian). Accessed 01 June 2020
2. Anedda, M., et al.: Heuristic optimization of DVB-T/H SFN coverage using PSO and SA algorithms. In: 2011 IEEE International Symposium on Broadband Multimedia Systems and Broadcasting (BMSB), pp. 1–5 (2011)
3. Aragon-Zavala, A., et al.: Radio propagation in terrestrial broadcasting television systems: a comprehensive survey. IEEE Access **9**, 34789–34817 (2021)
4. Bauschert, T., et al.: Network planning under demand uncertainty with robust optimization. IEEE Commun. Mag. **52**(2), 178–185 (2014)
5. Bertsimas, D., Brown, D., Caramanis, C.: Theory and applications of robust optimization. SIAM Rev. **53**(3), 464–501 (2011)
6. Bertsimas, D., Sim, M.: The price of robustness. Oper. Res. **52**(1), 35–53 (2004)
7. Blum, C., Puchinger, J., Raidl, G.R., Roli, A.: Hybrid metaheuristics in combinatorial optimization: a survey. Appl. Soft Comput. **11**, 4135–4151 (2011)
8. Büsing, C., D'Andreagiovanni, F.: New results about multi-band uncertainty in robust optimization. In: Klasing, R. (ed.) SEA 2012. LNCS, vol. 7276, pp. 63–74. Springer, Heidelberg (2012). https://doi.org/10.1007/978-3-642-30850-5_7
9. Büsing, C., D'Andreagiovanni, F.: A new theoretical framework for robust optimization under multi-band uncertainty. In: Helber, S., et al. (eds.) Operations Research Proceedings 2012. ORP, pp. 115–121. Springer, Cham (2014). https://doi.org/10.1007/978-3-319-00795-3_17
10. Büsing, C., D'Andreagiovanni, F., Raymond, A.: 0–1 multiband robust optimization. In: Huisman, D., Louwerse, I., Wagelmans, A.P.M. (eds.) Operations Research Proceedings 2013. ORP, pp. 89–95. Springer, Cham (2014). https://doi.org/10.1007/978-3-319-07001-8_13
11. Capone, A., Chen, L., Gualandi, S., Yuan, D.: A new computational approach for maximum link activation in wireless networks under the SINR model. IEEE Trans. Wirel. Commun. **10**, 1368–1372 (2011)
12. Chiaraviglio, L., et al.: Algorithms for the design of 5G networks with VNF-based reusable functional blocks. Ann. Telecommun. **74**, 559–574 (2019). https://doi.org/10.1007/s12243-019-00722-w
13. IBM ILOG CPLEX. http://www-01.ibm.com/software
14. D'Andreagiovanni, F.: On improving the capacity of solving large-scale wireless network design problems by genetic algorithms. In: Di Chio, C., et al. (eds.) EvoApplications 2011. LNCS, vol. 6625, pp. 11–20. Springer, Heidelberg (2011). https://doi.org/10.1007/978-3-642-20520-0_2

15. D'Andreagiovanni, F.: A hybrid exact-ACO algorithm for the joint scheduling, power and cluster assignment in cooperative wireless networks. In: Di Caro, G.A., Theraulaz, G. (eds.) BIONETICS 2012. LNICST, vol. 134, pp. 3–17. Springer, Cham (2014). https://doi.org/10.1007/978-3-319-06944-9_1
16. D'Andreagiovanni, F.: Revisiting wireless network jamming by SIR-based considerations and multiband robust optimization. Optim. Lett. **9**, 1495–1510 (2015)
17. D'Andreagiovanni, F., Lakhlef, H., Nardin, A.: A matheuristic for joint optimal power and scheduling assignment in DVB-T2 networks. Algorithms **13**(1), 27 (2020)
18. d'Andreagiovanni, F., Lakhlef, H., Nardin, A.: Green and robust optimal design of Single Frequency Networks by min-max regret and ACO-based learning. In: 2022 IEEE International Smart Cities Conference (ISC2), Pafos, Cyprus (2022)
19. D'Andreagiovanni, F., Mannino, C., Sassano, A.: Negative cycle separation in wireless network design. In: Pahl, J., Reiners, T., Voß, S. (eds.) INOC 2011. LNCS, vol. 6701, pp. 51–56. Springer, Heidelberg (2011). https://doi.org/10.1007/978-3-642-21527-8_7
20. D'Andreagiovanni, F., Mett, F., Nardin, A., Pulaj, J.: Integrating LP-guided variable fixing with MIP heuristics in the robust design of hybrid wired-wireless FTTx access networks. Appl. Soft Comput. **61**, 1074–1087 (2017)
21. Danna, E., Rothberg, E., Le Pape, C.: Exploring relaxation induced neighborhoods to improve MIP solutions. Math. Program. **102**, 71–90 (2005)
22. Dely, P., et al.: Fair optimization of mesh-connected WLAN hotspots. Wirel. Commun. Mob. Comput. **15**, 924–946 (2015)
23. DVB-T2. https://www.dvb.org/standards/dvb-t2. Accessed 01 June 2020
24. Georgiadias, A., Christodoulou, P., Zinonos, Z.: Citizens' perception of smart cities: a case study. Appl. Sci. **11**(6), 2517 (2021)
25. Liu, H., Wei, H.: Towards NR MBMS: a flexible partitioning method for SFN areas. IEEE Trans. Broadcast. **66**(2), 416–427 (2020)
26. Ismail, S., et al.: Recent advances on 5G resource allocation problem using PD-NOMA. ISNCC 2020, Montreal (2020). https://doi.org/10.1109/ISNCC49221.2020.9297208
27. International Telecommunication Union: Report BT.2140. https://www.itu.int/dms_pub/itu-r/opb/rep/R-REP-BT.2140-2008-PDF-E.pdf. Accessed 01 Feb 2023
28. International Telecommunication Union (ITU): DSB Handbook (2002). https://www.itu.int/pub/R-HDB-20. Accessed 01 Feb 2023
29. Kennington, J., Olinick, E., Rajan, D.: Wireless Network Design: Optimization Models and Solution Procedures. Springer, Heidelberg (2010). https://doi.org/10.1007/978-1-4419-6111-2
30. Koutitas, G.: Green network planning of single frequency networks. IEEE Trans. Broadcast. **56**(4), 541–550 (2010)
31. Jiang, H., Geertman, S., Witte, P.: The contextualization of smart city technologies: an international comparison. J. Urban Manage. **12**, 33–43 (2022)
32. Joshi, S., et al.: Developing smart cities: an integrated framework. Procedia Comput. Sci. **93**, 902–909 (2016)
33. Lanza, M., et al.: Coverage optimization and power reduction in SFN using simulated annealing. IEEE Trans. Broadcast. **60**(3), 474–485 (2014)
34. Li, C., et al.: Planning large single frequency networks for DVB-T2. IEEE Trans. Broadcast. **61**(3), 376–387 (2015)
35. Ligeti, A., Zander, J.: Minimal cost coverage planning for single frequency networks. IEEE Trans. Broadcast. **45**(1), 78–87 (1999)

36. Lomakin, A., et al.: Modeling and evaluation of intra-system interference in DVB-T2 single-frequency networks. In: FarEastCon 2019, Vladivostok (2019)
37. Mannino, C., Rossi, F., Smriglio, S.: The network packing problem in terrestrial broadcasting. Oper. Res. **54**, 611–626 (2006)
38. Nepal, S., et al.: Optimization of multi- frequency network with DVB-T2 services for regions with complex geographies: a case study of Nepal. IEEE Trans. Broadcast. **67**, 299–312 (2021)
39. Osenkowsky, T., et al.: National Association of Broadcasters Engineering Handbook. Routledge, New York (2018)
40. Rappaport, T.: Wireless Communications: Principles and Practices. Prentice Hall, Upper Saddle River (2001)
41. Richter, L., Reimers, U.H.: A 5G new radio-based terrestrial broadcast mode: system design and field trial. IEEE Trans. Broadcast. **68**, 475–486 (2022)
42. Sanchez-Corcuera, R., et al.: Smart cities survey: Technologies, application domains and challenges for the cities of the future. Int. J. Distrib. Sens. Netw. 15(6) (2019)
43. Shehab, M.J., et al.: 5G networks towards smart and sustainable cities: a review of recent developments. IEEE Access Appl. Future Perspect. **10**, 2987–3006 (2022)
44. United Nations - DESA. Around 2.5 billion more people will be living in cities by 2050 (2018). https://www.un.org/en/desa. Accessed 01 Feb 2023
45. Yang, C., et al.: Using 5G in smart cities: a systematic mapping study. Intell. Syst. Appl. **14**, 200065 (2022)

WanDa: A Mobile Application to Prevent Wandering

Berardina De Carolis, Vincenzo Gattulli(✉), Donato Impedovo, and Giuseppe Pirlo

Department of Computer Science, University of Bari Aldo Moro, Bari, Italy
vincenzo.gattulli@uniba.it

Abstract. The nature of Big Data tends to collect a huge quantity of useful information about human life. Implementing Artificial Life applications inherent to health could improve and sensitize individuals to the future. In fact, would be useful to implement an application that monitors people with neurodegenerative diseases when they are away from home to monitor Wandering. In this paper, an application called "WanDa" is proposed that monitors and prevents deviations from the usual path in real-time and, if wandering is detected, can guide the elderly person to a safe place and alert caregivers or relatives. The application uses the sensors and technologies of a generic Android smartphone and has a very simple interface to manage wandering behaviors. We tested the application from two perspectives: the accuracy of the algorithm in detecting wandering behaviors and the user experience. In both cases, WanDa was judged positively (Questionnaires performed by caregivers of patients who take the test), showing that it can be a useful support for managing, monitoring, and reporting wonder episodes.

Keywords: Health and Social Care · Smartphone · Wandering · Android Application · Automated Wandering Behavior · Behavioral Biometrics

1 Introduction

Artificial life is in overall evolution. In this area, it is possible to implement applications that succeed in improving the human standard of living. Behavioral Biometrics tends to fruit big data and patterns to be able to identify a specific behavior. Indeed, in this paper, we would like to monitor wandering. Wandering is often identified as a symptom of Alzheimer's disease or dementia. It is a widespread phenomenon in the elderly with dementia and occurs when a person roams around and becomes lost or confused about his location. People with dementia often wander because they are stressed, looking for something, engaging in past routines, or, other times, they may wander without aim at all. On the other hand, the sufferer tends to forget where they are going and what they were trying to do, wandering around and losing track of where they are. Wandering is also defined as "a dementia-related locomotor behavior syndrome having a frequent, repetitive, time-disordered and spatially disoriented nature that manifests in patterns of Lapping (more on that later), Random and Pacing" [1]. Wandering is also significantly common also among children with *ASD (Autism Spectrum Disorder)* and those

C. De Stefano et al. (Eds.): WIVACE 2022, CCIS 1780, pp. 232–244, 2023.
https://doi.org/10.1007/978-3-031-31183-3_19

with behavioral and developmental problems than among other children. Wandering among children with ASD tends to increase with increasing severity of ASD symptoms and decreasing developmental level and it is higher among males and children who have attention deficit hyperactivity disorder, anxiety, depression, or oppositional behavior [2]. Even wholly healthy, unimpaired people/children can find themselves at a loss, perhaps in times of high stress, which, induced by everyday life, can lead to a state of confusion at any time while they are headed toward a destination. Traditional methods impose physical restrictions to prevent wandering. However, physical, and psychological problems caused by physical restraints make traditional solutions impractical or ineffective in protecting vagrants. As an alternative, preventive measures have been recommended to maximize autonomy by minimizing the risk to older persons prone to wandering in home, community, and nursing facility settings. Wandering is a phenomenon that needs to be controlled because prohibiting patients from wandering could mean the loss of benefits associated with walking, including improved circulation and oxygenation and decreased risk of contractures. Interventions for wandering have prevented the behavior through physical and pharmacological restrictions. Apart from the known harmful effects of restraints, such as bedsores, anxiety, physical violence, falls, and high morbidity and mortality rates, the intervention is ineffective. Non-pharmacological interventions are a safer option and include electronic tagging and tracking devices, behavioral approaches, exercise, music therapy, aromatherapy, access camouflage, and environmental modifications [3]. This work aims at addressing one of the significant concerns of caregivers, which is to prevent and detect wandering episodes of individuals for whom they are in care. Thanks to smartphone technologies, it is possible to ease and support caregivers by assisting and monitoring the paths of assisted persons. In fact, in this work, we propose a solution offered through an Android application called WanDa which, thanks to a sought-after compromise between the efficiency of the techniques used and the accuracy of the results, succeeds in helping to manage the phenomenon of wandering. The application consists of three complementary software modules, thanks to the GPS hardware present in any Android smartphone, the app detects situations of wandering to present, when appropriate, quick solutions to the users to help them manage the situation of possible panic. The paper is structured as follows: the second section illustrates the state of the art of algorithms used by applications created for this purpose. Section three illustrates the design and implementation of the WanDa application. Then, in Section four tests in real-world contexts with experimental results will be illustrated. Finally, Section five illustrates the conclusions and planned future work.

2 Related Work

The design of a system depends substantially on the target scenario. As discussed by E. Batista et al. [4], there are two scenarios with a wide range regarding the location of roaming behavior: outdoor and indoor. The objective of this study is to provide a solution-oriented to the first category. To develop a solution that can prevent elderly people with dementia from transgressing the boundaries, which is often related to getting lost or other adverse events without timely care services, Q. Lin et al.[5] propose a Data Mining-based approach to build a personalized and secure outdoor geofence by extracting individuals' historical GPS trajectories, that is a kind of virtual boundary delineated around an area of interest that can be created with a variety of different techniques, such as Wi-Fi, cellular network, RFID and GPS. To build a secure external geofence customized for each elderly person, the method of geofencing is applied. It is proposed to model the movement of each elderly person as a graph, whose vertices refer to frequently visited places and the edges of this graph are paths between two adjacent locations. The qualitative evaluation results showed that the method is feasible for constructing customized secure geofences based on individuals' historical GPS trajectories. This secure geofence can be used to detect online whether an elderly person with dementia has moved across the boundary of the secure geofence [5]. Very similar work has been done by D. Zhang et al. [6] to provide appropriate real-time care services to elderly people who suffer from physical or cognitive disabilities and who often have difficulty navigating activities and remembering landmarks. They propose a disorientation detection method that detects outliers within GPS trajectories. Precisely, the first model of an individual's movement trajectories as a graph based on his historical GPS tracks by developing a method called iBDD that can detect two categories of peripheral trajectories in a uniform framework in real-time. To perform this detection, they project the historical movement trajectories of each elderly person onto a digital map and model the regular movement of each elderly person as a graph in which the vertices represent the places attended and the edges correspond to the paths between the places. Using the real-world GPS datasets of ten individuals, they demonstrated that iBDD could achieve a 95% disorientation detection rate [6]. Y. Chang [7] proposed real-time anomaly detection by considering user trajectories as input, and the anomaly is identified in output. Trajectories are modeled as a discrete-time series of axis-parallel edges ("*boxes*") in two-dimensional space. A trajectory is represented by a series of boxes, each with six attributes: The maximum longitude value, The minimum longitude value, The maximum latitude value, The minimum latitude value. A box is started with an acquired GPS point. When two boxes overlap, their similarity is equal to 1. The farther apart the two boxes are, the lower their similarity is. For 2D trajectories, the similarities can be decomposed into longitude and latitude similarity. Experimental results show that accuracy is 97.6% (*with Axis Adjustment*), and recall is 98.8% for participants with cognitive impairment [7]. The IRoute system invented by S. Hossain et al. [8] implements the Belief-Desire-Intention (BDI) architecture. It tracks the GPS position of the person with dementia in real-time and updates predictions based on changes in position. A deviation from the predicted path is considered abnormal behavior. As an intervention technique, a correct route is provided to induce the person with dementia to follow it. Failure to follow the guided route triggers the system to notify caregivers. Running on a GPS-equipped cell phone, the BDI agent is responsible for route prediction using user input (*list of travel locations and activities, frequency of events, start times, and destination locations*) and stored routes from previous trips (*a set of timestamps and GPS locations to a destination*) [8]. After showing the state of the art, some techniques chosen for the implementation of the WanDa application will be explained in more detail, namely the wandering detection algorithm called θ_WD and the *C-SIM* similarity measure. The θ_WD algorithm proposed by Lin et al. [9] is a wandering trajectory classification algorithm designed for the outdoor environment. They configured the algorithm by considering the angle between two trajectories considering lapping and pacing patterns as indicators. The authors [9] define the algorithm for wandering detection as "a ring trip, with each ring consisting of a series of track segments, blocked by two adjacent sharp points within a given distance interval." In summary, the algorithm has been implemented such that one loop, with four segments divided by sharp points, is sufficient to detect a wandering trace. According to the definition, repetitive loops are required to label a trace as stray. The algorithm attempts to identify only whether a trajectory contains loops and uses patterns to obtain a more generalized solution. The authors assumed that, in a wandering trajectory, there are points (those referred to as *"sharp points"*) with significant changes in direction, equal to vector angles equal to or greater than 90 degrees. Continuous acute points create at least one loop in the wandering trajectory. The angle between consecutive vectors (thus from three consecutive points) is an indicator of the direction of motion. If the angle between them is greater than 90° and this is repeated 4 times within the sub-segments (whose length is prefixed/parameter), then the algorithm returns the positive presence of a wandering situation [9]. Regarding the similarity between paths, the method illustrated in the article *R. Mariescu-Istodor* et al. [10] that illustrates its simplicity of implementation and at the same time, provides a fast algorithm with complexity in linear time. When overlapping two paths, the real path and the ideal path, the following measure is very useful. The amount of overlap measures how similar the paths are. Specifically, Cell Similarity (C-SIM) is a similarity measure that considers the area traversed by the two paths. It uses a grid to calculate a cell representation for the two trajectories and then measures how many cells are in common relative to the total number of distinct cells. In theory, an infinitely long path can exist in a single cell simply by moving in a circle within the cell. This type of behavior is sometimes noticed when the user is stationary, but the GPS signal fluctuates. Often points close to each other end up in different cells due to arbitrary grid division. This can produce errors when comparing routes. In principle, two people walking hand in hand can never share a cell. To resolve this and compensate for the arbitrary division of a grid, which can then allow points as far apart as 1 mm to lie in different cells, CSIM uses morphological dilation with a square (3 × 3) structural element. When using these grid-based approaches, the size (parameter L) of the cell acts as a distance threshold, and points mapped to the same cell are considered identical. The CSIM formula is essentially Jaccard's coefficient modified to handle dilated cells [11]:

$$S(C_A, C_B) = \frac{|(C_A \cap C_B) \cup (C_A \cap C_B^d) \cup (C_B \cap C_A^d)|}{|(C_A \cup C_B|} \quad (1)$$

From the point of view of effectiveness, C-SIM is the least affected by sampling rate variations and performs quite well under conditions of noise and point shift [10]. The θ_WD algorithm proposed by Lin et al. [9] will be used within one of the modules for wandering detection in the WanDa application. More precisely, it has been adapted in the component defined as "knowledgeless," which provides as input to the algorithm the GPS points of the user's walk that the smartphone acquires in real-time. While with regard to the path similarity measurement, the C-SIM algorithm [10] (https://github.com/uef-machine-learning/C-SIM) written in Java was chosen and after modifying it by implementing the sliding window method, it was included within the module called "with knowledge," precisely because it can detect wandering occasions thanks to the knowledge of the paths, the similarity of which with respect to the user's recent walk is carried out precisely thanks to the just-discussed C-SIM algorithm.

3 WanDa Mobile Application

The WanDa application is divided into two parts: the first intended for users and the second one intended for caregivers or parents in the case of child users. Users are provided with a simple interface consisting only of the central button for starting the Wandering service in the background, as well as related buttons for interacting with the map (focus and zoom in on the current location, zoom map, north up). Also designed for users is the alarm screen that appears on full screen whenever the system detects lost possibilities with appropriate buttons that can help them manage the situation. Also thought for users is the mode that makes Google Maps start navigation to one of the destinations previously stored in the settings and finally also the call to a preset phone number (relative, caregiver, rescue…). On the other hand, for the part of the application dedicated to caregivers, they can input settings (accessed from the button at the top right of the main screen), which will allow them to add information such as emergency numbers, user's usual routes, usual places, residence, and so on. Moreover, caregivers can use the app for localizing the assisted user in case of need (help request, wondering detection). The screens of the WanDa app will be illustrated below (Fig. 1).

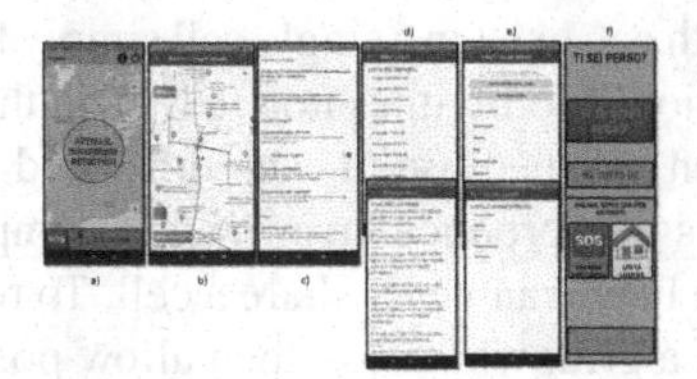

Fig. 1. a) Main screen, b) Usual routes entry screen, c) Settings, d) Report screen, e) Screen for entering usual places, f) Alarm (six),

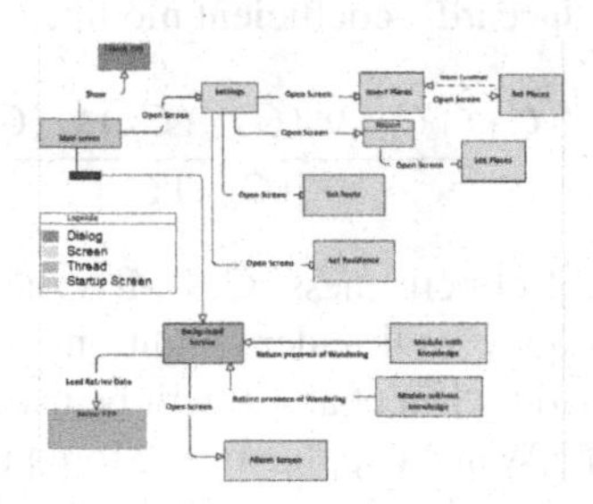

Fig. 2. WanDa Application Class Diagram

The (Fig. 2) shows the diagram of the classes, that is, how the classes used for detection interact with each other.

3.1 Main Screen

The main screen contains simple graphical buttons with large print. The main screen includes a large central button that allows you to start the wandering detection service, a settings button, and an info button that disappears once the service is started (*This was done first and foremost so that the user cannot absentmindedly press it while walking*). On the first startup, you will be asked for geolocation permissions. After clicking the middle button, the Google map will be launched (Google Maps API with some simple features such as Zoom and map rotation) where some routes entered by the caregiver (*through the Gson class, which is an open-source Java library for serializing and deserializing Java objects to JSON*) and stored locally (*3.3 Usual routes entry screen*) will be present in memory. Concerning the bottom section, there will be two buttons: "SOS" is used to initiate a phone call to the number set for emergencies; "routes icon" which allows to open another screen that contains the list of saved places and if an item is clicked on it, it makes the Google Maps app open with already set up walking navigation via to the relevant coordinates that were previously saved in the places entry section.

3.2 Settings Screen

The settings screen is accessible only when the app is not in the Wandering Detection state; it is divided into several sections. The first section allows for the entry of both an emergency phone number and the maximum minute by which you want the user to return to the home residence. The second section relates to habitual "*Places and Routes,*" the first item is a link that leads to another screen used for entering the user's habitual routes, and the second item is a link that leads to a screen where the name and coordinates of places can be recorded, and the third item provides the option of setting the residence. The third section contains a link that leads to the user's report.

3.3 Usual Routes Entry Screen

The screen for entering routes features a Google map in the center and buttons about the features implemented for entering the user's usual routes. First, it is a + button to be able to add a new route, the starting point of which can be entered by clicking on the desired map point. In addition, there is a "Cancel" button that allows the user to delete the entire set of various points entered. These, once saved by clicking "Save Route," will be automatically overlaid on the map present on the main screen. This step is used for caregivers.

3.4 Alarm Screens

The alarm screens appear one consequent to the other. When the system detects a wandering state, a screen appears with two buttons. The first one is used to say that "*you are actually lost*" and immediately opens the management screen of the dangerous situation, while the one further down, implying that it is a false positive wandering situation, closes the alarm screen. The action of the first button will open another screen with three buttons. The first one immediately initiates a call to the preset emergency telephone number. The second one refers to the screen containing the usual locations to initiate the relevant navigations. Finally, the third one at the bottom is used to close the screen. When launched, the smartphone's vibration is activated for four seconds, as well as playing a ringing sound to call the user's attention. The management of feedback given to the user in case of wandering has been studied by experts in the field. In fact, the caregiver must make sure that the sound is bailed on the device and that it is loud. In addition to the loud sound emitted, vibration of the device has also been added and in addition a design with large, colorful words to alert caregivers or the patient himself if he is still conscious.

3.5 Screen for Entering Usual Places

This screen consists of a field to enter the name you want to give to the usual place (*to perhaps help better remind the user which place it is*), a button that redirects to another where you center the place with the cursor placed in the middle. Finally, a button to store the place with the information entered just described. This, once clicked, in addition to storing the various information locally, causes the name of the place just added to appear in a list placed at the bottom. If the relevant item in the list is clicked on, it is deleted both from it and from the local memory of the smartphone.

3.6 Report Screen

The report screen, accessible through the settings section, contains a list of all reports related to walks started by the user, such that it has as headers the date and time related to the time when that tracking session was started. After clicking on an item, a screen will open regarding the session and contain information. More specifically, it will be indicated, with date and time, when the user departed from the usual routes and whether he returned, when he made abnormal movements, and when he was around the residence. Finally, if you want to delete a report, hold down on the one you want to delete from the list.

3.7 Starting Wandering Detection

When you click on the center green button on the main screen, a popup appears asking the user to enable it if GPS is not active. If yes, the application switches to the active detection state. Mobile devices have several techniques for keeping tabs on their geographical location; in addition to GPS, they can take advantage of some information about the Wi-Fi network to which they are connected. However, if you have no special requirements, the easiest way to monitor the user's location is to use the LocationManager service. One of the service tasks that will run in the background (dedicated to wandering detection) is to communicate with this service, activating it during initialization and listening for its updates. For wandering detection, two modules have been designed that work consequently with each other. The first is called the "Module with knowledge," which manages to detect wandering possibilities by already knowing the routes that the user usually takes (Entered previously by the caregiver); the second is called the "Module without knowledge," that is, the one that works without the knowledge of the usual routes that the user takes. The "Module with knowledge" first calculates the length of the recent route taken by the user, which corresponds to the set of GPS points acquired from the smartphone that is stored (*always and only the last x points, where x is a predefined parameter*). Then, each route stored in the habitual will be overlaid with the user's route in real-time (*3.8 Sliding Window Algorithm*). Suppose none of the maximum similarities obtained from this comparison is greater than or equal to the value set as a threshold (decided after several tests). In that case, a timer of 1 min (the time the user must re-enter the routes entered) is started. If one of the similarities has a value greater than the set threshold during the expiration of the time, then the timer will be canceled, asserting that the user has re-entered the set route. The alarm screen will appear if the minute is exceeded, and the user has not re-entered the preset routes. After this first step is carried out by the "Module with knowledge," further verification will be carried out by the "Module without knowledge." It can detect, based on the "*morphology*" of the user's route, possible wandering. The system would report wandering if the angles formed by the GPS positions include circular-type trajectories or anomalies within the path. The detection of these unusual movements is left to a special algorithm that, if successful, triggers the wandering alarm with the consequent appearance of the alarm screen. This algorithm is referred to as θ*_WD*. In addition, the calculation of the scalar product and the norm useful for calculating acute angles has also been included, which, if greater than three, cause the function to return a positive outcome. Finally, a third module that works by simply setting a timer whose duration is set from the application settings has also been implemented. The timer starts as soon as the user initiates Wandering detection from the button on the main screen. The alarm, in this case, is triggered only if the user does not return to the residence within the time set by the caregiver. The WanDa detector is activated by the caregivers with the purpose of monitoring from the starting point the entire pathway the patient performs. The WanDa detector will always be turned off by the caregivers.

3.8 Sliding Window Algorithm

By testing the C-SIM algorithm first among samples of sets of GPS points and then among data received by the application, it was found that the algorithm, in returning the percentage of equality between two routes, considers their length of them. However, the latter feature made the similarity untrue since the length of the user's last steps was often somewhat shorter than the length of the various stored routes. Therefore, the C-SIM similarity algorithm was adapted to the "sliding window" approach. This means that the algorithm compares the user's last steps and all the various possible fragments for each of the paths. In addition, as a matter of efficiency, this will be done only with those that are longer, or slightly shorter, than the length of the walk. Each comparison made with each portion then returns an index concerning its similarity and the last steps taken. Finally, the highest similarity that was obtained is retained for each path. All this is then done to proportionally scale the comparison between the various routes concerning the distance of the walk taken by the user. Otherwise, an incorrect similarity would be obtained due precisely to the different lengths of both in terms of meters.

4 Testing and Results

The test of the application was divided into five parts. The first three experiments were performed by domain experts (*Heuristic Tests*) with the purpose of testing the modules with knowledge and without through the simulator offered by Android Studio, while the last two tests were performed with real users. The tests are named *"Heuristic Tests on the modules with and without knowledge"*, *"Tests with different parameters"*, *"Tests on two GPS track datasets "*, *"Real-world Testing"* and *"Questionnaire results"* with SUS, UEQ and NPS Questionnaire to evaluating WanDa usability. Heuristic tests were conducted for the first three tests (4.1, 4.2, 4.3), they were carried out by a group of system experts that do not have any wandering pathology. The tests 4.4 were performed in real contexts by 12 patients that have wandering pathology. The 4.5 test was performed by caregivers of patients. The first three experiments were carried out by experts, they are analyzing that do not consider saved datasets.

4.1 Heuristic Tests on Forms with and Without Knowledge

These tests, for both modules were successful, it was noticed both numerically and visually that the evaluator was proceeding in the route showing green dots in real-time. In Fig. 2 (Module with knowledge - left) we can see that the user followed the route precisely and starts to deviate from it by turning left for a few meters. In the upper left corner, the similarity of 1 is shown, a value that will remain for a few more steps if he continues that route. In Fig. 2 (Form with knowledge - right), on the other hand, the user has deviated quite a bit from the previously entered route (colored yellow), and in fact the similarity is very close to the default 0.6 set threshold (Fig. 4).

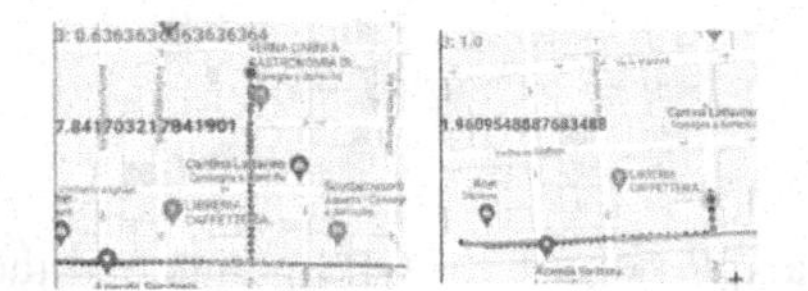

Fig. 3. Example Module with knowledge

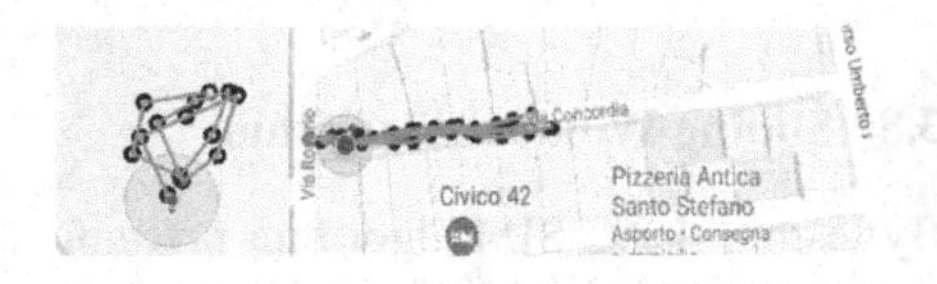

Fig. 4. Example Module without knowledge

In Fig. 3 (Module without knowledge - left) since the repetition of the path occurred relative to only the first 50 m, the second module gave wandering alerts given the sudden change of course that occurred several times. In Fig. 3 (Module without knowledge - right) there is a circle-shaped path forming some rather sharp angles that triggered the wandering alert.

4.2 Testing with Different Parameters

The evaluators experimentation will set parameters that will be evaluated in the actual tests. The first parameter concerns the number of GPS positions (steps). Several tests were carried out with 30, 40, 50 and 60 steps. The results shown in Table 1:

Table 1. Number of tracks correctly recognized per parameter GPS positions to be kept

Parameter Values				
Recognized traces	30	40	50	60
	8/10	9/10	8/10	7/10

The value 40, through experimental traces fed into the Android emulator, is the value that gave the best results, and this is due to the fact that if the number of steps is too short, especially below 30 steps. With too high a value would have the opposite effect of taking too many past points making the similarity untrue, in addition to the fact that it would make the calculation operated by the various modules more onerous. The second parameter relates to the size of the grid cells used by the "*knowledgeless module*." Detecting this parameter experimentally was difficult since it will depend on it how far the user can stray from the predetermined routes (Table 2).

Table 2. Number of traces recognized correctly regarding the width of grid cells

Parameter Values				
Recognized traces	20	30	35	40
	1/10	7/10	9/10	8/10

The value of 37 m is ideal, since in practice it would correspond to allowing parallel routes even up to almost 80 m away. Parallel routes are still considered viable and accepted as like the route set by the caregiver. As can be seen from the table above, even values close to 37 (30 and 40). In the case of small villages, this metric corresponds to one or even two parallel routes (albeit close to each other) that still allow some freedom of decision in choosing an alternative route but at the same time prevent going very far. Finally, regarding the similarity threshold value, during the various tests it was set equal to 0.6 given parameter 37 for the width of each cell.

4.3 Testing on Two GPS Track Datasets

In this subchapter, paths were simulated thanks to the emulation of GPS locations in Android Studio with the purpose of verifying the reactions and outputs of the application. The purpose of this test is to ascertain the behavior of the application with complete paths and in different situations. Two datasets will be extracted for this test:

- *Dataset1:* 10 GPS routes where there is no presence of wandering;
- *Dataset2:* 10 GPS routes with wandering situations.

Table 3 shows the number and relative percentage of tracks recognized correctly by the app. An accuracy of 85% was obtained with this test:

Table 3. Trace set confusion matrix

	Dataset1	Dataset2
Recognized	8	9
Not Recognized	2	1

Regarding the first dataset there were two traces that were not recognized exactly. This happened because at the moment when the user was re-entering the path (module with knowledge), the similarity did not go back up correctly (which it would have safely done a few steps further, however) causing the one-minute timer to start and expire. Regarding the second dataset, it was the second module that sensed the wandering at the moment when the user slowed down right under a balcony that also made the points less accurate thus forming "*sharp corners.*" The datasets referencing this experiment are not public. They were created by a group of experts in the evaluation of the heuristic test (the same experiments considered for experiments 4.1 and 4.2). This group of people can test the application before using the patients in the real world experiment (4.4).

4.4 Real-World Testing

Single test: In this subchapter, the WanDa application was tested with 12 real users in real contexts. The 12 users were made to perform tasks such as, for example, following certain set routes or parallel routes (*Task without wandering*), and tasks such as getting lost on the way or making abnormal movements such as turning around on a point or stopping for a long time (*Task with wandering*) with the purpose of testing the application. The results obtained from this test have been summarized in Table 4. An accuracy of 96% was obtained from this real-world test.

Table 4. Confusion matrix on real-world testing

	Task without wandering	Task with wandering
Recognized	11	12
Not Recognized	1	0

As can be seen from Table 4, the real-world test found very good results, confirming the validity of the parameters set in the previous tests (*number of steps to keep in memory, cell width, and threshold value (4.2 Tests with different parameters)*). The only unrecognized path is inherent in the first module, which incorrectly recognizes a path without wandering as wandering. Analyzing the data on the server remotely, a slight shift caused the similarity to decrease, which triggered the alarm screen. *Group test*: This test was carried out which consists of performing the same tests as the single test but moving as a group (*like a group of students to school*). The purpose of this experiment is to test whether the distance between users, a few meters apart, could affect the detection of wandering. The results highlighted how, due in part to the accuracy of the various smartphones, that the points detected when stopping, especially in areas with interference, were 3 or 4 m apart between the various smartphones of the group members. However, this equally allowed wandering to be detected only when appropriate, achieving 96% accuracy concerning the first module. On the other hand, when it comes to the second module, depending on the angles in the walk, wandering was still detected by the various smartphones, although on a few occasions there happened to be a few tens of seconds delay in detection from one device to another.

4.5 Questionnaire Results

The SUS, UEQ, and NPS questionnaires were administered to each user to measure the users' subjective impression of using the app and to understand how effective the GUI was. Regarding the *UEQ (User Experience Questionnaire)* the app presented some problems related to perspicuity for some users. But for most of them, on the other hand, it was found to be efficient (it responds abruptly to detection). Finally, it was found to have high originality about how it works. Next, the *SUS (System Usability Scale)* questionnaire was submitted, finding an overall score of 73. Looking at the rankings of the SUS questionnaire scores, the application ranks on grade B, thus with the relative adjective of "Good," presenting a usable interface. Finally, the *NPS (Net Promoter Score)* questionnaire was also submitted, and by calculating the score based on the number of detractors, passives, and promoters, it turns out that the score is 75. Regarding the questions related to originality and innovation, it was inferred that the application was also found to be very original and proposed a solution that many users had not seen or otherwise heard about before.

5 Conclusions and Future Work

The study enabled the implementation and testing of the WanDa application that was able to detect the presence of wandering by individuals with neurovegetative diseases. The application, after being carefully set up and adjusted by the caregiver, provides high accuracy in detecting wandering, and this can help and support both actors, with all the benefits that come with it. The solution, observing the various tests and questionnaires, presents some limitations. In fact, as discussed in the experimentation chapter, some situations were found to be mistakenly recognized as pertaining to wandering episodes. In addition, through SUS, NPS, and UEQ questionnaires administered to users, it was found that the application interface could be improved for some aspects. This, no doubt, indicates that the application could be further improved in the future not only from the UX point of view but also as far as functionalities are concerned. To this aim, we will conduct a study for a longer period in cooperation with a local association that assists people with dementia. From its results we will refine the application to better meet the needs of final users. Moreover, we will investigate also how the application could take advantage of using smart bands and smartwatches. The WanDa application is not yet available, some parts of it will be improved and later other modules will be added such as the fall detection module.

Acknowledgments. This work is supported by the Italian Ministry of Education, University and Research within the PRIN2017 - BullyBuster project - A framework for bullying and cyberbullying action detection by computer vision and artificial intelligence methods and algorithms.

References

1. Lin, Q., Zhang, D., Chen, L., Ni, H., Zhou, X.: Managing elders' wandering behavior using sensors-based solutions: a survey. Int. J. Gerontol. **8**, 49–55 (2014). https://doi.org/10.1016/J.IJGE.2013.08.007
2. Wiggins, L.D., et al.: Wandering among preschool children with and without autism spectrum disorder. J. Dev. Behav. Pediatr. **41**, 251–257 (2020). https://doi.org/10.1097/DBP.0000000000000780
3. Adekoya, A.A., Guse, L.: Wandering behavior from the perspectives of older adults with mild to moderate dementia in long-term care. Res. Gerontol. Nurs. **12**, 239–247 (2019). https://doi.org/10.3928/19404921-20190522-01
4. Batista, E., Casino, F., Solanas, A.: On wandering detection methods in context-aware scenarios. In: IISA 2016 - 7th International Conference on Information, Intelligence, Systems and Applications. (2016). https://doi.org/10.1109/IISA.2016.7785349
5. Qiang, L., Xinshuai, L., Weilan, W.: GPS trajectories based personalized safe geofence for elders with dementia. In: 2018 IEEE SmartWorld, Ubiquitous Intelligence & Computing, Advanced & Trusted Computing, Scalable Computing & Communications, Cloud & Big Data Computing, Internet of People and Smart City Innovation, pp. 505–514 (2018). (SmartWorld/SCALCOM/UIC/ATC/CBDCom/IOP/SCI). https://doi.org/10.1109/SmartWorld.2018.00111
6. Lin, Q., Zhang, D., Connelly, K., Ni, H., Yu, Z., Zhou, X.: Disorientation detection by mining GPS trajectories for cognitively-impaired elders. Pervasive Mob. Comput. **19**, 71–85 (2015). https://doi.org/10.1016/J.PMCJ.2014.01.003
7. Chang, Y.J.: Anomaly detection for travelling individuals with cognitive impairments. ACM SIGACCESS Accessibility Comput. **97**, 25–32 (2010). https://doi.org/10.1145/1873532.1873535
8. Hossain, S., Hallenborg, K., Demazeau, Y.: iRoute: cognitive support for independent living using BDI agent deliberation. Adv. Intell. Soft Comput. **90**, 41–50 (2011). https://doi.org/10.1007/978-3-642-19931-8_6
9. Lin, Q., Zhang, D., Huang, X., Ni, H., Zhou, X.: Detecting wandering behavior based on GPS traces for elders with dementia. In: 2012 12th International Conference on Control, Automation, Robotics and Vision, ICARCV 2012, pp. 672–677 (2012). https://doi.org/10.1109/ICARCV.2012.6485238
10. Mariescu-Istodor, R., Fränti, P.: Grid-based method for GPS route analysis for retrieval. ACM Trans. Spat. Algorithms Syst. **3**, 1–28 (2017). https://doi.org/10.1145/3125634
11. Kumar, A., Lau, C.T., Chan, S., Ma, M., Kearns, W.D.: A unified grid-based wandering pattern detection algorithm. In: Proceedings of the Annual International Conference of the IEEE Engineering in Medicine and Biology Society, EMBS. 2016-October, pp. 5401–5404 (2016). https://doi.org/10.1109/EMBC.2016.7591948

Modeling of Ferrite Inductors Power Loss Based on Genetic Programming and Neural Networks

Giulia Di Capua[1], Mario Molinara[1], Francesco Fontanella[1](✉), Claudio De Stefano[1], Nunzio Oliva[2], and Nicola Femia[2]

[1] University of Cassino and Southern Lazio, Cassino (FR), Italy
{giulia.dicapua,m.molinara,fontanella,destefano}@unicas.it
[2] University of Salerno, Fisciano (SA), Italy
{noliva,femia}@unisa.it

Abstract. This work compares two behavioral modeling approaches for predicting AC power loss in Ferrite-Core Power Inductors (FCPIs), normally used in Switch-Mode Power Supply (SMPS) applications. The first modeling approach relies on a genetic programming algorithm and a multi-objective optimization technique. The resulting AC power loss model uses the voltage and switching frequency imposed on the FCPI as input variables, whereas the DC inductor current is used as a parameter expressing the impact of saturation on the magnetic device. A second modeling approach involves a Multi-Layer Perceptron, with a single hidden layer. The resulting AC power loss model uses the voltage, switching frequency and DC inductor current as input variables. As a case study, a 10 μH FCPI has been selected and characterized by a large set of power loss experimental measurements, which have been adopted to obtain the training and test data. The experimental results confirmed the higher flexibility of the FCPI behavioral modeling based on genetic programming.

1 Introduction

A Switch-Mode Power Supply (SMPS) is an electronic circuit that integrates a switching regulator for efficient electrical power conversion. Magnetic components (like power inductors and transformers) are the biggest and heaviest components in an SMPS, limiting its overall power density [1]. Ferrite-Core Power Inductors (FCPIs), whose magnetic cores are made of ferrite materials, are extensively employed in SMPSs, due to their low cost and low power loss. However, their inductance drops sharply as the current increases because of magnetic core saturation. Accordingly, it is considered a good practice to have FCPIs operating in the so-called *weak saturation* region (within 10–20% inductance drop, [2]). Nevertheless, it has been proved that saturation is not an issue, neither for the inductor nor for the whole power converter, if the current ripple, the power loss, and the temperature rise of the FCPI fall within given limits, acceptable for

C. De Stefano et al. (Eds.): WIVACE 2022, CCIS 1780, pp. 245–253, 2023.
https://doi.org/10.1007/978-3-031-31183-3_20

the device and the application. Recent studies have proved that FCPIs working in sustainable saturation conditions can achieve SMPSs with high power density [3]. Accordingly, the interest in SMPSs with partially saturating FCPIs has greatly increased in recent years.

Enhanced inductance models and power loss models have been then proposed for reliable current ripple and power loss prediction in partial saturation conditions. Several works discuss the recent advances in nonlinear inductance modeling for power inductors to properly reproduce the inductance versus current curve (or *L versus i* curve, [2]). A complete survey with comparison among nonlinear behavioral inductance models, including different types of mathematical functions (polynomial, arctangent, piecewise-affine, etc.) and the nature of the optimization problem (quadratic programming or nonlinear programming), is provided in [4]. Conversely, just a few works deal with power loss modeling of power inductors operating in partial saturation conditions. Many loss formulations consider total inductor power losses split into two main contributions: power loss in the magnetic core (p_{core}) and power loss in the windings (p_{wind}). Unfortunately, these loss formulations involve the evaluation of not easily measurable magnetic quantities (like the permeability, the magnetic flux density, etc.), and rely on the knowledge of geometrical and physical properties of the device, such as core material, winding turns number and cross-sectional area, which are not disclosed by manufacturers for off-the-shelf FCPIs. For this reason, as an alternative, the power loss can conveniently be split into the DC component (p_{DC}) and the AC component (p_{AC}). These separate contributions can be computed based on the DC and AC components of the inductor voltage and current, which can be easily measured through experimental voltage and current measurements. The DC component of the power loss is simply due to the inductor mean current value and expressed through the inductor DC resistance R_{DC} as $P_{DC} = R_{DC} i_{DC}^2$. Some behavioral models involving easy-to-measure electrical quantities have been instead developed to evaluate the AC power loss term [5,6]. Some simplified models like [5] are valid for a limited set of operating conditions and express the P_{AC} contribution in terms of the RMS value of the inductor current and/or the duty-cycle of the SMPS. In [6], a new behavioral AC loss model of FCPIs in SMPS applications, based on a Genetic Programming (GP) algorithm [7,8]. It includes the effects of saturation and relies on SMPS operating conditions (duty-cycle and switching frequency) and electrical quantities (current and voltage).

On this premise, we propose the use of a supervised learning technique for modeling power losses in FCPIs. The AC power loss contribution is herein modeled by means of a Neural Network (NN). To allow the full exploitation of sustainable saturation operation for FCPIs, training and test data have been built by including partially-saturated operating conditions.

2 Genetic Programming, Neural Networks, and FCPIs

The *AC* loss prediction based on NN ($P_{AC,nn}$) will be compared to the *AC* power loss model based on GP ($P_{AC,gp}$), which was proposed in [6] and whose main details are provided in Subsect. 2.2, for the sake of completeness.

2.1 Case Study

The Coilcraft MSS1260-103 part has been adopted as a case study, whose main characteristics are listed in Table 1. The experimental characterization of this part has been realized by means of the MADMIX system [9]. The MADMIX emulates the operation of a real open-loop *DC-DC* power converter. This system imposes desired SMPS operating conditions (input voltage V_{IN}, duty-cycle D, switching frequency f_s and output current I_{OUT}), which result in a certain equivalent voltage $V_{eq} = V_{IN} \times D \times (1 - D)$ and current $I_L = I_{OUT}$ upon the power inductor, and realizes power inductor measurements, including *DC* and *AC* inductor power losses.

Table 2 lists all the operating conditions tested by the MADMIX system for the Coilcraft MSS1260-103 part. The 1040 experimental data point making up our training data was generated by combining all these operating conditions. The inductance de-rating achieved a maximum of 50% with respect to the nominal inductance L_{nom}, with an average current of 7.25 A. Finally, Table 3 shows the operating conditions combined to generate the additional 94 test points used for the model verification.

Table 1. Coilcraft MSS1260-103 and its main characteristics.

L $[\mu H]$	Dimensions $[mm^3]$	DCR $[m\Omega]$	$I_{sat}[A]$ 10% drop	$I_{sat}[A]$ 20% drop	$I_{rms}[A]$ 40°C rise
10	12 × 12x6	24	6.18	6.92	4.00

Table 2. Training Set Operating Conditions.

f_s $[kHz]$	V_{IN} $[V]$	D	I_L $[A]$
200; 300; 400; 500	6; 8; 10; 12	0.2; 0.35; 0.5; 0.65; 0.8	3; 3.5; 4; 4.5; 5; 5.5; 5.75; 6; 6.25; 6.5; 6.75; 7; 7.25

Table 3. Test Set Operating Conditions.

f_s $[kHz]$	V_{IN} $[V]$	D	I_L $[A]$
250	7; 11	0.2; 0.35; 0.5; 0.65; 0.8	4.2; 5.2; 6.2; 7.2
350	7; 11	0.3; 0.5; 0.7	4; 5; 6; 7
450	7; 11	0.2; 0.35; 0.5; 0.65; 0.8	4.2; 5.2; 6.2

2.2 GP-Based Loss Model

The FCPIs power loss model presented in [6] was obtained by means of a GP algorithm, in combination with a multi-objective optimization approach, aiming to realize an optimal trade-off between the accuracy and complexity of the resulting formula. Differently from previously proposed approaches, this GP-based model considers the applied voltage, the duty cycle, and the switching frequency as main variables of the power loss formula, whereas the *DC* current is adopted as a *biasing* parameter, which influences the model coefficients to take into account the saturation impact.

For the selected FCPI, a set of m average current conditions I_{Lj} have been considered ($j = 1, ..., m$). For each current value, n couples of frequency and equivalent voltage conditions (f_{si}, V_{eqi}) have been adopted ($i = 1, ..., n$). For each of the resulting $n \times m$ test conditions, we created a data point, including the test values (f_{si}, V_{eqi}) and the corresponding experimental *AC* power loss $y_{ij} = P_{ac,exp}(f_{si}, V_{eqi}, I_{Lj})$. We used these samples as the training set. The goal is to identify the following behavioral model:

$$P_{ac,gp} = \mathbf{F}\left(f_s, V_{eq}, \mathbf{p}(I_L)\right) \tag{1}$$

such that the value of the function $\mathbf{F}$ computed for each test condition of the training data set is as close as possible to the corresponding experimental value y_{ij}, $\forall i \in \{1, ..., n\}$ and $\forall j \in \{1, ..., m\}$.

The proposed GP algorithm starts from a population of 500 individuals and evolves over 300 generations, with a maximum tree size of 50 nodes. The crossover and mutation probabilities have been fixed to 0.8 and 0.2, respectively, through sub-tree crossover and sub-tree and node mutation mechanisms. We used the tournament selection strategy (tournament size = 2).

A set of elementary functions have been assumed for the *non-terminal set*, whereas constant coefficients and input variables have been adopted for the *terminal set*. Complexity factors cf have been assigned to all the elements of these sets. In particular, the following values have been assumed for the terminal set: $cf = 0.6$ for the multiplication of input variables, $cf = 1$ for all the other operations between input variables and for constant coefficients. Instead, the following values have been considered for the non-terminal set: $cf = 1$ for the sum and/or multiplication of elementary functions, $cf = 1.5$ for more complex functions (e.g., logarithms, exponentials, powers, arctangents, hyperbolic tangents, etc.)

Fixed the input variables (f_s and V_{eq}) of the model, the structure of the behavioral function $\mathbf{F}$ has to be the same for all the average current conditions, whereas the coefficients $\mathbf{p}$ are functions of I_L. A non-linear least squares method has been used to determine the coefficients $\mathbf{p}$ valid for each current I_{Lj} ($j = 1, ..., m$). In particular, the values of coefficients $\mathbf{p}$ have been determined by minimizing the χ-squared error, evaluated between the experimental loss y_{ij} and the GP-predicted loss $F\left(f_{si}, V_{eqi}, \mathbf{p}\left(I_{Lj}\right)\right)$, as given in (2):

$$\chi_j^2 = \frac{1}{n} \sum_{i=1}^{n} \left\{ F\left(f_{si}, V_{eqi}, \mathbf{p}\left(I_{Lj}\right)\right) - y_{ij} \right\}^2 \tag{2}$$

Accuracy and *complexity* of each GP-based model have been evaluated to select the best AC loss model among all the discovered ones.

Models accuracy has been determined by means of the Root Mean Square Error ($RMSE$) between the experimental loss and the GP-predicted loss over the whole training data set, calculated as follows:

$$RMSE = \sqrt{\frac{1}{m}\sum_{j=1}^{m}\chi_j^2} \tag{3}$$

Models complexity has been determined by means of the complexities of the elementary functions in the model structure **F**. A global term $F_{complexity}$ has been eventually evaluated, according to the following rules:

- *if* a function is the argument of another function, *then* the complexity factors cf of the two functions are multiplied;
- *if* two functions are multiplied or summed, *then* their complexity factors cf are summed and subsequently multiplied by the complexity factor of a sum or a product, respectively.

An elitist Non-dominated Sorting Genetic Algorithm (NSGA-II) has been adopted to discover the behavioral model (6) offering a good trade-off between $RMSE$ and $F_{complexity}$, which have been selected as objective functions to be minimized in a Multi-Objective Optimization (MOO) approach. Thirty independent runs of the resulting GP-MOO algorithm have been executed to check the repeatability of the resulting behavioral models. In order to select an optimal AC loss model among all the obtained Pareto-optimal solutions, some metrics have been adopted to classify each GP-based model, like the following ones:

- the number of GP runs during which the algorithm has discovered the model;
- the average number of generations during which the model exists within the population;
- the mean value, standard deviation, and maximum value of the percentage error distribution of the model over the training set;
- the average number of intervals over which the model coefficients change their monotonicity with respect to the DC current I_L.

Only the non-dominated Pareto-optimal solutions occurring for at least eight runs have been selected and considered for further comparison. Among these solutions, the following model presents an optimal trade-off among all such metrics:

$$P_{ac,gp} = p_0 exp\left(-p_1 f_s\right) V_{eq}^{p_2} + p_3 V_{eq}^2 \tag{4}$$

with maximum percent error $err_{max} = 25\%$, mean percent error $\mu_{err} = 1\%$ and standard deviation $\sigma_{err} = 5\%$. Interpolating functions $\mathbf{p}(I_L)$ have been determined, and the coefficients $\{p_0, ..., p_3\}$ have been modeled as given in (5):

$$\mathbf{p}\left(I_L\right) = \mathbf{a}_0 exp\left(\mathbf{a}_1 I_L\right) + \mathbf{a}_2 I_L + \mathbf{a}_3 \tag{5}$$

where the vector coefficients $\{\mathbf{a}_0, ..., \mathbf{a}_3\}$, determined by means of a non-linear least square method (Levenberg-Marquardt algorithm), are listed in Table 4.

Table 4. Behavioral model coefficients for Coilcraft MSS1260-103 part.

coefficients	a_0	a_1	a_2	a_3
p_0	5.76E−04	1.70E+00	−2.71E+00	3.86E+01
p_1	2.20E−06	1.23E+00	−4.16E−04	6.78E−03
p_2	9.65E−14	4.12E+00	−5.39E−03	2.05E+00
p_3	2.36E+01	1.36E−01	−5.23E+00	−1.78E+01

2.3 NN-Based Loss Model

The wide range of knowledge developed in the Machine Learning (ML) field allows us to obtain excellent results in the use of learning techniques to approximate functions of great complexity. The ability of NN to approximate a function has been theoretically demonstrated and verified in numerous application contexts, even for complex cases. For instance, pioneering results on the modeling ability of neural systems are reported in [10,11], which proved that a single hidden layer network can properly approximate any function to an arbitrary degree of precision. NN topologies are usually organized in *layers*. The patterns are presented to the network via the "input layer", whereas the final answer is provided through an "output layer". A Multi-Layer Perceptron (MLP) is a particular Neural Network (NN). Once the network topology has been chosen, the NN must be trained by providing a set of labeled samples as input. We used a feed-forward completely connected network, trained using the back-propagation algorithm [12]. In this work, we present an application of these learning techniques to the evaluation of the power losses of FCPIs. We search for an approximation of the following function:

$$P_{ac,nn} = \mathbf{F}\left(f_s, V_{eq}, I_L\right) \tag{6}$$

We used the Weka MLP implementation [13], with three inputs, one output, and sixteen neurons in the hidden layer. After preliminary experiments, we set the hyperparameter values as shown in Table 5. The global relation between input and output can be summarized as in:

$$P_{ac,mlp} = \sum_{i=1}^{16} w_{i,o}\sigma\left(b_i + w_{i,f_s} f_s + w_{i,V_{eq}} V_{eq} + w_{i,I_L} I_L\right) + b_o \tag{7}$$

Table 5. Hyper-parameters values of the NN.

Hyper-parameter	Value
Learning rate	0.3
Momentum	0.2
Batch size	10
Number of epochs	5000
Activation function	Sigmoid

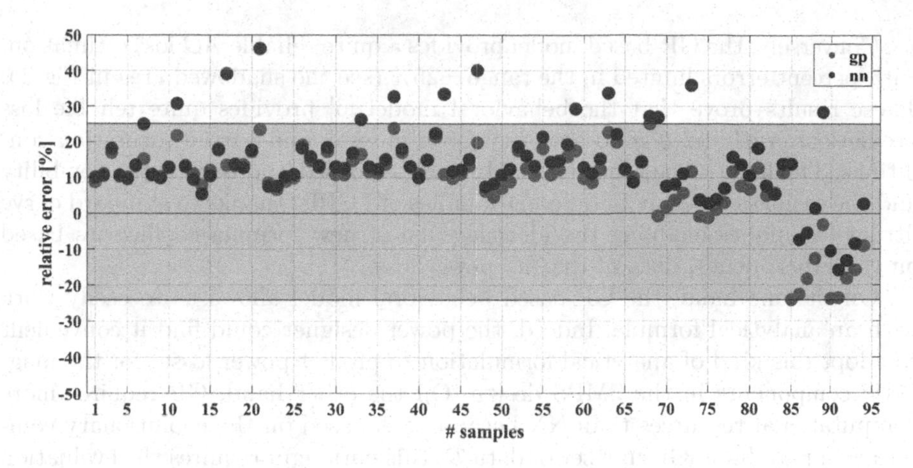

Fig. 1. MSS1260-103: relative percent errors of GP-based model (red) *versus* NN-based model (blue) in weak-saturation (samples from #1 to #53) and partial-saturation conditions (samples from #54 to #94) (Color figure online).

where σ is the sigmoid function:

$$\sigma(x) = \frac{1}{1 + e^{-x}} \tag{8}$$

$w_{i,*}$, $w_{i,o}$ are the weights of the hidden and output layers, respectively, and b_* the biases.

3 Results and Discussion

The *AC* power loss for the Coilcraft MSS1260-103 part has been evaluated for all the operating conditions given in the test set (see Table 2). Figure 1 shows the errors in the *AC* loss prediction of the GP-based model $P_{ac,gp}$ (filled red circles) and of the NN-based model $P_{ac,nn}$ (filled blue circles). Samples are sorted according to increasing values of the current (from 4.0 A to 7.2 A).

Table 6 shows the maximum, the mean, and the standard deviation values of the two models' percent error distribution over the test set of Table 3. The NN-based model overestimates the *AC* power loss in some operating conditions distributed over the entire test set, with a high maximum percent error and slightly higher mean percent error.

Table 6. Comparison of the models' percent error distribution.

metrics	$P_{AC,gp}$	$P_{AC,mlp}$
err_{max}[%]	25.1%	46.2%
μ_{err}[%]	8.8%	13.9%
σ_{err}	10.6	10.6

Conversely, the GP-based model provides a quite reliable AC loss estimation, with percent errors limited in the range $\pm 25\%$ (see the shadowed area in Fig. 1). These results prove that the behavioral model [6] provides quite reliable loss predictions, with quite good accuracy, both in weak and partial saturation conditions. The huge advantage ensured by our GP-based approach is the flexibility and the easiness to apply to different families of FCPIs, thanks to standard curve fitting techniques enabling the identification of new formula coefficients based on the experimental data of the AC power loss.

On the one hand, the GP-based behavioral model allows us to easily work with an analytical formula. Indeed, the power designer could find it convenient to adopt this kind of analytical formulation to predict power losses for the magnetic components in the SMPS design. On the other hand, GP requires more computational resources than NN because it is based on the evolutionary computation paradigm. Given a set of data $\mathcal{D}$, this paradigm requires the evaluation of $N_T = N_p \times N_g$ solutions on $\mathcal{D}$, with N_p and N_g representing the number of individuals in the population and the number of generations, respectively. Typically, these N_T evaluations require more computational resources than those required to train a NN.

References

1. Milner, L., Rincon-Mora, G.A.: Small saturating inductors for more compact switching power supplies. IEEJ Trans. Electr. Electron. Eng. **7**(1), 69–73 (2012)
2. Di Capua, G., Femia, N.: A novel method to predict the real operation of ferrite inductors with moderate saturation in switching power supply applications. IEEE Trans. Power Electron. **31**(3), 2456–2464 (2016)
3. Di Capua, G., Femia, N., Stoyka, K.: Switching power supplies with ferrite inductors in sustainable saturation operation. Int. J. Electr. Power Energy Syst. **93**, 494–505 (2017)
4. Oliveri, A., Lodi, M., Storace, M.: Nonlinear models of power inductors: a survey. Int. J. Circuit Theory Appl. **50**(1), 2–34 (2022)
5. Lodi, M., Bizzarri, F., Linaro, D., Oliveri, A., Brambilla, A., Storace, M.: A nonlinear behavioral ferrite-core inductance model able to reproduce thermal transients in switch-mode power supplies. IEEE Trans. Circuits Syst. I Regul. Pap. **67**(4), 1255–1263 (2020)
6. Stoyka, K., Di Capua, G., Femia, N.: A novel ac power loss model for ferrite power inductors. IEEE Trans. Power Electron. **34**(3), 2680–2692 (2019)
7. De Falco, I., Tarantino, E., Cioppa, A.D., Fontanella, F.: A novel grammar-based genetic programming approach to clustering. In: Proceedings of the 2005 ACM Symposium on Applied Computing, SAC 2005, pp. 928–932. Association for Computing Machinery New York, NY, USA (2005)
8. De Falco, I., Tarantino, E., Cioppa, A., Fontanella, F.: An innovative approach to genetic programming-based clustering. In: Abraham, A., de Baets, B., Köppen, M., Nickolay, B. (eds.) Applied Soft Computing Technologies: The Challenge of Complexity. AINSC, vol. 34, pp. 55–64. Springer, Heidelberg (2006)
9. Wens, M., Thone, J.: MADMIX: The standard for measuring SMPS inductors. Bodo's Power Systems, pp. 52–54 (2015)

10. Hornik, K., Stinchcombe, M., White, H.: Multilayer feedforward networks are universal approximators. Neural Netw. **2**(5), 359–366 (1989)
11. Hornik, K.: Approximation capabilities of multilayer feedforward networks. Neural Netw. **4**(2), 251–257 (1991)
12. Kelley, H.J.: Gradient theory of optimal flight paths. Ars J. **30**(10), 947–954 (1960)
13. Frank, E., Hall, M.A., Witten, I.H.: The WEKA Workbench. Online Appendix for "Data Mining: Practical Machine Learning Tools and Techniques", Morgan Kaufmann, Fourth Edition (2016)

Evolutionary Stitching of Plot Units with Character Threads

Pablo Gervás(✉)

Facultad de Informática, Universidad Complutense de Madrid,
Madrid 28040, Spain
pgervas@ucm.es
http://nil.fdi.ucm.es

Abstract. Ideation of novel narrative plots is often achieved by recombination of elements from prior plots. This approach presents two challenges: what elements to select from prior narratives, and how to find a new combination that makes sense without replicating any of the originals. The present paper relies on prior results for representing elementary narrative units for plot representation that both respect connectivity across a plot and allow a high degree of abstraction from previous instances of plot. An evolutionary procedure is applied to search the space of possible combinations, driven by metrics on coherent connectivity across plot units.

Keywords: story generation · evolutionary approach · metrics on story cohesion · character threads

1 Introduction

Stories are often built by reusing material from existing stories, from full reuse of the actual plot – popular in the film industry in the form of *remakes* of earlier films – to reusing relevant episodes, situations or plot lines. Many examples of these reusable elements for building stories are reviewed in the *TVTropes*[1] web site, a very impressive crow-sourced compilation [2]. In computational terms, the task of deconstructing existing stories, selecting elements at the right level of granularity, adapting them to a different context, and recombining them into a new story appears to be just as complex as inventing a new story from scratch.

The present paper considers some existing work on potential means of representing plot at different levels of granularity and explores the possibility of applying an evolutionary search procedure, informed by fitness functions designed to capture common sense connectivity across the constituent elements, to generate outlines of story plots.

[1] https://tvtropes.org/.

This paper has been partially funded by the project CANTOR: Automated Composition of Personal Narratives as an aid for Occupational Therapy based on Reminescence, Grant. No. PID2019-108927RB-I00 (Spanish Ministry of Science and Innovation) and the ADARVE (Análisis de Datos de Realidad Virtual para Emergencias Radiológicas) Project funded by the Spanish Consejo de Seguridad Nuclear (CSN).

C. De Stefano et al. (Eds.): WIVACE 2022, CCIS 1780, pp. 254–265, 2023.
https://doi.org/10.1007/978-3-031-31183-3_21

2 Related Work

Three topics are considered relevant for this paper: representing plot at the right granularity, story construction by combination of plot-relevant units, and evolutionary approaches to creation of narratives of some kind.

Propp's morphology of the Russian folk tale [12] postulates the concept of a *character function* as an abstract representation of the action of a character – characters meeting, misbehaving, fighting, travelling – that is relevant to the plot of the story, with particular types of characters play certain types of roles in the story: the villain misbehaves, the victim suffers as a consequence, the hero takes action to right the wrongs of the villain... When recombining plot elements at this level of granularity the achievement of an interesting overall structure becomes a significant challenge. Other approaches attempt to capture the structure of plot from beginning to end [1,3], in structures that may serve as templates to instantiate to achieve new plots. The *axes of interest* postulated in the PlotAssembler system [7] constitute a representation at an intermediate level of granularity – but provided with means for establishing links across characters dynamically during combination – that has the potential of addressing the difficulties of low granularity and high granularity representations. As axes of interest are chosen for the present paper, they are reviewed below.

The idea of building stories by combining fragments that already hold a partial representation of plot has been explored to build stories with more than one plot line. In these efforts, a *plot line* is a sequence of plot-relevant elements or *scenes* that make sense in the order in which they appear in the story, and linked by at least a shared set of protagonist and secondary characters. Fay [5] starts from a set of narrative threads for particular types of characters, and constructs multi-plotline stories in response to a given story request by selecting threads that match the type of characters in the request, ordering the events in each thread into a consistent overall timeline and binding together some characters in different threads. The PlotAssembler system [7] – mentioned above – interweaves the scenes in the axes of interest provided as input based on information on probabilities of character continuity across scenes – as mined from a corpus of prior stories. The work of Concepción et al. [4] operates on a set of plot templates and proposes procedures for weaving them together, including both know literary techniques and basic computational alternatives proposed as baselines for the task.

The use of evolutionary algorithms for story generation has been considered in the past. McIntyre and Lapata [10] use genetic algorithms to combine plot lines represented as a partially ordered graph of events associated to the entity, with a fitness function aimed to maximise story coherence and story interest. Gómez de Silva and Pérez y Pérez [13] use an evolutionary approach to refine an initial population built using knowledge-based heuristics. Fredericks and DeVries [6] generate small fragments of narratives applying an evolutionary solution driven by novelty. The work of Gervás et al. [9] explores an evolutionary solution for the combination of plot templates as described in [4]. Two relevant subtasks were identified: discourse planning – which involves sequencing decisions on the order in which elements from different plot lines are combined – and character fusion – which involves decisions on when/whether to assign the same character of the overall story to different narrative roles in more than one of the plot lines being combined. A genetic representation was proposed for a combination of plot lines, including genes for which plotline to start with, when to switch from the plotline being followed onto a new one, which plotline to jump to on switching, and which story characters to assign to the set of narrative roles in the plot lines being combined. The evolutionary algorithm relies on metrics for story consistency from a semantic point of view [8] based on the formative qualitative evaluation of multi-plotline stories that was reported in [4].

3 Evolutionary Combination of Plot Units Driven by Consistency Metrics

The solution described in this paper explores the application of the genetic representation presented in [9] for combining plot templates to the task of combining into a simple story a set of axes of interest, the much smaller plot-relevant units described in [7].

3.1 Knowledge Representation for Plot

A fundamental aspect of the solution presented here is the use of a knowledge representation that allows decomposition of stories into fragments small enough to allow variety in the set of constructed stories, but also captures connections across the different elements at several levels: in terms of sequences of events that are connected – departure on a journey is related to return from a journey – and in terms of participating characters – the character on a journey may be kidnapped, or may decide to kidnap someone, or may be called upon to rescue someone who has been kidnapped. The set of representation used to achieve this is presented here.

The plots considered for the present paper are represented in terms of structured compositions of basic units called *plot atoms*. A plot atom is a unit similar to a character function which explicitly holds additional information to indicate how the roles specific to the plot element (kidnapper, kidnapped) are filled in by roles that are relevant to the plot (villain, victim). This refinement allows for interesting articulation between roles specific to a plot element and roles more general to the plot at large.

The plot atoms in a plot are organised in a complex structure that combines different sequences of plot atoms. This is required to allow for the concepts of plots that relate actions that take place at non-contiguous points in time (villainy early in the story, revenge at the end of it, with other unrelated events happening in between), or plots that combine more than one subplot (each subplot is a different sequence of plot atoms which may be interleaved with the other subplots by breaking their sequence down into smaller subsequences that constitute different scenes of the subplot).

These are called *axes of interest* (AoIs), which constitute intermediate units of plot structure that involve a set of plot atoms related by a long range dependency. For example, an AoI representing an *Abduction* as it features in classic stories would include the actual kidnapping (which would happen somewhere towards the start of the story) and the corresponding *Release* (which would happen somewhere towards the end of the story), but these two plot atoms are structurally connected. An axis of interest has a set of narrative roles – those of its constituent plot atoms – that are initially free variables but which can be instantiated to specific constants when the axes of interest is combined into larger structures.

An example of axis of interest is shown in Table 1. To assist in the process of combining them into more elaborate structures, each axis of interest specifies what the roles relevant to the axis of interest are.

Table 1. The Axis of Interest for ABDUCTION: co-occurrence in variable names is relevant, so the character that plays `abducted` in the `Kidnapping` must be the same as the one that plays `abducted` in the `Rescue`.

AXISofINTEREST =	ABDUCTION
Kidnapping	abductor = x, abducted = y
Rescue	abducted = y, rescuer = z

The set of plot atoms and AoIs used in the present paper corresponds to the 34 plot atoms over the 19 axes of interest presented in [7]. These resulted from a knowledge engineering effort informed by a number of sources in the literature, including Propp's character functions [12], Booker's seven basic plots [1] and Polti's situations [11]. Interested readers are referred to the original paper for details.

3.2 Discourse Planning and Character Fusion over AoIs

Axes of interest can be combined together, weaving their corresponding sequences of atoms with those of other axes of interest, to form story drafts. In a story draft, the plot atoms from the axes of interest that have been combined appear in an ordered sequence that corresponds to the discourse for the story draft, but each plot atom is labelled to indicate which axes of interest it corresponds to. Additionally, the story draft lists for each plot atom how the roles specific to the various axes of interest are instantiated in terms of the set of constants that encode the overall set of narrative roles involved in the story draft.

An example of story draft is presented in Table 2, which shows how the *Abduction* and *CallToActionReward* axes of interest are interleaved to form the basic story draft. It also shows how the narrative roles for the story draft (*hero*, *villain*, *victim*, *sender*) are mapped to the roles specific to the plot atoms of the constituent axes of interest (*abductor*, *abducted*, *rescuer* for the *Abduction* axis of interest and *called*, *caller*, *rewarded* for the *CallToActionReward* axis of interest). This ensures that the various plot atoms in the plot are instantiated in a manner coherent with the narrative roles that the characters play in the overall story draft.

Table 2. Example of story draft for a basic plot combining axes of interest for ABDUCTION and CALLTOACTIONREWARD. The first column shows the interweaving of the axes of interest. The co-occurrence of constants in the final column – shown in **bold** – indicates an instance of character fusion between the two axes.

ABDUCTION	Kidnapping	(abductor=villain,abducted=victim)
CALLTOACTIONREWARD	Call	(called=**hero**,caller=sender)
ABDUCTION	Rescue	(abducted=victim,rescuer=**hero**)
CALLTOACTIONREWARD	Reward	(rewarded=**hero**)

For a story draft to be considered a valid narrative plot, the variables in the axes of interest that compose it must be instantiated in a coherent manner. For the example in Table 2, the required restriction can be expressed by requiring that the character *abducted* in the *Kidnapping* be the same as the *abducted* in the *Rescue*, and that the character who is *called* in the *Call* be the same one who acts as *rescuer* in the *Rescue* and as *rewarded* in the *Reward*. Of these, the relevant connection is the one that relates plot atoms across the different axes of interest being combined. Note that unless the *called/rewarded* and the *rescuer* are the same character the proposed story draft makes no sense. The restrictions holding for characters within an axis of interest are captured in its definition. The inclusion of this type of connection in a story constitutes an instance of character fusion. These connections that relate plot atoms across the different axes of interest being combined are going to be used to build the metrics that will be used as fitness functions in the evolutionary procedure.

3.3 Evolutionary Construction of AoI-based Stories

The adaptation of the genetic representation presented in [9] to the combination of AoIs rather than templates for complete plots is described here. As in the original approach, the representation covers two different aspects: a part of the representation is intended to address discourse planning decisions and different part addresses decisions concerning character fusion. Because both plot templates and AoIs are in essence a sequence of plot units related by a set of characters that instantiates specific roles in the plot units, the actual computational representation is mostly equivalent. To ensure that the present paper is understandable as a self-contained unit, the representation is nevertheless summarised below.

The problem of constructing a story as a combination of AoIs is determined by the set of AoIs taken as input. This implies that, for a particular problem of combining N AoIs, the length of the final discourse is determined by the total number of scenes in the AoIs being considered, and the maximum number of possible characters featuring in the story is determined by the union of the sets of characters in the AoIs being considered.

The caracterisation of the discourse plan for a given story candidate requires: establishing which AoI to start on, defining the specific points in the final sequence in which the discourse switches to a different AoI, and defining to which of the other available AoIs the discourse is switching when it does. This is encoded into a genetic vector of integers as follows: (1) a single digit (0 or 1) defines which AoI the final discourse starts with, (2) a sequence of digits (0 or 1) defines for the total number of scenes in the final discourse whether the next scene follows on with the prior AoI or it switches to a different AoI, (3) a sequence of digits (ranging between 1 and N-1, where N is the total number of plots being combined) defines how many of the available AoIs are skipped whenever the discourse switches to a different AoI.

The information on character fusion is represented in terms of a vector that defines how the character roles for the different AoIs are instantiated by the character names that appear in the final story draft. The variable names need to be distinct across the different AoIs to avoid confusion when different AoIs have similar variables names – say both having a `hero`. This is ensured by assigning a prefix with the AoI name to all the variable names that feature in any given AoI.

The caracterisation of the choices for character fusion for a given story candidate requires an assignment of character names to each of the variables in the joint set of variables for the story. For simplicity, the set of potential character names of each sort for the story is defined to be the set of integers from 0 to C, with C being the cardinality of the joint set of character variables for the story. This is sufficient to represent any choices made in terms of entity fusion (with variables in two different positions in the name-assignment vector being assigned to the same integer).

An initial population of story candidates is built by assigning values to the representation described. The parts of the genetic representation that rely on integers simply need a random value in the appropriate range. For the assignment of story draft character names to the character variables of each AoI, care must be taken to ensure that variables from the same AoI should not be assigned the same character name, to avoid fusing in the same character roles that are distinct in the context of the specific AoI – say fusing villain and victim for a given kidnapping. Details of the procedures may be found in [9].

Once a population has been constructed, mutation and cross over operators are applied to it. Because of the different nature of the various parts of the representation, specific operators of each kind are applied to the different parts. Again, details of the specific procedures may be found in [9].

3.4 Fitness Functions Based on Story Cohesion Metrics

The quantitative evaluation of the narrative quality of a given combination of AoIs would have to consider a large number of aspects that are very difficult to quantify. The perception of the quality of narrative is known to hinge on many subjective matters such as emotion, attribution of motivation, empathy... that are not only complex in themselves but also insufficiently known for successful formalisation. Additionally, none of these aspects is captured in the representation being considered. For these reasons, the evaluation of the quality of the constructed stories is restricted in the present paper to acceptability of the two aspects that are represented: a correct sequence of plot atoms as a result of the discourse planning task, and acceptable occurrence of characters sharing roles across AoIs.

To capture these aspects in our fitness functions, we developed a set of metrics. For simplicity, a different metric is defined for each potential pairwise combination of two AoIs. An overal fitness functions for the system is defined by assigning as a final score to a given draft the average of the scores given by the metrics for each of the possible pairwise combinations of the AoIs included in it.

The metric for a particular pairwise combination of AoIs is defined in terms of the following aspects:

- *role-sharing constraints* on a particular character playing a role X in one of the AoIs and a role Y in the other (say, the traveller in a journey becoming the victim of a kidnapping)
- particular *sequencing constraints* on the atoms for the AoIs involved, possibly arising from a particular shared role (for instance, a kidnapped traveller should return only after he has been released from his kidnapping)

The particular constraints for a pairwise combination of AoIs is registered as an entry of the form:

```
Abduction+CallToActionReward =
hero + hero =
Abduction + CallToAction / CallToAction + Rescue / Rescue + Reward
```

where the second line describes role-sharing constraints and the third line describes sequencing constraints – in each case with each individual constraint separated by a `/`.

This expresses that fact that, for a combination of a kidnapping (`Abduction`) and a hero called to action (`CallToActionReward`), the `hero` of one should be the `hero` of the other, the kidnapping should take place before the call to action, the call to action before the rescue, and the reward after the rescue.

A particular pairwise combination of AoIs is assigned a numerical score over a total of 100. Of that score, 20 points are assigned based on role-sharing constraints. Each role-sharing constraint present is scored 100 if met and 0 otherwise, and the average value of all role-sharing constraints taken as the total role-sharing score (normalised to 20). The remaining 80 points are computed by:

- assigning 100 points to any precedence constraint that is met (for A + B, A appears before B in the discourse sequence)
- if a required precedence constraint is not met, a partial score over 100 is assigned corresponding to the number of positions that one of the elements

would need to shift for the constraint to hold (normalised over the length of the sequence)
- the average of all sequencing constraints taken as the total sequencing score (normalised to 80).

These metrics provide a progressive scoring, so that drafts where the sequencing constraints are not met are scored relative to how far they need to be modified for the constraints to be met. This allows mutations that modify the sequence in the right direction to be scored progressively higher, allowing evolution to converge towards optimal solutions.

4 Discussion

The proposed system is run in each case with an initial population of 100 individuals generated at random, with the described operators for mutation (probability of mutation set to 0.2) and crossover (probability of cross over set to 0.05), for 20 generations. At each generation populations are culled by selecting the next generation using a best scoring criterion.

The evolutionary procedure is tested by comparison with individuals produced at random without any evolution. To achieve this, in each case a sample population of 10 individuals was created by instantiating the genetic representation at random for a given combination of AoIs, and the individuals in this population were then scored using the same metrics used for the fitness function of the evolutionary procedure. The results are compared with the best scoring individuals produced by applying the evolutionary procedure 10 different times to the same combination of AoIs. In each case, the average score, the highest and the lowest scores are recorded. Results for four different combinations of AoIs are presented in Table 3.

Table 3. Table of comparative scoring for random baseline (**bl**) and evolutionary (**ev**) story draft generation over four different combinations of 5 AoIs. The label of the combination in each case encodes the initials of the AoIs involved.

AoIs comb.	Sol	Av	Hi	Lo
AB-CO-JO-SL-VA	bl	45	56	35
	ev	77.4	81	63
RR-CD-RL-RI-TA	bl	55	79	43
	ev	92.7	93	91
CA-HL-PU-RI-VP	bl	43	51	39
	ev	79.1	80	78
RE-CO-DO-JO-TA	bl	42	56	32
	ev	86.3	87	86

The results support three different conclusions. First, that different combinations of AoIs establish a different maximum threshold for overall score. This is due to the way the constraints applicable in each case restrict the potential of achieving optimal scores. Second, that for each specific combination, the evolutionary procedure consistenly achieves higher scores. This shows that the procedure does achieve an improvement in the quality of the story drafts. Third, that for all the combinations the results for the random baseline show a wide spread of scores, while the results of the evolutionary procedure are more tightly grouped close to the optimal scores achievable for the given combination.

In terms of the observable quality of the output story drafts, some examples of results are discussed in more detail below (Table 4).

Table 4. An example of output story combining the AoIs Relenting Guardian (UG) and abduction (Abd). Columns show: description of the *discourse plan* indicating for each plot atom which AoI it comes from, description of the cast indicating the roles they play in the respective plot lines and what names have been assigned to them – which reflect any *character fusions* present. Numbers used internally as character names have been replaced with capital letters for ease of reading. Columns 1-3 are alligned by scenes, column 4 applies to the story draft as a whole. The story corresponds to a suitor imprisoned by their beloved's guardian and rescued by the beloved. The allignments between suitor and victim and between guardian and kidnapper are favoured by the metrics, while the allignment between beloved and rescuer arises freely due to evolution.

#	AoI	Plot Atom	Variable	Role	Name
0	RelentingGuardian	CoupleWantsToMarry	UG-0	lover	A
1	RelentingGuardian	UnrelentingGuardian	UG-1	beloved	B
2	Abduction	Abduction	UG-2	guardian	C
3	Abduction	Rescue	Abd-0	abducted	A
4	RelentingGuardian	RelentingGuardian	Abd-1	abductor	C
5	RelentingGuardian	Wedding	Abd-2	rescuer	B

Fitness Scoring

Role-sharing	Abduction victim is also RelentingGuardian hero	100 %	20
	Abduction villain is also RelentingGuardian obstacle	100 %	
Sequencing	UnrelentingGuardian precedes Abduction	100 %	80
	Rescue precedes RelentingGuardian	100 %	

To illustrate the operation of the specific part of the fitness function dealing with discourse planning and sequencing constraints, examples of the structure of different stories at different levels of conformance with sequencing constraints are shown in Table 5.

Given that the constraint for this particular pairwise combination of AoIs is registered as an entry of the form:

```
Abduction+RelentingGuardian =
victim + hero / villain + obstacle =
UnrelentingGuardian + Abduction / Rescue + RelentingGuardian
```

Table 5. Structures for stories at different levels of conformance of discourse structure with sequencing constraints, for combinations of the AoIs Relenting Guardian (UG) and abduction (Abd) used above – for ease of comparison. In each case, the score assigned by the sequencing component of the scoring function is shown at the bottom. For ease of understanding of the structure of the discourse plans, plot atoms corresponding to the Abd AoI are highlighted in bold.

	CoupleWantsToMarry **Abduction** UnrelentingGuardian RelentingGuardian **Rescue** Wedding	CoupleWantsToMarry **Abduction** **Rescue** UnrelentingGuardian RelentingGuardian Wedding	CoupleWantsToMarry UnrelentingGuardian **Abduction** RelentingGuardian **Rescue** Wedding	CoupleWantsToMarry UnrelentingGuardian **Abduction** **Rescue** RelentingGuardian Wedding
UnGuard < Abd	83	66	100	100
Resc < RelGuard	66	100	83	100
	74.5	83	91.5	100

Table 6. Examples to illustrate options on character fusion. Names of character variables for different AoIs whose values have been fused are highlighted in bold.

abducted	**abducted = lover**	abducted	**abducted = lover**
abductor	abductor	**abductor = guardian**	**abductor = guardian**
rescuer	rescuer	rescuer	rescuer
lover		lover	
beloved	beloved	beloved	beloved
guardian	guardian		
0	50	50	100

the operation of the specific part of the fitness function dealing with character fusion, examples of different stories can be illustrated by the examples shown in Table 6.

The procedure has been illustrated this far for combinations of only two AoIs, but it clearly has more advantages when a larger number of AoIs is combined. Table 7 shows examples of stories obtained for combinations of three AoIs. In these cases, the interactions between the metrics on pairwise combinations lead to innovative sequences while ensuring that the resulting story makes sense.

One important shortcoming is the need for a significant knowledge engineering effort for the construction of the metrics. As each pairwise combination of AoIs requires only a single entry to define the corresponding metrics – symmetric combinations can be handled by a single metric – the number of metrics required is not as high as it might have been.

The procedure originally presented for combining axes of interest in [7] exhaustively generated all combinations deemed to be valid in terms of whether they matched the probabilities of character continuity across scenes as obtained from a prior corpus. This criterion ensures local consistency, but it it does not take into consideration long ranging connections across non-contiguous scenes. The new evolutionary solution, by relying on a fitness function based on metrics

Table 7. Examples of stories obtained for combinations of three AoIs. To facilitate understanding a textual rendering of the gist of the resulting story is added after each story.

AoIs	Score	PlotAtom	Character assignment
Abduction RelentingGuardian Task	100	CoupleWantsToMarry UnrelentingGuardian DifficultTask Solution Abduction Rescue RelentingGuardian Wedding	lover=C-3, beloved=C-0 lover=C-3, guardian=C-6, beloved=C-0 setter=C-6, solver=C-3 solver=C-3 abducted=C-0, abductor=C-1 abducted=C-0, rescuer=C-2, abductor=C-1 lover=C-3, guardian=C-6, beloved=C-0 lover=C-3, beloved=C-0
C-3 wants to marry C-0. C-6 opposes the idea. C-6 sets C-3 a task. C-3 solves it. C-1 kidnaps C-0. C-2 rescues C-0. C-6 relents. C-1 marries C-0.			
Abduction RelentingGuardian ShiftingLove	100	CoupleWantsToMarry UnrelentingGuardian RelentingGuardian Wedding BoyMeetsGirl LoveShift Reconciliation Abduction Rescue	lover=C-0, beloved=C-4 lover=C-0, guardian=C-1, beloved=C-4 lover=C-0, guardian=C-1, beloved=C-4 girl=C-0, boy=C-7 lover=C-7, beloved=C-0, rival=C-8 lover=C-0, beloved=C-4 lover=C-7, beloved=C-0 abducted=C-0, abductor=C-1 abducted=C-0, rescuer=C-2, abductor=C-1
C-0 wants to marry C-4. C-1 opposes the idea. C-1 relents. C-0 marries C-4. C-0 meets C-7. C-7 spurns C-0 for C-8. C-7 returns to C-0. C-1 kidnaps C-0. C-2 rescues C-0.			
Abduction RelentingGuardian VillainyPunishment	90	CoupleWantsToMarry UnrelentingGuardian Abduction Rescue Villainy Punishment RelentingGuardian Wedding	lover=C-3, beloved=C-1 lover=C-3, guardian=C-2, beloved=C-1 abducted=C-3, abductor=C-1 abducted=C-3, rescuer=C-2, abductor=C-1 villain=C-2 punished=C-2 lover=C-3, guardian=C-2, beloved=C-1 lover=C-3, beloved=C-1
C-3 wants to marry C-1. C-2 opposes the idea. C-1 kidnaps C-3. C-2 rescues C-1. C-2 commits a villainy. C-2 is punished. C-2 relents. C-1 marries C-3.			
Abduction RelentingGuardian Validator	87	CoupleWantsToMarry UnrelentingGuardian Abduction Tested Character'sReaction Validation ValidationRecognised Rescue RelentingGuardian Wedding	lover=C-2, beloved=C-4 lover=C-2, guardian=C-1, beloved=C-4 abducted=C-0, abductor=C-1 tested=C-1, tester=C-7 tested=C-1, tester=C-7 validated=C-1, validator=C-7 validated=C-1 abducted=C-0, rescuer=C-2, abductor=C-1 lover=C-2, guardian=C-1, beloved=C-4 lover=C-2, beloved=C-4
C-2 wants to marry C-4. C-1 opposes the idea. C-1 kidnaps C-0. C-1 tested by C-7. C-1 reacts to test. C-1 validated. C-2 rescues C-0. C-1 relents. C-2 marries C-4.			

built heuristically to capture common sense connections across AoIs, improves upon the original on this aspect.

5 Conclusions

The evolutionary approach to constructing plot outlines for stories by combining axes of interest based on metrics for common sense connections between them provides efficient means for building a population of drafts that satisfy constraints on semantic validity over the final linear discourse for the story. Due to the progressive nature of the metrics used as fitness function the population converges reasonably quickly for a low number of constituent axes of interest. It remains to be seen whether the solution will scale well towards higher numbers of constituents. The proposed solution improves upon prior work in [7] by replacing random choice over statistically probable connections with an evolutionary procedure, and it improves upon [9] in that it relies on a representation of narrative of lower granularity – patterns of connection between events rather than templates for full plot lines – which increases it ability to produce new stories.

References

1. Booker, C.: The Seven Basic Plots: Why We Tell Stories (2004). Continuum. http://books.google.pt/books?id=qHJj9gOl0j8C
2. Brehob, E.: Online Academics: The Wiki TV Tropes as a Community of Pseudo-Academic Producers. Ph.D. thesis, University of Michigan (2013)
3. Campbell, J., Cousineau, P., Brown, S.: The Hero's Journey: Joseph Campbell on His Life and Work. Collected works of Joseph Campbell. New World Library, Novato (1990). https://books.google.co.uk/books?id=0LIxpikJraoC
4. Concepción, E., Gervás, P., Méndez, G.: Exploring baselines for combining full plots into multiple-plot stories. N. Gener. Comput. **38**(4), 593–633 (2020). https://doi.org/10.1007/s00354-020-00115-x
5. Fay, M.P.: Driving story generation with learnable character models. Ph.D. thesis, Massachusetts Institute of Technology (2014)
6. Fredericks, E.M., DeVries, B.: (Genetically) improving novelty in procedural story generation. arXiv preprint arXiv:2103.06935 (2021)
7. Gervás, P.: Generating a search space of acceptable narrative plots. In: 10th International Conference on Computational Creativity (ICCC 2019). UNC Charlotte, North Carolina, USA (2019)
8. Gervás, P., Concepción, E., Méndez, G.: Assessing multiplot stories: from formative analysis to computational metrics. In: 12th International Conference on Computational Creativity (ICCC 2019). Mexico City, Mexico (2021)
9. Gervás, P., Concepción, E., Méndez, G.: Evolutionary construction of stories that combine several plot lines. In: Martins, T., Rodríguez-Fernández, N., Rebelo, S.M. (eds.) EvoMUSART 2022. LNCS, vol. 13221, pp. 68–83. Springer, Cham (2022). https://doi.org/10.1007/978-3-031-03789-4_5
10. McIntyre, N., Lapata, M.: Plot induction and evolutionary search for story generation. In: Proceedings of the 48th Annual Meeting of the Association for Computational Linguistics, pp. 1562–1572. Association for Computational Linguistics, Uppsala, Sweden (2010). https://aclanthology.org/P10-1158
11. Polti, G., Ray, L.: The Thirty-Six Dramatic Situations. Editor Company, New Jersey (1916)
12. Propp, V.: Morphology of the Folktale. University of Texas Press, Austin (1968)
13. de Silva Garza, A.G., y Pérez, R.P.: Towards evolutionary story generation. In: Colton, S., Ventura, D., Lavrac, N., Cook, M. (eds.) Proceedings of the Fifth International Conference on Computational Creativity, ICCC 2014, Ljubljana, Slovenia, 10–13 June 2014, pp. 332–335 (2014). https://computationalcreativity.net/

Impact of Morphology Variations on Evolved Neural Controllers for Modular Robots

Eric Medvet[1(✉)] and Francesco Rusin[2]

[1] Department of Engineering and Architecture, University of Trieste, Trieste, Italy
emedvet@units.it
[2] Department of Mathematics and Geosciences, University of Trieste, Trieste, Italy

Abstract. Modular robots, in particular those in which the modules are physically interchangeable, are suitable to be evolved because they allow for many different designs. Moreover, they can constitute ecosystems where "old" robots are disassembled and the resulting modules are composed together, either within an external assembling facility or by self-assembly procedures, to form new robots. However, in practical settings, self-assembly may result in morphologies that are slightly different from the expected ones: this may cause a detrimental misalignment between controller and morphology. Here, we characterize experimentally the robustness of neural controllers for Voxel-based Soft Robots, a kind of modular robots, with respect to small variations in the morphology. We employ evolutionary computation for optimizing the controllers and assess the impact of morphology variations along two axes: kind of morphology and size of the robot. Moreover, we quantify the advantage of performing a re-optimization of the controller for the varied morphology. Our results show that small variations in the morphology are in general detrimental for the performance of the evolved neural controller. Yet, a short re-optimization is often sufficient for aligning back the performance of the modified robot to the original one.

Keywords: Embodied cognition · Soft robotics · Adaptation

1 Introduction and Related Works

Fully autonomous robotic systems require to be adaptable to environmental changes without an external intervention. One main path toward adaptation of an entire robotic ecosystem, instead of the single robot, consists in having a population of robots that are built, "live", and are disposed in such a way that their robotic material can be reused for building other robots [3]. Modular robots are particularly suitable to form such ecosystems, because their modularity eases the building and disposal phases [4,12]. Moreover, modular robots also favor adaptation, in particular through evolution [2], because they allow for great expressiveness for the morphology and the controller [1,9].

However, actually realizing the scenario of real (i.e., with hardware robots) evolution of a robotic ecosystem poses several challenges [2], ranging from the

C. De Stefano et al. (Eds.): WIVACE 2022, CCIS 1780, pp. 266–277, 2023.
https://doi.org/10.1007/978-3-031-31183-3_22

well-known reality gap problem [14,20], to the much longer times for evaluating candidate solutions [13], to the need of employing automatic assembly and disassembly procedures for robots, in order to make the system scalable and really autonomous.

The recent progresses on manipulators [12,13], as well as on self-assembly [17], make the automatic assembly of modular robots feasible. However, it can happen that the morphology obtained by unassisted assembly differs from the designed one, in particular for soft modular robots [5]. In such a case, the controller that was designed to be associated with the expected morphology might not be equally effective with the slightly different one. The drop in effectiveness might be large, according to the embodied cognition paradigm that states that the intelligent behavior of an embodied agent depends on the combined work of both its body and brain, since the body-brain misalignment might be detrimental.

In this paper, we experimentally investigate the impact of small variations in the morphology on the effectiveness of evolved neural controllers, i.e., of closed-loop controllers based on neural networks whose parameters are optimized, for a given morphology and task, by means of an evolutionary algorithm (EA). We consider the case of Voxel-based Soft Robots (VSRs), a kind of modular robots whose modules are soft cubes that can expand or contract individually based on signals dictated by the controller.

We perform a number of experiments with six morphologies, consisting in three base morphologies in small and large versions, with controllers evolved for the task of locomotion, a classic task of evolutionary robotics. We apply small random variations to each morphology and measure the impact on the degree to which the resulting VSRs solve the task (i.e., their velocity, for locomotion). We find that the decrease in robot velocity, i.e., the controller effectiveness, is large even for small modifications (one voxel for small original morphologies) and we explain this finding in terms of the great potential for morphological computation that VSRs offer. This potential results in the body having a key role in determining VSR behavior and hence make the body-brain misalignment particularly detrimental.

We also experimentally verify whether a re-optimization for the slightly modified morphology can make the controller back on par with the effectiveness of the one evolved for the original morphology. We found that (a) modified morphologies are not intrinsically worse and (b) seeding the re-optimization with the original controller makes the re-optimization efficient, besides effective.

We believe that our results contribute to strengthen the understanding of how body and brain interact in modular robots and, more broadly, constitute a further step toward autonomous and evolvable robotic ecosystems.

2 Evolutionary Optimization of Voxel-based Soft Robots

Voxel-based Soft Robots (VSRs) are a kind of modular robots where each module, called voxel, is a deformable cube that is attached to adjacent cubes at the vertexes. The volume of each voxel changes according to external forces acting on it, such as the gravity and those deriving from the contact with the ground, and to an internal force, itself modulated by a control value dictated by the controller of the robot. The way the controller varies the control value of each voxel over time determines, together with the robot-environment interaction, the behavior of the VSR.

Both the controller and the morphology of a VSR can be optimized in order to obtain a behavior that allows the robot to achieve a predefined task. For the purpose of this study, we are only concerned with the optimization of the controller; as an aside, previous works have shown that VSRs are particularly suitable for the concurrent optimization of controller and morphology [9,15].

In this study, we consider a 2-D variant of VSRs that can be simulated in discrete time and continuous space [8]. In the next sections, we describe in detail the morphology and the controller of our VSRs, as well as the way we optimize the latter by means of an EA.

2.1 VSR Morphology

A VSR *morphology* is unequivocally described by a 2-D grid in which each non-empty element describes a voxel. A *voxel* is modeled, in the simulation, as a compound of four rigid bodies (at the corners) and many spring-damper systems, for softness and elasticity, connecting the rigid bodies [8]. Adjacent voxels are glued together at the vertexes: that is, their rigid bodies at the corners are bound with joints that do not permit relative rotation, nor distance variation.

By varying the parameters of this mechanical model, the designer may vary the properties of the (simulated) material the voxel consists of, possibly impacting on the overall behavior of the VSR [11]. In this work, we assume that all the voxels consist of the same material, hence a morphology is unequivocally described by a Boolean grid (or matrix) $\boldsymbol{m} \in \{\mathrm{T}, \mathrm{F}\}^{w \times h}$, where the $m_{x,y}$ is set if there is a voxel at coordinates (x, y).

During the simulation, the area of voxels changes upon the combined effect of external forces and the voxel control value, dictated by the controller. For a voxel at (x, y), we denote the *control value* at time step k as $a_{x,y}^{(k)} \in [-1, 1]$, where -1 corresponding to maximum requested expansion, and 1 corresponding to maximum requested contraction. Contraction and expansion are modeled as linear variations of the rest-length of the spring-damper systems, proportional to value of $a_{x,y}^{(k)}$. We used the default values of the simulator 2D-VSR-Sim [8] for the minimum and maximum rest-length values.

2.2 VSR Controller

The controller of a VSR is in charge of determining the control value for each voxel of the VSR at each time step of the simulation.

In this study, we use a distributed neural controller [7] that determines the control values based on some sensory inputs acquired from the voxels, hence realizing a closed-loop control of the robot. Our controller is distributed in the sense that, for each voxel, the control value is determined locally based on local sensory inputs and information acquired from adjacent voxels. Moreover, it is neural because the processing of that information is performed with a feed-forward artificial neural network (NN) embedded in the voxel.

In detail, the distributed neural controller works as follows. At each time step k and for each voxel at (x, y), we collect a *sensor reading* $\boldsymbol{s}_{x,y}^{(k)} \in [-1,1]^4$ consisting of: (a) the ratio between the current area of the voxel and its rest area, (b) a binary value set to 1 if the voxel is in contact with the ground or -1 otherwise, and (c) the velocities of the center of mass of the voxel along the x- and y-axes. We normalize the values of the ratio and the velocities in order to ensure they are in the $[-1, 1]$ domain. Then, we use the NN located in the voxel to compute the control value $a_{x,y}^{(k)}$ and the information to be passed to adjacent voxels at the next time step:

$$\left[a_{x,y}^{(k)}\ \boldsymbol{i}_{x,y}^{\vartriangle(k)}\ \boldsymbol{i}_{x,y}^{\triangledown(k)}\ \boldsymbol{i}_{x,y}^{\triangleleft(k)}\ \boldsymbol{i}_{x,y}^{\triangleright(k)}\right] = \text{NN}_{\boldsymbol{\theta}}\left(\left[\boldsymbol{s}_{x,y}^{(k)}\ \boldsymbol{i}_{x,y-1}^{\vartriangle(k-1)}\ \boldsymbol{i}_{x,y+1}^{\triangledown(k-1)}\ \boldsymbol{i}_{x+1,y}^{\triangleleft(k-1)}\ \boldsymbol{i}_{x+1,y}^{\triangleright(k-1)}\right]\right),$$

where $\boldsymbol{i}_{x,y}^{\vartriangle(k)}, \boldsymbol{i}_{x,y}^{\triangledown(k)}, \boldsymbol{i}_{x,y}^{\triangleleft(k)}, \boldsymbol{i}_{x,y}^{\triangleright(k)} \in [-1,1]^{n_\text{comm}}$ is the information output by the voxel at (x, y) for its neighbors at $(x, y+1)$ ($\vartriangle$), $(x, y-1)$ ($\triangledown$), $(x-1, y)$ ($\triangleleft$), and $(x+1, y)$ ($\triangleright$), respectively, and $\boldsymbol{\theta} \in \mathbb{R}^p$ is the vector of the parameters (or weights) of the NN. The NN hence acts as a function $\mathbb{R}^{4+4n_\text{comm}} \to \mathbb{R}^{1+4n_\text{comm}}$.

The information propagation to and from adjacent voxels allows the distributed neural controller to realize a form of collective intelligence that results, upon optimization, in an emergent behavior of the VSR, despite being composed of independents controller modules (the NNs) [16]. Moreover, the one-step delay in propagation resulting from the fact that, at time k, a NN uses the information produced by adjacent NNs at time $k-1$, makes the distributed neural controller capable of exhibiting complex dynamical behaviors, despite being composed of static modules. However, it has been observed that this richness may often lead to vibrating behaviors [9] that would hardly be effective in reality (hence resulting in the so-called reality gap problem [14,20]). For this reason, we actually change the control value at a lower frequency, namely 5 Hz, than the one of the simulation, being 60 Hz.

Concerning the dimension of the information passed to and from NNs and the architecture of each NN, we here use the same architecture, with one inner layer with 8 neurons, and parameters $\boldsymbol{\theta}$ for each NN and $n_\text{comm} = 1$, as this setting has been proven to allow for effective control without increasing too much the search space [9,16]. We hence have $\boldsymbol{\theta} \in \mathbb{R}^p$, with $p = (4+4+1) \cdot 8 + (8+1) \cdot (1+4) = 117$, where the trailing +1s are the bias. We use tanh as activation function.

A key property of the distributed neural controller that holds when, as in this case, all the NNs use the same parameters $\boldsymbol{\theta}$ is that a controller may be coupled with any morphology. This perfectly fits the scenario where voxels self-assemble to form a desired morphology on which a pre-trained controller is hence "installed". In our work, we exploit this this property to evolve a controller for one morphology and then couple it with slightly different morphologies to study its effectiveness in the new conditions.

2.3 Evolution of VSR Controller

In order to evolve the controller for a VSR, namely its parameters $\boldsymbol{\theta} \in \mathbb{R}^p$, we employ a standard $\mu + \lambda$ EA, as follows.

We iteratively evolve a population of n_{pop} solutions, i.e., numerical vectors, until we have performed n_{evals} fitness evaluations. At each iteration, we build the offspring by repeating n_{pop} times the following steps: (1) we randomly select the crossover (with probability p_{xover}) or the mutation (with probability $1 - p_{\text{xover}}$) genetic operator; (2) we select one or two parents (depending on the chosen operator) with a tournament selection of size n_{tour}; (3) we apply the operator to the parents obtaining a new individual for which we evaluate the fitness. Then, we merge the parents and the offspring, we retain only the n_{pop} best individuals, and proceed to the next generation. At the end of the evolution, we pick the solution in the population with the best fitness as the evolved controller.

For initializing the population, we build each initial solution $\boldsymbol{\theta}$ by sampling each θ_i from $U(-1, 1)$. Concerning the genetic operators, we use the extended geometric crossover, where the child $\boldsymbol{\theta} \in \mathbb{R}^p$ is obtained from the parents $\boldsymbol{\theta}_1, \boldsymbol{\theta}_2 \in \mathbb{R}^p$ as $\boldsymbol{\theta} = \boldsymbol{\theta}_1 + \boldsymbol{\alpha}(\boldsymbol{\theta}_2 - \boldsymbol{\theta}_1) + \boldsymbol{\beta}$, where $\boldsymbol{\alpha} \in \mathbb{R}^p$ is sampled as $\alpha_i \sim U(-0.5, 1.5)$, and $\boldsymbol{\beta} \in \mathbb{R}^p$ is sampled as $\beta_i \sim N(0, \sigma_{\text{xover}})$. As mutation, we use the Gaussian mutation, where $\boldsymbol{\theta} = \boldsymbol{\theta}_1 + \boldsymbol{\beta}$, with betas sampled from $N(0, \sigma_{\text{mut}})$.

3 Experiments and Discussion

Our broad aim is to study the impact of small variations in the morphology on the effectiveness of evolved neural controllers. More precisely, we aim at answering the following research questions:

RQ1 To which degree do small morphology variations affect the controller effectiveness? Do overall robot size and initial morphology have an impact on how variations affect effectiveness?

RQ2 Is it possible to mitigate the decrease in effectiveness of an evolved controller by re-optimizing it on the varied morphology?

In order to answer these questions, we performed a large number of experiments in which we (i) evolved the controllers for a few pre-defined morphologies, taking note of their effectiveness upon evolution on a pre-defined task, (ii) modified the morphologies to different extents, and (iii) measured the effectiveness on the task of the evolved controller applied to the modified morphology. Moreover, for the purpose of addressing RQ2, we performed a further re-optimization of the evolved controller on the modified morphology, taking note of its effectiveness.

We experimented with the task of locomotion, in which the robot is placed on a flat surface and has to move the farthest possible towards the right (i.e., along the positive direction of the x-axis). For this task, we measure the effectiveness as the average velocity v_x along the x-axis computed by considering the x-position of the center of mass of the VSR at the beginning and at the end of a simulation lasting $t_{\text{sim}} = 30$ s. We use v_x as the fitness of the robot during the evolution: clearly, the larger, the better.

Concerning the pre-defined morphologies, i.e., the ones on which we evolved the controllers, we experimented with 6 different cases, that we show in Fig. 1 (in darker green). They are a biped, a comb, and a worm, each one in two sizes (small and large): the biped consists of an horizontal body and two legs, the comb extends this concept to four legs, resembling indeed a comb, and the worm is just a full rectangle. They are composed by 10, 11, and 10 voxels (biped, comb, worm), for small morphologies, and 40, 44, and 40 voxels, for large ones.

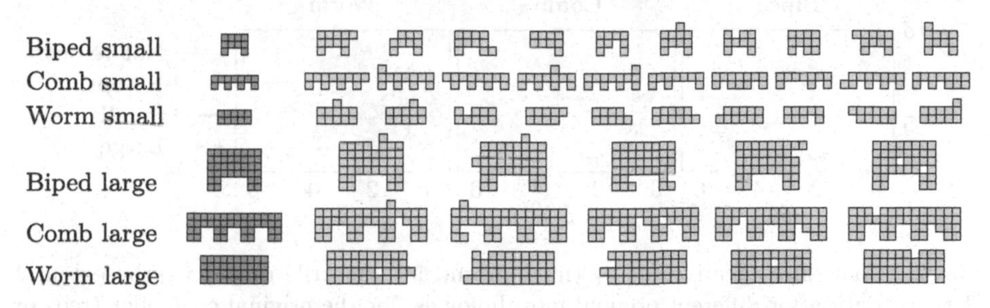

Fig. 1. The 6 morphologies of our experiments (in darker green) with 10 (small, with $\delta = 1$) or 5 (large, with $\delta = 2$) example variations (in lighter green). (Color figure online)

For producing the variations in the morphology, that are a key component of this study, we proceeded as follows. Given a morphology $\boldsymbol{m}$ and a target extent δ of the variation, in number of voxels, we iteratively and randomly removed or added voxels, i.e., we randomly changed the Boolean values of $\boldsymbol{m}$, until the Hamming distance between the varied morphology $\boldsymbol{m}'$ and the original one was $d_{\text{hamming}}(\boldsymbol{m}, \boldsymbol{m}') = \delta$. At each random addition or removal of a voxel from a morphology, we ensured that it remained connected, i.e., a single body; that is, we always dealt with proper polyominoes. Figure 1 shows, for each original morphology and in lighter green, some example variations obtained with $\delta = 1$ (small) or $\delta = 2$ (large).

We performed our experiments with 2D-VSR-Sim [8], for simulating the VSRs, and with JGEA [10], for the evolutionary optimization. Concerning the former, we set all the parameters to their default values. For the latter, we set $n_{\text{pop}} = 100$, $n_{\text{evals}} = 10000$, $p_{\text{xover}} = 75\%$, $n_{\text{tour}} = 5$, and $\sigma_{\text{mut}} = \sigma_{\text{xover}} = 0.35$. We used $\delta \in \{1, 2, 3\}$ for small morphologies and $\delta \in \{2, 4, 6\}$ for large ones.

We performed statistical significance tests with the Mann-Whitney U rank test with the null hypothesis of equality of the median, after having verified all other relevant hypotheses, and with $\alpha = 0.05$.

3.1 Results and Discussion for RQ1: Impact of Variations

Figure 2 shows the results concerning the impact on controller effectiveness v_x (on the y-axis) of morphology variations of extent δ (on the x-axis) applied after the evolution of the neural controller (red line) and without any re-optimization, for the six considered morphologies—the blue line is discussed in the next section as it pertains to RQ2. The value of v_x for $\delta = 0$ corresponds to the fitness at the end of the evolution on the original morphology, i.e., without any variation, and is computed as the median across 10 independent evolutionary runs. The value of v_x for $\delta > 0$ is the median of the velocities of the 10 controllers obtained with the original morphology applied each one to 10 variations, for each δ, of the original morphology; i.e., for each $\delta > 0$, there are 100 values.

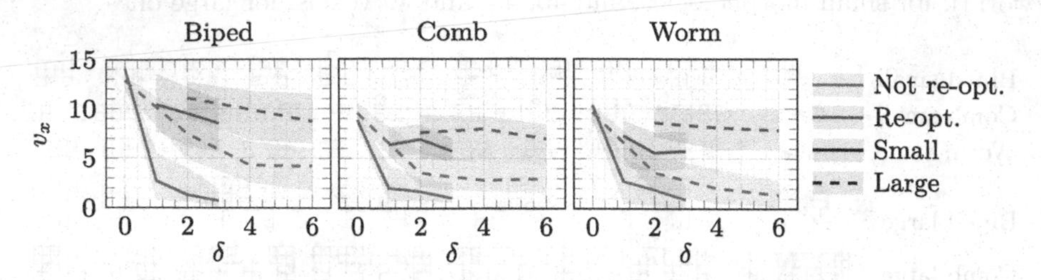

Fig. 2. Controller effectiveness v_x (median and interquartile range) vs. the extent δ of the variation for different original morphologies, for the original controller (red) or with re-optimization (blue). (Color figure online)

The main finding we infer from Fig. 2 is that there is a clear decrease in the effectiveness of the neural controller once the morphology it has been evolved on changes. The decrease is apparent for all the original morphologies in both sizes and it is statistically significant for any $\delta > 0$. The difference between the v_x on the original morphology and the one on the modified morphology ranges from ≈ 10 ($\approx 70\%$) for the small biped to ≈ 5 ($\approx 55\%$) for the large comb.

Interestingly, the extent δ of the morphology variation appears to play a minor role in the decrease of effectiveness, in particular for the small original morphologies—the differences for pairs of $\delta > 0$ are statistically significant only in a few cases. The drop in effectiveness looks smoother, with respect to δ, only for the large biped.

For understanding the reason for such a drastic decrease in the effectiveness, even when the variation in the morphology consists in just one voxel, we analyzed many simulations visually, i.e., we observed the behaviors in locomotion of the corresponding VSRs. In general, the degree to which the behavior appears successful in the task of locomotion depends on the original morphology, the small biped being the best one. For this case, it is sharply clear that the gait is very negatively affected by addition of removal of just one voxel: we made a video for the gait of the VSR with the original morphology and with $\delta = 1$ publicly available at https://youtu.be/bB1u3Yj6FTo. The gait for the original small biped is effective because the controller evolved to master the dynamics of the body: the latter has a peculiar periodicity that can be exploited to obtain an effective gait. When just one block is removed or added, the periodicity changes and the controller is no more able to exploit it. More broadly, the VSR is a dynamical system whose attractor in the space of poses is cyclic and highly functional to the locomotion: when some properties of the morphology change, the attractor becomes much less effective.

We believe that these experiments further corroborate the validity of the embodied cognition paradigm: the ability of the VSR to perform a task is hosted jointly in the VSR brain and body, not just, nor mostly in the former. This is particularly true for VSRs because their body, being an aggregation of many soft components, each corresponding to a simple dynamical system, offers a great potential for performing morphological computation [18,22].

3.2 Results and Discussion for RQ2: Re-Optimization

With the experiments discussed in the previous section, we found that the small variations in the morphology are greatly detrimental for the effectiveness of the evolved neural controllers. We also hypothesized that the decrease in effectiveness is rooted in the misalignment between body and brain that arises from the morphology variation. Here we wonder if another, not necessary alternative, motivation is in the fact that the modified morphologies are intrinsically less suitable for performing the task. For investigating this scenario, we considered, for each original morphology and each value of $\delta > 0$, the $10 \cdot 10$ varied morphologies and performed one evolutionary optimization of a neural controller for each of them. We used the same EA, with the same parameters, we used for the previous experiments.

Figure 2 presents the results of this experiment. It shows, through the blue lines, the controller effectiveness v_x (on the y-axis) obtained at the end of the re-optimization vs. the morphology variations of extent δ (on the x-axis), for the six considered morphologies.

The foremost finding is that the controller effectiveness v_x for the modified morphologies is, upon re-optimization, almost on par with the effectiveness of the original controller on the original morphology—the small differences are always statistically significant, with the exception of the large comb and worm. From another point of view, if a new neural controller is evolved for the modified morphology, it is clearly better than the non re-optimized controller that was obtained for the original morphology.

A second observation concerns the variability of v_x for the re-optimized controllers– we recall that both red and blue lines of Fig. 4 are, for $\delta > 0$, computed on 100 values. The figure suggests that, for half of the morphologies, the interquartile range for re-optimized controllers is larger than the one for the original controllers. We looked at a few behaviors of the VSRs with re-optimized controllers and we found that, indeed, some of the morphologies appeared more suitable for locomotion than others. While it has already been showed that "regular", hand-designed morphologies are not, in general, better than evolved ones [9,21], we were not able to identify a single criterion of improvement. In particular, it was not the size of the VSR, i.e., bigger (and hence stronger) robots were not in general faster—we discuss this analysis more deeply in Sect. 3.2.

Summarizing, this experiment showed that the modified morphologies are not, in general clearly worse than the original ones and that a evolving a new controller for them can make them work.

Smarter Re-Optimization. Re-optimizing a new neural controller for a modified morphology proved to be effective, according to the experiment discussed above. In a practical realization of the scenario considered in this study, however, this finding would hardly be exploitable. The requirement of fully evolving from scratch a neural controller for a morphology resulting from self-assembly that just slightly differ from the expected one would make self-assembly a fragile component of the adaptation of the robotic ecosystem. On the other hand, every skill acquired by optimizing the controller for the original morphology would be dropped entirely, hence wasting previous optimizations effort.

To address this point, we explored the possibility of starting the re-optimization from the neural controller evolved for the original morphology instead of starting from scratch. In detail, for each one of the 10 controllers evolved for each original morphology, we took each of the 10 corresponding variations and performed an evolutionary optimization in which the initial population was not entirely random, but partly built based on the original controller $\boldsymbol{\theta}^\star$. In particular, we built $\frac{1}{2}n_\text{pop}$ individuals randomly, i.e., by sampling $U(-1,1)$, $\frac{1}{2}n_\text{pop}-1$ individuals by applying the Gaussian mutation to $\boldsymbol{\theta}^\star$, and finally included $\boldsymbol{\theta}^\star$ itself—that is, we *seeded* the initial population with $\boldsymbol{\theta}^\star$ as done in [7,19]. For reducing the computational effort, we considered only the small morphologies with $\delta = 1$.

Figure 3 shows the results of this experiments. It shows how v_x on the modified morphologies changes during the evolution when starting from a random initial population (blue line) or from a population seeded with $\boldsymbol{\theta}^\star$ (green line). For comparison, the figure also shows the evolution of v_x on the original morphology (red line). This is computed on 10 values, differently than the other two lines that are computed on 100 values, and results hence less smooth.

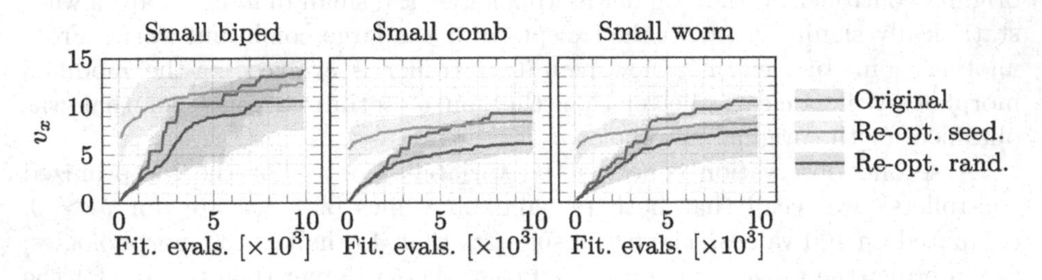

Fig. 3. Controller effectiveness v_x (median and interquartile range) during the evolution with the original morphology (red), a modified morphologies with $\delta = 1$ with re-optimized controller with random (blue) or seeded (green) population. (Color figure online)

The main finding arising from Fig. 3 is that re-using the controller evolved for the original morphology is very effective. For the biped and worm morphologies, v_x for the best individual at the first iteration of the EA with the seeded initial population is very close to the random initial population case after 5000 fitness evaluations. For the comb, the initial seeded best is on par with the final best of the random case. In other words, with the seeded population, one can obtain an effective neural controller even with a very short re-optimization, i.e., much more efficiently.

A secondary, yet important finding, is that not only the seeded population positively impacts on efficiency of the optimization, but it also allows to obtain more effective controllers. For the biped and comb morphologies, the re-optimized controllers on the modified morphologies are not statistically significantly worse than the original controller on the original morphology.

Overall, this experiments show that a short re-optimization of the controller is enough to re-align it to the modified body, hence making the VSR ecosystem truly capable of adaptation and robustness.

Discovery of More Effective Morphologies. While analyzing the raw results for the discussion of Sect. 3.2, we discovered that for some modified morphologies, the neural controllers resulting from re-optimization (from random initial population) achieved sharply better v_x than those obtained with the original morphology.

In Fig. 4 we present the v_x obtained for all the variants of the small biped morphology, including the original, unmodified one. We chose this case because it was the one resulting, in general, in the most effective gaits. The figure shows one point for each of morphology, with coordinates given by its v_x (upon re-optimization, on the x-axis) and its size (number of voxels, on the y-axis) and color based on the morphology: green for the original one, gray for other cases, and blue and red for the Talos and horse variants.

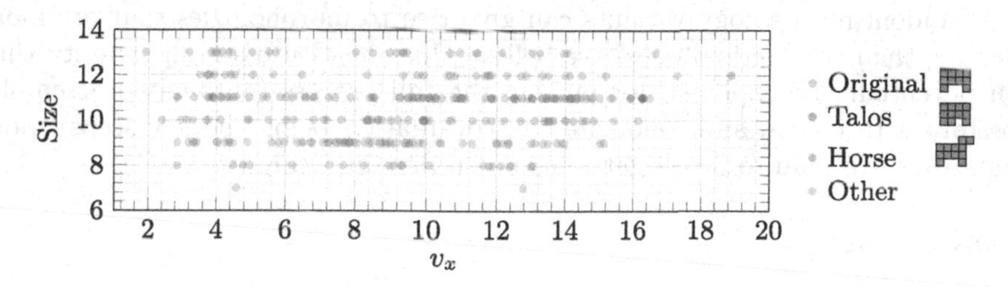

Fig. 4. Controller effectiveness v_x vs. size (num. of voxels) for all the 100 small biped morphologies, each with a re-optimized controller.

The latter two morphologies exhibited larger v_x values that were easily explainable by visually inspecting their behavior. The Talos variant, which we named after the main character of the book "Lo scudo di Talos" [6], had a thicker rear leg that conferred it greater strength for hopping faster. The horse variant used its "head", moved in counter-phase with other parts of its body, to balance the overall movement and hence advance faster.

In order to validate this casual observation, we performed 10 evolutionary runs for each of the two variants and found that their suitability for locomotion was systematic. The evolved controllers obtained a better v_x than the one for the original biped, with statistically significant differences. We deem this finding particularly promising, in perspective. If new, more effective morphologies can be discovered "by chance" by re-optimizing existing controllers from morphologies that present small, erratic variations, then there is an opportunity for a further adaptation that propagates these small, yet positive advancements, hence making the robotic ecosystem, as a whole, more adaptive. We plan to investigate this possibility more deeply in future works.

Finally, Fig. 4 also shows that there is no correlation between the size of the VSR and v_x ($R^2 \approx 0.005$). In other words, for the biped variants, it does not hold that bigger morphologies result in faster robots, despite having, in principle, greater available strength.

4 Concluding Remarks

We considered the case of Voxel-based Soft Robots (VSRs) and experimentally characterized the impact of small morphology variations on the effectiveness of evolved neural controllers. We found that even small variations are highly detrimental for controller effectiveness, i.e., robots with a controller evolved for a morphology are much smaller if applied on morphologies differing in one or more voxels. By analyzing robot behaviors, we motivated this finding in terms of body-brain misalignment: since VSRs are soft, their bodies exhibit a rich dynamics which plays a key role in determining the behavior.

We also experimented with the re-optimization of controllers for the modified morphologies and found that it is both efficient and effective if it is seeded with the controller evolved for the original morphology. As an aside, we discovered that random morphology variants can give rise to morphologies that are more effective than the original ones. We believe this constitute an opportunity that can potentially be exploited for making the full process (automated assembly, possibly with errors, and robot life with evaluation being the key steps) more adaptable. We plan to investigate this possibility in future works.

References

1. Corucci, F., Cheney, N., Giorgio-Serchi, F., Bongard, J., Laschi, C.: Evolving soft locomotion in aquatic and terrestrial environments: effects of material properties and environmental transitions. Soft Rob. **5**(4), 475–495 (2018)
2. Faiña, A.: Evolving modular robots: challenges and opportunities. In: ALIFE 2021: The 2021 Conference on Artificial Life. MIT Press (2021)
3. Hale, M., et al.: The are robot fabricator: how to (re) produce robots that can evolve in the real world. In: International Society for Artificial Life: ALIFE2019, pp. 95–102. York (2019)
4. Li, S., et al.: Scaling up soft robotics: a meter-scale, modular, and reconfigurable soft robotic system. Soft Rob. **9**(2), 324–336 (2022)
5. Malley, M., Haghighat, B., Houe, L., Nagpal, R.: Eciton robotica: design and algorithms for an adaptive self-assembling soft robot collective. In: 2020 IEEE International Conference on Robotics and Automation (ICRA), pp. 4565–4571. IEEE (2020)

6. Manfredi, V.M.: Lo Scudo di Talos. Edizioni Mondadori, Milan (2013)
7. Medvet, E., Bartoli, A., De Lorenzo, A., Fidel, G.: Evolution of distributed neural controllers for voxel-based soft robots. In: Proceedings of the 2020 Genetic and Evolutionary Computation Conference, pp. 112–120 (2020)
8. Medvet, E., Bartoli, A., De Lorenzo, A., Seriani, S.: 2D-VSR-Sim: a simulation tool for the optimization of 2-D voxel-based soft robots. SoftwareX **12**, 100573 (2020)
9. Medvet, E., Bartoli, A., Pigozzi, F., Rochelli, M.: Biodiversity in evolved voxel-based soft robots. In: Proceedings of the Genetic and Evolutionary Computation Conference, pp. 129–137 (2021)
10. Medvet, E., Nadizar, G., Manzoni, L.: JGEA: a modular java framework for experimenting with evolutionary computation. In: Proceedings of the Genetic and Evolutionary Computation Conference Companion (2022)
11. Medvet, E., Nadizar, G., Pigozzi, F.: On the impact of body material properties on neuroevolution for embodied agents: the case of voxel-based soft robots. In: Proceedings of the Genetic and Evolutionary Computation Conference Companion (2022)
12. Moreno, R., Faiña, A.: EMERGE modular robot: a tool for fast deployment of evolved robots. Front. Robot. AI **8**, 198 (2021)
13. Moreno, R., Faiña, A.: Out of time: on the constrains that evolution in hardware faces when evolving modular robots. In: Jiménez Laredo, J.L., Hidalgo, J.I., Babaagba, K.O. (eds.) EvoApplications 2022. LNCS, vol. 13224, pp. 667–682. Springer, Cham (2022). https://doi.org/10.1007/978-3-031-02462-7_42
14. Mouret, J.B., Chatzilygeroudis, K.: 20 years of reality gap: a few thoughts about simulators in evolutionary robotics. In: Proceedings of the Genetic and Evolutionary Computation Conference Companion, pp. 1121–1124 (2017)
15. Nadizar, G., Medvet, E., Miras, K.: On the schedule for morphological development of evolved modular soft robots. In: Medvet, E., Pappa, G., Xue, B. (eds.) Genetic Programming. EuroGP 2022. Lecture Notes in Computer Science (Part of EvoStar), vol. 13223, pp. 146–161. Springer, Cham (2022). https://doi.org/10.1007/978-3-031-02056-8_10
16. Nadizar, G., Medvet, E., Nichele, S., Pontes-Filho, S.: Collective control of modular soft robots via embodied Spiking Neural Cellular Automata. arXiv preprint arXiv:2204.02099 (2022)
17. Peck, R.H., Timmis, J., Tyrrell, A.M.: Self-assembly and self-repair during motion with modular robots. Electronics **11**(10), 1595 (2022)
18. Pfeifer, R., Gómez, G.: Morphological computation-connecting brain, body, and environment. In: Creating Brain-Like Intelligence, pp. 66–83. Springer, Cham (2009)
19. Pigozzi, F., Tang, Y., Medvet, E., Ha, D.: Evolving modular soft robots without explicit inter-module communication using local self-attention. In: Proceedings of the Genetic and Evolutionary Computation Conference (2022)
20. Salvato, E., Fenu, G., Medvet, E., Pellegrino, F.A.: Crossing the reality gap: a survey on sim-to-real transferability of robot controllers in reinforcement learning. IEEE Access **9**, 153171–153187 (2021)
21. Talamini, J., Medvet, E., Nichele, S.: Criticality-driven evolution of adaptable morphologies of voxel-based soft-robots. Front. Robot. AI **8**, 673156 (2021)
22. Zahedi, K., Ay, N.: Quantifying morphological computation. Entropy **15**(5), 1887–1915 (2013)

EGSGP: An Ensemble System Based on Geometric Semantic Genetic Programming

Liah Rosenfeld and Leonardo Vanneschi(✉)

NOVA Information Management School (NOVA IMS), Universidade Nova de Lisboa, Campus de Campolide, 1070-312 Lisbon, Portugal
{lrosenfeld,lvanneschi}@novaims.unl.pt

Abstract. This work is inspired by the idea of seeding Genetic Programming (GP) populations with trained models from a pool of different Machine Learning (ML) methods, instead of using randomly generated individuals. If one considers standard GP, tackling this problem is very challenging, because each ML method uses its own representation, typically very different from the others. However, the task becomes easier if we use Geometric Semantic GP (GSGP). In fact, GSGP allows us to abstract from the representation, focusing purely on semantics. Following this idea, we introduce EGSGP, a novel method that can be seen either as a new initialization technique for GSGP, or as an ensemble method, that uses GSGP to combine different Base Learners (BLs). To counteract overfitting, we focused on the study of elitism and Soft Target (ST) regularization, studying several variants of EGSGP. In particular, systems that use or do not use elitism, and that use (with different parameters) or do not use ST were investigated. After an intensive study of the new parameters that characterize EGSGP, those variants were compared with the used BLs and with GSGP on three real-life regression problems. The presented results indicate that EGSGP outperforms the BLs and GSGP on all the studied test problems. While the difference between EGSGP and GSGP is statistically significant on two of the three test problems, EGSGP outperforms all the BLs in a statistically significant way only on one of them.

Keywords: Genetic Programming · Geometric Semantic Genetic Programming · Initialization · Ensemble methods · Meta-learning

1 Introduction

Ensemble Methods (EMs) focus on combining the knowledge obtained from multiple models resulting from different algorithms, with the goal of achieving a final model that is better than any of its singular components [1]. More specifically, EMs form their final models by constructing an ensemble, which is simply a bundle of learners – commonly known as Base Learners (BLs) – whose individual decisions are combined in some way [2]. EMs are therefore anchored in the

C. De Stefano et al. (Eds.): WIVACE 2022, CCIS 1780, pp. 278–290, 2023.
https://doi.org/10.1007/978-3-031-31183-3_23

Aristotelian idea that the whole should be better than the sum of its parts, relying on the "power of combination" to achieve a more accurate final result [1]. Meta-learning is a specific field within EMs that, as its name indicates, focuses on raising the level of abstraction a step further, seeking for the ability of "learning to learn" [3]. More precisely, while a ML algorithm (commonly denoted as the first-level or base learner) learns to identify patterns in the data in order to make predictions, a meta-learning algorithm (often denoted as the second-level learner) learns how to combine these first-level predictions [1]. Hence, in meta-learning an algorithm learns from previous experience in a data-driven way [4] by taking the output of different base learners as input [5].

This paper proposes Ensembled Geometric Semantic Genetic Programming (EGSGP) as a meta-learning technique, that uses Geometric Semantic Genetic Programming (GSGP) to combine the information obtained from a heterogeneous set of ML algorithms, with the goal of using the power of combination for a better predictive performance. GSGP [6] is an extension of Genetic Programming (GP) [7], that has the goal of evolving individuals while directly searching in the semantic space [8]. In this context, the semantic of an individual refers to its predictions on the training data or, in other words, the vector of its output values [9]. This ability to directly search in the semantic space is possible thanks to the replacement of standard crossover and mutation, typically based on individuals' syntax, with geometric semantic operators (GSOs) [10]. GSOs bestow GSGP with the property of inducing a unimodal error surface [10] (i.e., a fitness landscape characterized by the lack of locally optimal solutions). Besides the introduction of GSOs, several other contributions focused on facilitating GP's search using semantic awareness [8]. For instance, Pawlak and Kraweic introduced semantic geometric initialization [11], opening gates to other GSGP focused initialization methods (e.g., the EDDA initialization [12]), recognizing the importance of an appropriate population initialization and how it may play a crucial role for GSGP.

The proposed technique (EGSGP) is contextualized in this research track. In fact, it can be seen as an ensemble initialization method that utilizes the outputs obtained from a heterogeneous set of ML algorithms, the BLs, as the initial population for GSGP. This is possible because, as recognized in [6], GSGP can abstract from representation, i.e. from the genotype of the individuals, focusing solely on their semantics, a property that distinguishes GSGP from traditional GP. So, even though the different BLs have very diverse representations, the evolution can be based only on their output values, calculated on the training observations. One can perceive EGSGP as either a novel initialization method for GSGP, where the initial population is no longer composed of random individuals, but rather of the predictions (i.e., the semantics) of the BLs, or as a meta-estimator that utilizes GSGP as a trainable combiner (i.e., the entity responsible for combining the knowledge obtained from the different models [13]). Additionally, an analysis of the effects of elitism and of the incorporation of a regularization method called Soft Target (ST) [14] in the models, with the goal of limiting overfitting and dealing with potential diversity issues, was performed.

This document is organized as follows: in Sect. 2, we present GSGP; Sect. 3 introduces ST regularization; Sect. 4 discusses the proposed EGSGP technique; Sect. 5 presents our experimental study, including a discussion of the test cases, of the used experimental settings and of the obtained results; finally, Sect. 6 concludes the work and proposes ideas for future research.

2 Geometric Semantic Genetic Programming

Let $X = \{\overrightarrow{x_1}, \overrightarrow{x_2}, ..., \overrightarrow{x_n}\}$ be the set of input data (training instances, observations or fitness cases) of a supervised learning problem, and $\overrightarrow{t} = [t_1, t_2, ..., t_n]$ the vector of the respective expected output or target values (in other words, for each $i = 1, 2, ..., n$, t_i is the expected output corresponding to input $\overrightarrow{x_i}$). A GP individual (or program) P can be seen as a function that, for each input vector $\overrightarrow{x_i}$ returns the scalar value $P(\overrightarrow{x_i})$. Following [6], we call *semantics* of P the vector $\overrightarrow{s_P} = [P(\overrightarrow{x_1}), P(\overrightarrow{x_2}), ..., P(\overrightarrow{x_n})]$. This vector can be represented as a point in a n-dimensional space, the *semantic space*. As explained above, GSGP is a variant of GP, where the traditional crossover and mutation are replaced by new operators, the Geometric Semantic Operators (GSOs). The objective of GSOs is to define modifications on the syntax of GP individuals, that have a precise effect on their semantics. One of the reasons for the success of GSOs is because they induce an unimodal error surface (on training data) for any supervised learning problem, where fitness is calculated using an error measure between outputs and targets. In other words, using GSOs the error surface on training data is guaranteed to not have any locally optimal solution, except for the global optima. The definitions of the GSOs are, as given in [6], respectively:

Geometric Semantic Crossover (GSC). Given two parent functions $T_1, T_2 : \mathbb{R}^n \to \mathbb{R}$, the geometric semantic crossover returns the real function $T_{XO} = (T_1 \cdot T_R) + ((1 - T_R) \cdot T_2)$, where T_R is a random real function whose output values range in the interval $[0, 1]$.

Geometric Semantic Mutation (GSM). Given a parent function $T : \mathbb{R}^n \to \mathbb{R}$, the geometric semantic mutation with mutation step ms returns the real function $T_M = T + ms \cdot (T_{R1} - T_{R2})$, where T_{R1} and T_{R2} are random real functions.

Even though this is not in the original definition of GSM, later contributions [10] have clearly shown that limiting the codomain of T_{R1} and T_{R2} in a predefined interval (for instace $[0, 1]$, as it is done for T_R in GSC) helps improving the generalization ability of GSGP. As in several previous works [10], here we constrain the outputs of T_R, T_{R1} and T_{R2} by wrapping them in a logistic function. Only the definitions of the GSOs for symbolic regression problems are given here, since they are the only ones used in this work. For the definition of GSOs for other domains, the reader is referred to [6].

Reference [10] contains a detailed and intuitive explanation of the reason why the error surface is unimodal when GSOs are used. However, even without dwelling with the formal details, that can be found in that paper, one can have an intuition of why GSO induce a unimodal error surface, looking at the GSM definition. GSM corresponds to box mutation in the semantic space, which means that if we mutate an individual x, it is possible to create an individual with any semantic included in a box centered in the semantic of x. As such, and recalling that also the target vector is a point in the semantic space, with a GSM step applied to any x it is always possible to get closer than x to the global optimum. One important notation is that, even though having a unimodal error surface significantly simplifies the learning process, it does not "trivialize" the problem. If fact, it is important to understand that we are facing supervised machine learning problems here, and not pure optimization problems. As such, what is interesting for us is the model accuracy on *unseen* data, while the unimodality of the error surface can be proven only on training data.

Furthermore, using GSOs the semantics of the offspring is completely defined by the semantics of the parents: the semantics of an offspring produced by GSC will lie on the segment joining the semantics of the parents (geometric crossover), while GSM defines a mutation such that the semantics of the offspring lies within the ball of radius ms centered in the semantics of the parent (geometric mutation, or "ball mutation"). As discussed in [6,10], GSOs always generate larger offspring than the parents, and this entails a rapid growth of the size of the individuals in the population. To counteract this problem, in [10] an implementation of GSOs is presented that makes GSGP not only usable in practice, but also significantly faster than standard GP. This is possible by means of a particular representation of GP individuals, that allows us to not store their genotypes during the evolution. This point, i.e. not having to store the genotype of the evolving individuals, is key for our work, since the initial individuals are models coming from the training of very diverse BLs. So, the implementation presented in [10] is the one we used here.

3 Soft Target Regularization

Considering that EGSGP constructs an initial population for GSGP using pretrained BL models, a potential tendency for overfitting, due to the lack of initial diversity, could imaginably become a problem. For this reason, regularization may be an essential step for EGSGP. The choice of implementing Soft Target (ST) (first proposed by Aghajanyan in [14]) and not another regularization method was mainly anchored in the fact that ST was already successfully employed with GP [15]. ST works by maintaining information from the models' outputs in the early phases of the training process, as their early mistakes can be useful and provide information regarding what Aghajanyan called the "co-label effects" [15]. More specifically, ST redefines the traditional loss function as a quantification of the distance between the calculated outputs and the soft target. The soft target is a weighted average between the expected outputs and some information concerning the past models' outputs. At each generation t, this information consists in a moving average of the past models' outputs $\hat{y}^t$, defined as:

$$\begin{cases} \hat{y}^0 = f^0(X) \\ \hat{y}^t = \beta \cdot \hat{y}^{t-1} + (1-\beta) \cdot f^t(X), \quad \text{if } t > 0 \end{cases} \tag{1}$$

where X is the set of input vectors, $X = \{\overrightarrow{x_1}, \overrightarrow{x_2}, ..., \overrightarrow{x_n}\}$. Consequently, we update the value of the soft target $T^t = \{t_1^t, t_2^t, ..., t_n^t\}$ through a weighted average of $\hat{y}^t$ and the true expected output $y = \{y_1, y_2, ..., y_n\}$, as follows:

$$\begin{cases} t^0 = y \\ t^t = \gamma \cdot \hat{y}^t + (1-\gamma) \cdot y, \quad \text{if } t > 0 \end{cases} \tag{2}$$

In the previous two equations, β and γ are parameters of the algorithm. The former (β) allows us to tune the importance of the past models' outputs compared to the current one, while the latter (γ) allows us to tune the importance of the moving average of the past models, compared to the true expected output. Finally, for each iteration t, the loss function $\mathcal{L}$ is defined in the following way:

$$\mathcal{L}(f^t) = \left(\frac{1}{n} \sum_{i=1}^{n} \mathcal{L}_i(f^t(x_i), t_i^t) \right)^m \tag{3}$$

where, in the case of our work, we used $\mathcal{L}_i(a, b) = (a - b)^2$ and $m = 1/2$, thus turning $\mathcal{L}$ into a Root Mean Square Error (RMSE), as it is customary for GP [15]. ST regularization requires two parameters in addition to β and γ: the burning period (*bp*) and the frequency of update (*fu*), corresponding to the number of initial generations in which the algorithm is executed without any ST regularization and the frequency of update of the soft target, respectively. Thus, setting the ST parameters allows us to tune how much importance is given to the early stages of the learning. The next section provides further details regarding how ST regularization was used by the EGSGP algorithm.

4 Ensembled Geometric Semantic Genetic Programming

Simply put, EGSGP consists in a novel initialization for GSGP. In EGSGP, the population is no longer seeded with random individuals. Instead, the initial individuals are generated using the semantics of the BL models, that had been previously trained. Six different methods were constructed in order to assess the performance of our proposed technique. These methods are characterized by the possible combinations of elitism and ST regularization, that we identify as two important elements to counteract overfitting. As previously discussed, the ST regularization depends on different parameters, that allow us to tune the importance given to the previous models during the learning. Thus, when ST regularization is used, two different sets of ST parameter combinations are employed. For simplicity, we call the two variants ST_{pa} and ST_{pb}.

ST_{pa} consists of ST regularization with a parameter set that gives high importance to past models' outputs (and therefore to the original BLs), starting the regularization at an early stage of the learning process. The parameters that characterize ST_{pa} version are: $\beta = 0.95, \gamma = 0.95, bp = 0, fu = 2$.

ST_{pb}, instead, applies ST regularization with a parameter set that is more balanced, giving importance both to the original BLs and to the changes that happen to them throughout the learning process. The parameters that characterize ST_{pb} are: $\beta = 0.5, \gamma = 0.5, bp = 5, fu = 5$.

Furthermore, EGSGP with no regularization was also studied. In this fashion, the combination of the whether Elitism or ST regularization (and which ST regularization parameters) were or not employed results in six EGSGP models: EL_NO_ST, that uses elitism and no ST; NO_EL_NO_ST, that uses no elitism and no ST; EL_STA, that uses elitism and ST_{pa}; NO_EL_STA, that uses no elitism and ST_{pa}; EL_STB, that uses elitism and ST_{pb}; and NO_EL_STB, that uses no elitism and no ST.

5 Experimental Study

Base Learners. In this work, EGSGP seeds the initial population of GSGP with the semantics of 15 different BL pre-trained models, that underwent no parameter optimization, since their outputs will later be further trained by GSGP. Hence, the default setting of the Python libraries' functions were kept. The used BLs were: Decision Trees, Random Forests, Extra Tree Regressor, AdaBoost, Gradient Boosting, Lasso Regression, Ridge Regression, Support Vector Machines, K-Nearest Neighbours, Elastic Net, Multilayer Perceptron, ADR Regressor, Kernel Ridge and two variants of XgBoost.

Test Cases. Three real-life regression problems were used to assess the performance of EGSGP. The first one, the Concrete dataset, has the objective to predict the strength of high perfomance concrete given its components. The remaining two datasets are problems from the biomedical area; the Plasma Protein Binding (PPB) levels dataset aims at predicting the value of a pharmacokinetic parameter (the PPB), based on a set of molecular descriptors of potential new drugs, and the Unified Parkinson's Disease Rating Scale (UPDRS) dataset seeks to predict the UPDRS assessment, in order to aid the automatization of the tracking of the development of Parkinson's disease symptoms. These three datasets have often been used as benchmarks for GP. The Concrete dataset is discussed in [16], and it is characterized by 8 features and 1030 observations. The PPB dataset is described in [17], and it is characterized by 626 features and 131 observations. The Parkinson's dataset was introduced in [18], and it is characterized by 18 features and 5875 observations.

Experimental Settings. We performed the Monte-Carlo cross-validation with 30 independent runs, using a different random seed for each one of them. At each run, 30% of the data was used for testing. The remaining portion of the data (70%) was split in two parts: training and validation, with 70% and

30% of that remaining portion respectively. For each dataset, the six different EGSGP models were tested. The GSGP parameters, like the probability of mutation (PMut), the probability of crossover (PXover), the mutation step (*ms*), and others, were determined via a grid-search, where the parameter combination with the best median validation error was chosen. The parameter values tested in the grid search were: crossover and mutation rates equal to 0.25, 0.5, 0.75 and 1; mutation step equal to 0.001, 0.01, 0.1 and 1; tournament size equal to 2 and 10; population size equal to 25, 50, 100, 200, 250; maximum number of generations equal to 25, 50, 100, 200, 250 (plus the additional values of 650 and 1250, that were tested only for GSGP). We gave GSGP the possibility of evolving for a larger number of generations than EGSGP (in fact, for GSGP also the values 650 and 1250 were tested), because the initial population of EGSGP is composed of pre-trained models. Hence, we considered that allowing GSGP to train for more generations is important for having a fair comparison. More specifically, the value 1250 was chosen in view of the fact that the highest number of maximum iterations for the training stage of the BLs was 1000. Therefore, EGSGP can be seen as having an additional 1000 iterations when compared to GSGP, unless we allow GSGP to run for 1000 more generations.

Finally, an assessment of the models was carried out by executing a comparison between EGSGP, the best BL for each test case, and traditional GSGP. The validation dataset was not solely used for parameter optimization. Additionaly, BLs that were deemed as "overfitters" based the on validation dataset were not included in the initial population. For the Concrete and Parkinson's case studies, we considered that a BL overfits if its median validation error was more than four times its median training error (given that amount was far from the "normally expected" seen to unseen error worsening in our particular case studies). For the PPB dataset, given its noticeable tendency to overfit, BLs with a median training error below 0.15 and with a median validation error above two times the median validation error of all the BLs, were excluded. Finally, in the Concrete test case, no BLs were excluded from the initial EGSGP population. For the PPB dataset, the BL models that were excluded were the ones generated by Extra Tree Regressor, XgBoost, Gradient Boosting, Lasso, Ridge, K-Nearest Neighbours, Elastic Net and Multilayer Perceptron. For the Parkinson's test case, only one model was excluded: the one generated by AdaBoost.

Experimental Results. The obtained experimental results are reported in Tables 1, 2 and 3 (for the Concrete, PPB and Parkinson's case studies, respectively). The first column of these tables reports the studied methods (either one of the EGSGP variants, or traditional GSGP, or the best BL). The remaining columns present the median RMSE of the models for both training and unseen (i.e., testing) data. Furthermore, to assess the statistical significance of the results, a set of tests were performed. Their outcomes are presented in Tables 4, 5 and 6 (for the Concrete, PPB and Parkinson, respectively). Since the results of the Kolmogorov-Smirnov tests revealed that the data are not normally distributed, the non-parametric Mann-Whitney U-statistic was chosen for our

Table 1. Results of the median training and unseen error obtained on the Concrete case study over 30 independent runs for the EGSGP models, GSGP and the best BL. The best training and unseen error are highlighted in bold.

Method	Median Training Error	Median Unseen Error
EL_NO_ST	3.290	5.462
NO_EL NO_ST	**3.317**	5.495
EL_STA	3.434	5.468
NO_EL_STA	3.421	5.461
EL_STB	3.332	**5.442**
NO_EL_STB	3.339	5.491
Best BL	3.455	5.477
GSGP	13.325	13.627

Table 2. Results of the median training and unseen error obtained on the PPB case study over 30 independent runs for the EGSGP models, GSGP and the best BL. The best training and unseen error are highlighted in bold.

Method	Median Training Error	Median Unseen Error
EL_NO_ST	0.188	33.378
NO_EL NO_ST	1.535	33.280
EL_STA	**0.186**	33.366
NO_EL_STA	0.650	**32.915**
EL_STB	0.260	33.042
NO_EL_STB	1.536	33.282
Best BL	0.189	33.378
GSGP	36.890	39.749

Table 3. Results of the median training and unseen error obtained on the Parkinson's case study over 30 independent runs for the EGSGP models, GSGP and the best BL. The best training and unseen error are highlighted in bold.

Method	Median Training Error	Median Unseen Error
EL_NO_ST	2.607	3.221
NO_EL NO_ST	**2.587**	**3.218**
EL_STA	2.594	3.227
NO_EL_STA	2.623	3.284
EL_STB	2.607	3.223
NO_EL_STB	2.588	3.220
Best BL	2.641	3.309
GSGP	7.699	7.721

Table 4. p-values of the Mann-Whitney test for the Concrete dataset.

Method	Method vs GSCP	Method vs BL
EL_NO_ST	$< 10^{-4}$	0.17774
NO_EL NO_ST	$< 10^{-4}$	0.33137
EL_STA	$< 10^{-4}$	0.35859
NO_EL_STA	$< 10^{-4}$	0.32071
EL_STB	$< 10^{-4}$	0.18554
NO_EL_STB	$< 10^{-4}$	0.34210

analyses. As we can see from the tables, EGSGP is able to outperform traditional GSGP in all the studied test problems, both on training and unseen (i.e., testing) data. The difference between the two methods is statistically significant in all case studies, with the exception of PPB. However, when we observe the same tables and compare EGSGP with the best BL, we conclude that the frequently observed improvement on both training and unseen error is only enough to yield statically significant results on the Parkinson's case study.

Table 5. *p*-values of the Mann-Whitney test for the PPB dataset.

Method	Method vs GSCP	Method vs BL
EL_NO_ST	0.08343	0.47642
NO_EL NO_ST	0.08573	0.47937
EL_STA	0.08806	0.48526
NO_EL_STA	0.05599	0.38656
EL_STB	0.07472	0.4676
NO_EL_STB	0.08573	0.47937

Table 6. *p*-values of the Mann-Whitney test for the Parkinson's dataset.

Method	Method vs GSCP	Method vs BL
EL_NO_ST	$< 10^{-4}$	0.00031
NO_EL NO_ST	$< 10^{-4}$	0.0002
EL_STA	$< 10^{-4}$	0.00026
NO_EL_STA	$< 10^{-4}$	0.10310
EL_STB	$< 10^{-4}$	0.00036
NO_EL_STB	$< 10^{-4}$	0.00019

For each one of the studied test problems, Fig. 1 reports the error evolution on the test set, against generations, for the best EGSGP model, the best BL (represented as an horizontal line in the plots) and GSGP, while Fig. 2 reports an overall comparative performance assessment of the different EGSGP models. For the Concrete dataset, we can conclude that the EGSGP models are able to improve the fitness of the best BL without, however, yielding significant statistical results. This is true with the exception of methods NO_EL_NO_ST and NO_EL_STB. Note that both of these methods do not use elitism. Interestingly, adding a ST regularization that gives great importance to previous models' outputs (method NO_EL_STA) clearly reduces overfitting. Also, EGSGP was able to improve the fitness of the best BL, regardless of the hyper-parameter model combination, in both the PPB and Parkinson's case studies. Finally, we point out that in the Parkinson's case study, EGSGP significantly outperformed the best BL (with the exception of method NO_EL_STA).

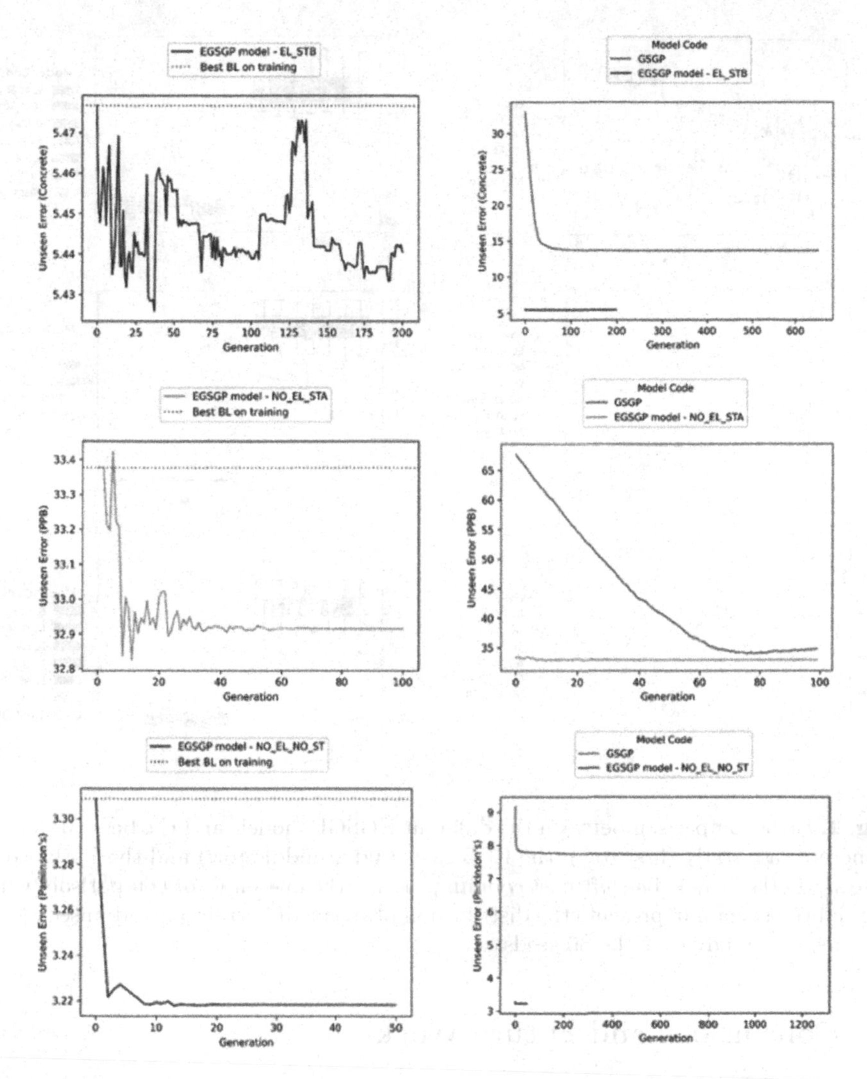

Fig. 1. Unseen error comparison between the best EGSGP method and the best BL (on the left) and the best ESGP method and GSGP (on the right) in the Concrete case study (first row), the PPB case study (second row) and the Parkinson's case study (third row).

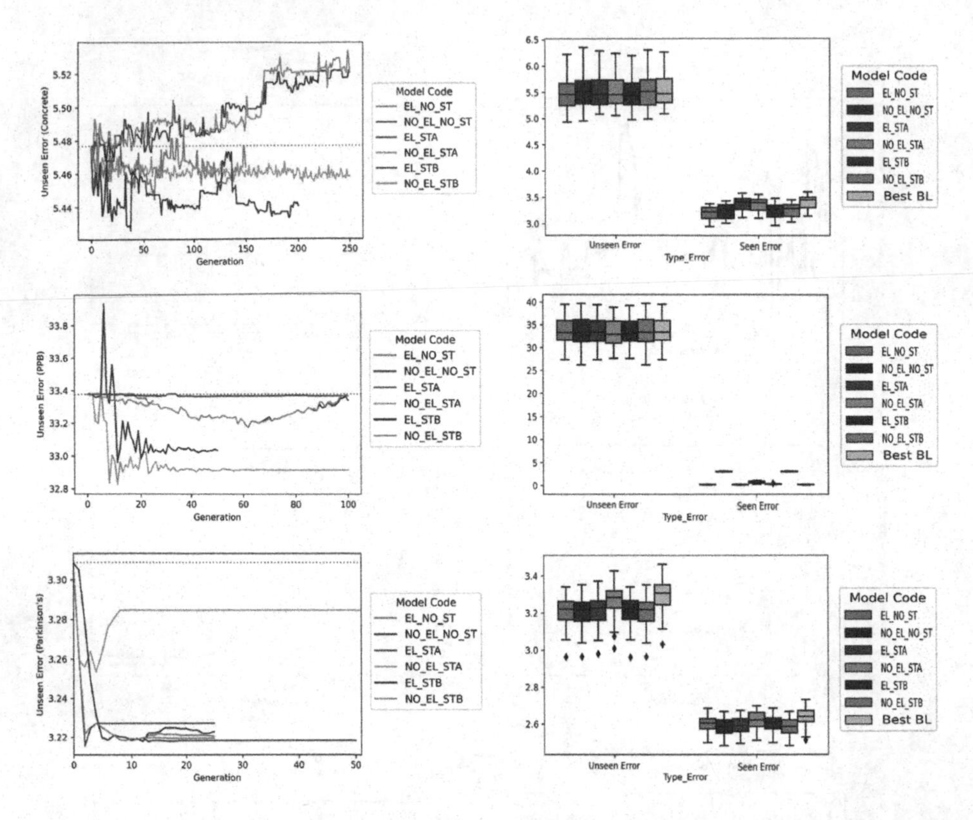

Fig. 2. Error comparison between the different EGSGP models and the best BL for the Concrete case study (first row), the PPB case study (middle row) and the Parkinson's case study (last row). The leftmost column presents the unseen error comparison, while the rightmost column presents the distribution of the median training and unseen error for each case study over the 30 seeds.

6 Conclusions and Future Work

This work exploited the idea of seeding a Genetic Programming (GP) population with models generated by other Machine Learning (ML) methods, instead of using randomly generated individuals. While tackling this problem with standard GP would have been extremely challenging, because each ML method uses its own representation, typically very different from the others, the task becomes straightforward if we use Geometric Semantic GP (GSGP). In fact, GSGP allows us to abstract from the representation of the evolving individuals, focusing purely on their semantics (i.e., the vector of their output values on the training observations). Following this idea, in this paper we have presented EGSGP, a novel method, that can be seen either as a new intialization technique for GSGP or as an ensemble method, that uses GSGP to combine different Base Learners (BLs). Given that the initial population of EGSGP is composed by pre-trained models, it is easy to expect that a further optimization of those models, given by the GSGP evolution, would lead the system to overfit training data. For this reason, we identified elitism and Soft Target (ST) regularization as elements that, if appropriately tuned, could help us limit overfitting and so several variants of EGSGP were studied. In particular, variants that use or do not use elitism, and variants that use (with different parameters) or do not use ST were studied. Those variants were compared with the best BL and with GSGP on three real-life regression problems (Concrete, PPB and Parkinson's datasets). An analysis of the presented results indicates that there is no EGSGP hyper-parameter combination that outperforms the others in all case studies. For the Concrete and PPB case studies, ST yielded the best results (with Elitism and without elitism, respectively) while the best model for the Parkinson's case study was obtained without ST and without elitism. Furthermore, the best models were, in two of the three case studies, the ones with no elitism. In general, having elitism or not affects the models' results more when the ST regularization process is started at an earlier stage of the evolution, giving more importance to the original BLs. With respect to its competitors, EGSGP was able to not only outperform GSGP, but also the best BL. This improvement yielded statistically significant results in one of the three case studies. Nonetheless, we consider the performance of EGSGP to be overall positive, given the fact that it was able to improve the best BL in all case studies, even when not enough to be deemed statistically significant. These results reflect GSGP's potential to become a powerful meta-learning combiner that is able to integrate the knowledge obtained from a set of different ML models, in a way that outputs results that are better than the ones individually obtained. Future work should focus on the reconstruction of the genotype of the final model returned by EGSGP. The task is ambitious, because it implies the definition of a middle layer of representation, common to all the ML methods used as BLs.

References

1. Zhou, Z.H.: Ensemble Methods: Foundations and Algorithms. Chapman and Hall/CRC, Boca Raton (2019)
2. Dietterich, T.G.: Ensemble methods in machine learning. In: Kittler, J., Roli, F. (eds.) MCS 2000. LNCS, vol. 1857, pp. 1–15. Springer, Heidelberg (2000). https://doi.org/10.1007/3-540-45014-9_1
3. Thrun, S., Pratt, L.: Learning to learn: introduction and overview. In: Learning to Learn, pp. 3–17. Springer, Heidelberg (1998). https://doi.org/10.1007/978-1-4615-5529-2_1
4. Vanschoren, J.: Meta-learning. In: Hutter, F., Kotthoff, L., Vanschoren, J. (eds.) Automated Machine Learning. TSSCML, pp. 35–61. Springer, Cham (2019). https://doi.org/10.1007/978-3-030-05318-5_2
5. Rokach, L.: Pattern classification using ensemble methods, vol. 75. World Scientific (2010)
6. Moraglio, A., Krawiec, K., Johnson, C.G.: Geometric semantic genetic programming. In: Coello, C.A.C., Cutello, V., Deb, K., Forrest, S., Nicosia, G., Pavone, M. (eds.) PPSN 2012. LNCS, vol. 7491, pp. 21–31. Springer, Heidelberg (2012). https://doi.org/10.1007/978-3-642-32937-1_3

7. Koza, J.R.: Genetic Programming: On the Programming of Computers by Means of Natural Selection, vol. 1. MIT press, Cambridge (1992)
8. Bakurov, I., Vanneschi, L., Castelli, M., Fontanella, F.: EDDA-V2 – an improvement of the evolutionary demes despeciation algorithm. In: Auger, A., Fonseca, C.M., Lourenço, N., Machado, P., Paquete, L., Whitley, D. (eds.) PPSN 2018. LNCS, vol. 11101, pp. 185–196. Springer, Cham (2018). https://doi.org/10.1007/978-3-319-99253-2_15
9. Vanneschi, L., Castelli, M., Silva, S.: A survey of semantic methods in genetic programming. Genet. Program. Evol. Mach. **15**(2), 195–214 (2014)
10. Vanneschi, L.: An introduction to geometric semantic genetic programming. In: Schütze, O., Trujillo, L., Legrand, P., Maldonado, Y. (eds.) NEO 2015. SCI, vol. 663, pp. 3–42. Springer, Cham (2017). https://doi.org/10.1007/978-3-319-44003-3_1
11. Pawlak, T.P., Krawiec, K.: Semantic geometric initialization. In: Heywood, M.I., McDermott, J., Castelli, M., Costa, E., Sim, K. (eds.) EuroGP 2016. LNCS, vol. 9594, pp. 261–277. Springer, Cham (2016). https://doi.org/10.1007/978-3-319-30668-1_17
12. Bakurov, I.: An initialization technique for geometric semantic genetic programming based on demes evolution and despeciation: machine learning for rare diseases: a case study. PhD thesis, NOVA IMS (2018)
13. Fan, D.W., Chan, P.K., Stolfo, S.J.: A comparative evaluation of combiner and stacked generalization. In: Proceedings of AAAI-96 Workshop on Integrating Multiple Learned Models, pp. 40–46 (1996)
14. Aghajanyan, A.: Soft target regularization: an effective technique to reduce overfitting in neural networks. In: 2017 3rd IEEE International Conference on Cybernetics (CYBCONF), pp. 1–5. IEEE (2017)
15. Vanneschi, L., Castelli, M.: Soft target and functional complexity reduction: a hybrid regularization method for genetic programming. Expert Syst. Appl. **177**, 114929 (2021)
16. Castelli, M., Vanneschi, L., Silva, S.: Prediction of high performance concrete strength using genetic programming with geometric semantic genetic operators. Expert Syst. Appl. **40**(17), 6856–6862 (2013)
17. Vanneschi, L.: Improving genetic programming for the prediction of pharmacokinetic parameters. Memetic Comput. **6**(4), 255–262 (2014)
18. Castelli, M., Vanneschi, L., Silva, S.: Prediction of the unified parkinson's disease rating scale assessment using a genetic programming system with geometric semantic genetic operators. Expert Syst. Appl. **41**(10), 4608–4616 (2014)

Real-Time Monitoring Tool for SNN Hardware Architecture

Mireya Zapata[1(✉)], Vanessa Vargas[2], Ariel Cagua[2], Daniela Alvarez[2], Bernardo Vallejo[3], and Jordi Madrenas[3]

[1] Centro de Investigación en Mecatrónica y Sistemas Interactivos - MIST, Universidad Tecnológica Indoamérica, Quito, Ecuador
mireyazapata@uti.edu.ec

[2] Universidad de las Fuerzas Armadas ESPE, Av. General Rumiñahui S/N, 17000 Sangolquí, Ecuador
vcvargas@espe.edu.ec

[3] Department of Electronic Engineering, Universitat Politècnica de Catalunya, Jordi Girona 1, 08034 Barcelona, Spain
{bernardo.javier.vallejo,jordi.madrenas}@upc.edu

Abstract. Spiking Neural Networks (SNN) are characterized by their brain-inspired biological computing paradigm. Large-scale hardware platforms are reported, where computational cost, connectivity, number of neurons and synapses, speed, configurability, and monitoring restriction, are some of the main concerns. Analog approaches are limited by their low flexibility and the amount of time and resources spent on prototype development design and implementation. On the other hand, the digital SNN platform based on System on Chip (SoC) offers the advantage of the Field-programmable Gate Array (FPGA) technology, along with a powerful Advanced RISC Machine (ARM) processor in the same chip, that can be used for peripheral control and high-bandwidth direct memory access.

This paper presents a monitoring tool developed in Python that receives spike data from a large-scale SNN architecture called Hardware Emulator of Evolvable Neural System for Spiking Neural Network (HEENS) in order to on-line display in a dynamic raster plot in real-time. It is also possible to create a plain text file (.txt) with the entire spike activity with the aim to be analyzed offline. Overall, the monitoring tool and the HEENS functionalities working together show great potential for an end-user to bring up a neural application and monitor its evolution introducing a low delay, since a FIFO is used to temporarily store the incoming spikes to give the processor time to transmit data to the PC through Ethernet bus, without affecting the neural network execution.

Keywords: Real-time neural monitoring · Spiking Neural Network · Raster plot · FPGA

C. De Stefano et al. (Eds.): WIVACE 2022, CCIS 1780, pp. 291–302, 2023.
https://doi.org/10.1007/978-3-031-31183-3_24

1 Introduction

The human brain is a source of inspiration for the development of efficient technological advances. Tasks such as speech recognition, object classification, machine translation, etc., have been implemented based on complex learning algorithms with high processing costs and high energy consumption compared to the brain [1,2]. Thus, new approaches are required for applications with size and power consumption limitations. On the one hand, different neural network compression techniques and machine learning algorithms have been developed in software-based solutions, which allow accelerating their processing, however, given their sequential nature, a consequent time delay cannot be avoided. On the other hand, hardware solutions and algorithms have been developed for neural networks that emulate physiological and structural characteristics of the brain at various levels of plausibility, called neuromorphic computing. Under this scheme, information is transmitted through patterns of action potentials or spikes under parallel processing, where the information is also time-locked. The latter are the pillars of the 3rd generation of neural networks called spiking neural networks (SNN).

In neuromorphic computing, neural spiking algorithms have been developed with different degrees of biological realism, such that the more biological features they can reproduce, the higher their computational cost will be. Among the best known are Leaky Integrate and Fire (LIF) and Izhikevich, Hodgking - Huxley, among others. Literature reports several neuromorphic architecture proposals for running spiking neural algorithms. These proposals differ in the number of neurons and synapses they can implement, the flexibility of implementation of neural and synaptic models, the type of technology, etc. Some of the most relevant architectures are BrainScale [4], Neurogrid [5], TrueNorth [6], SpiNNNaker [7], Loihi2 [8] among others. For large-scale SNN platforms such as those mentioned above, monitoring the complete network activity of all the chips that comprise it is essential to study its behavior, patterns, and evolution, especially when it is required to do so in a single terminal and without affecting its execution within the 1 ms window considered as real-time.

This paper proposes an alternative for spike real-time monitoring in an SNN implemented on the architecture called HEENS (Hardware Emulator of Evolvable Neural System for Spiking Neural Network) which is described in more detail in the following section. The interface presented consists of a hardware-software co-design that allows capturing the spikes generated from a neural network to be visualized in a raster plot, which also captures all the information in a plain text file for subsequent offline analysis. As a proof of concept, a network with 496 excitatory neurons was implemented using LIF as the neural algorithm.

In the following section, HEENS architecture is briefly explained. Section 3 details the neural algorithms and the topology implemented. Section 4 presents the methodology to develop the monitoring interface. Experimental results are shown in Sect. 5. Finally, Sect. 6 provides conclusions and future work.

2 HEENS Architecture

HEENS architecture is a flexible neuromorphic proposal that allows emulating spiking neural networks (SNN) and monitoring spike activity, both in real-time. Thanks to its reconfigurable characteristics, this architecture allows to easily scale. Its multi-chip architecture allows interconnecting of up to 127 chips in a ring topology through a high-speed serial point-to-point communication (see Fig. 1).

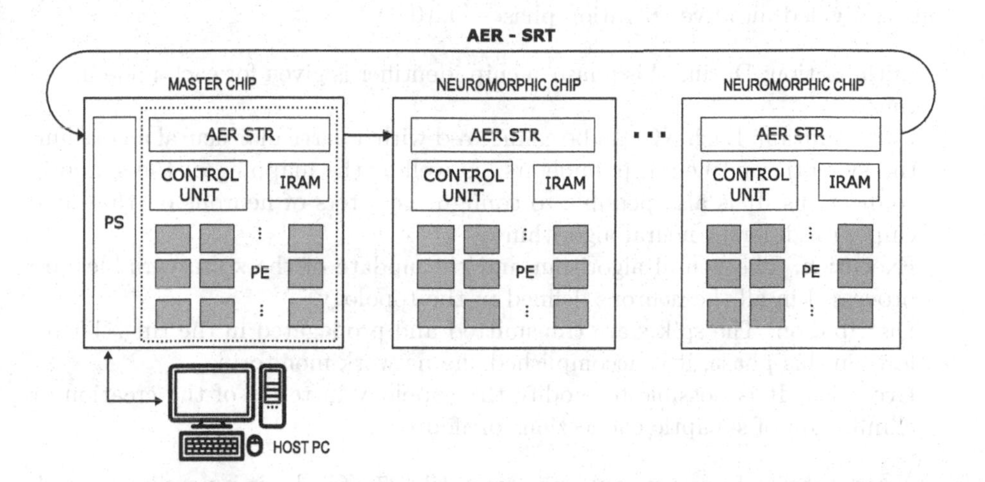

Fig. 1. HEENS architecture with one master chip and n neuromorphic chips (NC) connected as a ring.

The HEENS architecture follows a master-slave scheme, where the master chip (MC) has two main functions: a) serves as an interface between the host computer and the SNN; b) manages the configuration and the network execution. On the other hand, each neuromorphic chip (NC) is responsible for executing the neural algorithm and communicating to the network [3]. The MC consists of a processor system (PS) and a Programmable Logic (PL). The PS includes the ARM processor and the components to communicate to the PL through the Advanced eXtensible Interface(AXI) protocol. Also, the PS communicates to an external host computer(PC) by an ethernet interface. The PL includes the FPGA where neurons can be emulated, and the components to communicate to the ring Topology. It is important to highlight that the MC can operate standalone which means this can operate as a unique chip without being connected to any NC chips.

The NC run the neural algorithm through a 2D SIMD (Single Instruction Multiple Data) array of Processing Elements (PE). The spike distribution in the network is carry out through a communication controller with Address Event Representation (AER-SRT) as its communication protocol (see Fig. 1).

Each PE emulates individual biological neurons and it could emulate more than one neuron by means of time multiplexing. This multiplexing process, denoted as virtualization, does not consume extra area resources; however, it increases execution time. The architecture allows enabling until 8 virtual layers. Each PE has its own data memory where it stores the synaptic and neuronal parameters of each neuron. The control unit is in charge of managing the complete data flow and instructions of the PE array. Instructions are read from a single instruction memory block.

The HEENS processing sequence to emulate the biological behavior of neurons is divided into five operation phases [9,10]:

- Initialization: During this phase, a chip identifier is given for each node in the ring.
- Configuration: Each NC can be configured with a particular neural algorithm, the synaptic and neural parameters, as well as the mapping of the synaptic connections. It is also possible to configure clusters of neurons on the same chip with different neural algorithms.
- Execution: The neural algorithm and the update of the state variables are processed in all the neurons defined in the topology.
- Distribution: The spikes are transmitted and propagated in the ring. Therefore, in this phase, it is accomplished the network monitoring.
- Evolution: It is possible to modify the topology in terms of the creation or elimination of synaptic connections or neurons.

Even if the architecture can configure until 127 NC, also it is possible to work only with the MC operating in standalone mode since the MC can implement a HEENS multiprocessor. The system has been described in VHDL (VHSIC Hardware Description Language) and it has been synthesized on FPGA of the Xilinx 7 for the NC and ZYNQ family for the MC. The connection between the host computer and the MC is done by using ethernet protocol while the network communication uses the AER bus.

3 Neural Network Topology

To model the biological neural behavior, this work uses the LIF neural algorithm that describes the membrane voltage as a function of a) the input current produced by an input pulse (spike) received from a different neuron; b) the membrane baseline state; and c) the leak of the energy produced by the diffusion of ions in the neural membrane. The LIF algorithm is described in equation (1).

$$V(t+1) = V_r + (V(t) - V_r)k_m + \sum_i \omega_i(t)S_i(t) + B(t) \tag{1}$$

Reset condition:

$$V(t) > V_{th} \rightarrow V(t) = V_r \tag{2}$$

$V(t)$: membrane potential
S_i: post-synaptic spikes
$\omega_i(t)$: synaptic weight = 2.5 mV
$B(t)$: background noise
V_{th}: threshold voltage (same for every neuron) = –55 mV
V_r: restoring voltage (same for every neuron) = –70 mV
$k_m = e^{\frac{-1}{\tau_m}}$: decay constant ($\tau_m = 20$)

In order to test the proposal, two types of topology were chosen. The first one implements the MC operating in standalone mode (one chip) with 496 neurons distributed on 10×16 PE array with 8 virtual layers. The neural network topology is illustrated in Fig. 2a, where all neurons are excitatory.

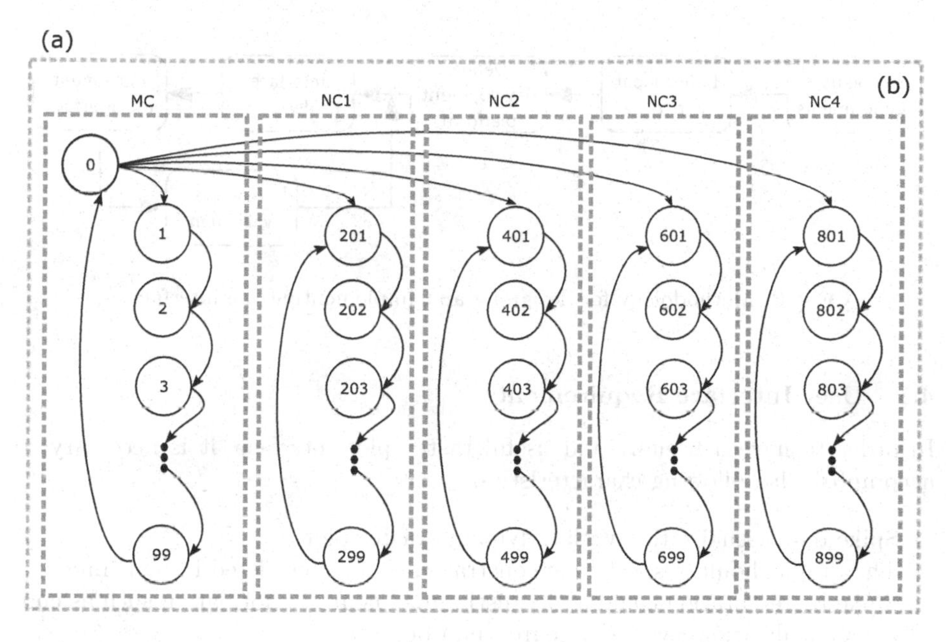

Fig. 2. Neural network topology for 496 neurons. (a) Entire neural topology implemented on MC with standalone mode (blue line), (b) The neural topology is implemented on 1 MC and 4 NC in a ring topology, and every chip is marked with a green line. (Color figure online)

The second one is a 496 neurons network distributed in 4 NC configuration with the same excitatory topology illustrated in Fig. 2b. In this case, each NPD is configured in a 5×5 PE array with 8 virtual layers. In both cases, no self-neural connections were considered. In addition, all neurons were defined in the same virtual layer with an initial membrane voltage $Vinit$. Also, a Linear Shift Feed Register (LSFR) circuit is implemented for each neuron to generate a random background noise source.

4 Neural Network Analysis and Monitoring

From the user point of view, an adequate human-machine interface (HMI) should simplify the comprehension and data analysis. When a considerable quantity of data must be analyzed at a glance, the use of a graphical interface could diminish the user effort. This work aims at proposing a friendly interface that allows to perform a static analysis of stored information about a neural activity, as well as having a real-time data chart. In the first case, the information is loaded from a database that stores neural network activity in plane text files. In the second case, the information comes from the computer host by an ethernet connection between the host and the master node of the neural network. Figure 3 illustrates the methodology to develop the monitoring interface.

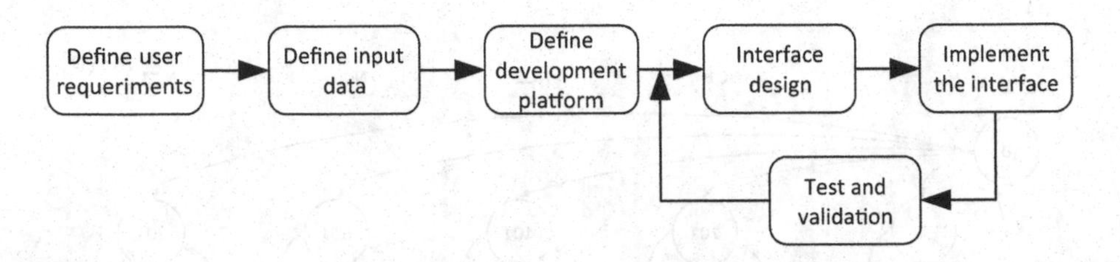

Fig. 3. Methodology for designing and implementing the interface.

4.1 User Interface Requirements

In order to have a friendly and useful raster plot interface, it is necessary to accomplish the following characteristics:

- Spike data visualization with a dynamic raster plot
- The graph should respect time constraints to be considered in real-time.
- Spike representation could easily distinguish neuron information such as chip ID, virtualization level, and neuron number.
- Display area should be customized in terms of execution cycles.
- Size of the spike representation in the plot should be customizable.
- Window message that shows neurons that have not been mapped.
- Start, pause and stop buttons to manipulate the dynamic graphic.
- Zoom and navigation buttons.

4.2 Neural Network Data

As it was aforementioned, this interface allows monitoring the neural activity in real-time or from activity saved in a plain text file. In both cases, the information has a pre-defined format established by HEENS architecture.

Format in Which Messages Are Sent from the PL to the PS in the Master Core. To transmit the neural activity in the ring node, a message consisting of 32 bits is used. A single message can send until two spikes of information of neural activity coming from the same chip id. The information is coded as follows: i)(bit_{31} to bit_{30}) indicates whether the transmitted spike number is even (10) or odd (01), ii)undefined bit (bit_{29}) iii) 7 bits for chip id (bit_{28} to bit_{22}) iv) 11 bits for information of the first spike (bit_{21} to bit_{11}), and v) 11 bits for information for the second spike (bit_{10} to bit_0). Figure 4 shows both possibilities. This information coming from the PL is sent to the ARM processor by using AXI protocol.

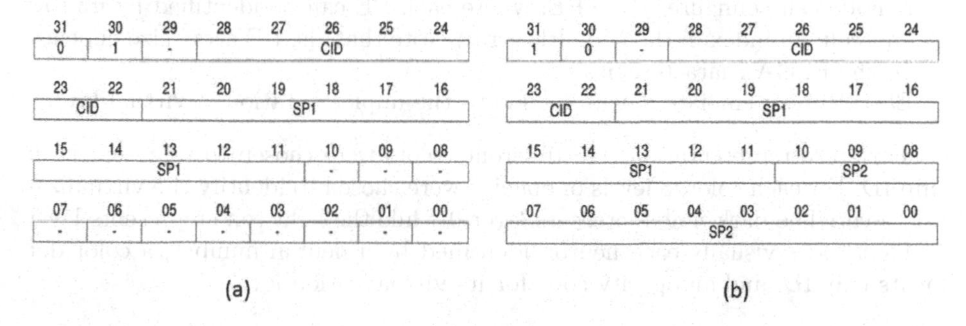

Fig. 4. (a) Spike transfer format - odd number of spikes, b) Spike transfer format - even number of spikes

Format in Which Messages Are Sent from the PS (ARM) to the PC. Before transmitting the neural information to the host computer, the MC generates a frame with a time stamp and encoded spike information. For each execution cycle (timestamp), a frame is sent through the ethernet connection. It includes the execution cycle (time stamp) of the neural activity, the chip id identifier (CID), the virtualization level (vir), row and column of the PE that fire the spike (SP). This information is stored in a text file for later analysis.

4.3 Development Platform

In order to accomplish the user interface requirements, the development platform selected for this proposal was a personal computer with a Linux Operating System by compatibility with the user toolset allowing the definition of applications (neural algorithm, topology, neural and synaptic parameters, etc.) [11].

The selected programming language was Phyton because of its portability and light execution response. In order to decode and graph the neural activity in a raster plot, it was necessary the use of different libraries such as: matplotlib.pyplot, numpy, array, sys, tkinter and FuncAnimation. The latter was initially selected because it is largely used for dynamic plotting in Phyton.

4.4 Graphical Interface Design

The design of the interface begins with decoding the stored data of the plain text file explained in Sect. 4.3. Then, to define the parameters to graph the neural activity in a raster plot, for that purpose of this proposal we consider the representation of neural activity as follows: only the spikes are represented; a different color is used for spikes coming from neurons located in each NC, and the opacity color represents the virtualization layer; dot size can be modified; the time stamp scale can be configurable. Under HEENS operating conditions the following restrictions are listed:

- The ring topology can have up to 128 NC (one MC and 127 NC)
- A node can configure $r \times c$ PE, where each PE can be identified by its row and column index in the node PE array. Note that the PE array size depends on the FPGA characteristics.
- Each PE can emulate 8 neurons due to the implementation of virtual layers.

For spike representation, 128 different colors were chosen to represent each chip ID. For each color, 8 levels of opacity were chosen to identify the virtualization layer: The dark tone represents layer 0 while the light tone represents layer 7. That codes visually each neuron identified by a neuron number, a color dot for its chip ID, and an opacity code for its virtualization level.

4.5 Raster Plot Implemented Interface

An important consideration during the development of the raster plot is the representation of several spikes in the same execution cycles. In this case, the use of FuncAnimation library tends to delay the graph. Therefore, it was necessary to develop a proprietary function to overpass this problem. Through the own implemented functions, this gap between each execution cycle was solved. Another difficulty was graphing the data coming from real-time activity. To solve this problem, a file was used to store data coming from the ethernet until data was decoded and graphed in the interface. To do this, certain programming flags were used to monitor the network in real-time.

The implemented graphical interface is shown in Fig. 5. It allows choosing the input data between real-time or stored data in a text file. The display can be scaled by selecting from 100 to 1000 timestamp units. Also, the size of the spike representation can be selected through the dot size bar. The interface also includes a message window displaying the neurons that have not been mapped fires due to some error or unwanted neural activity. Besides, to manipulate the dynamic plot there is a start, pause and stop buttons as well as zoom and navigation buttons.

5 Results

The implemented interface was tested in real-time monitoring mode, through ethernet communication with the Master Chip. The baud rate was limited to

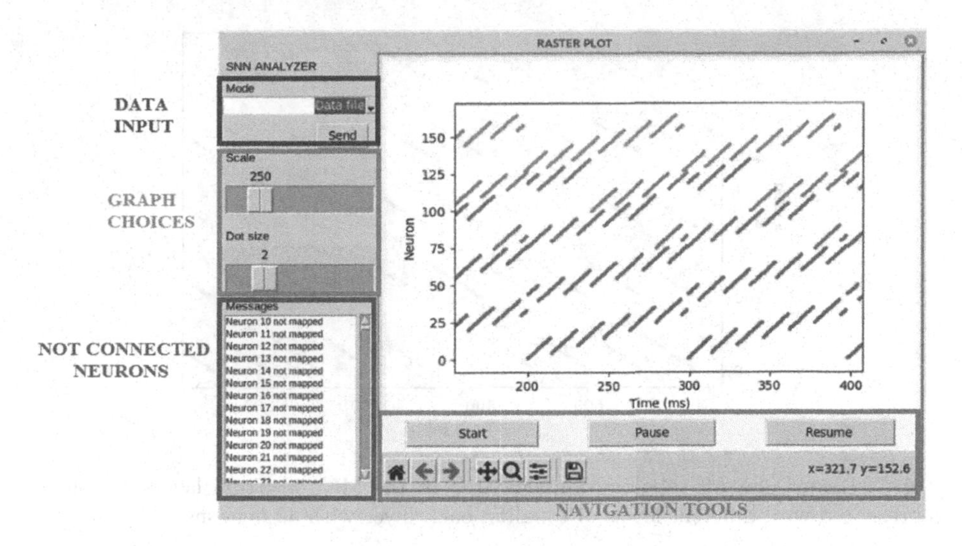

Fig. 5. Interface graphing an example file

100 Kbps due to a PS limitation. Also, a forward analysis option was proved by reading data in a plain text file (.txt). Figure 6 shows the hardware set up with Zedboard and Zynq ZC706, any of these boards can be used to implement the MC.

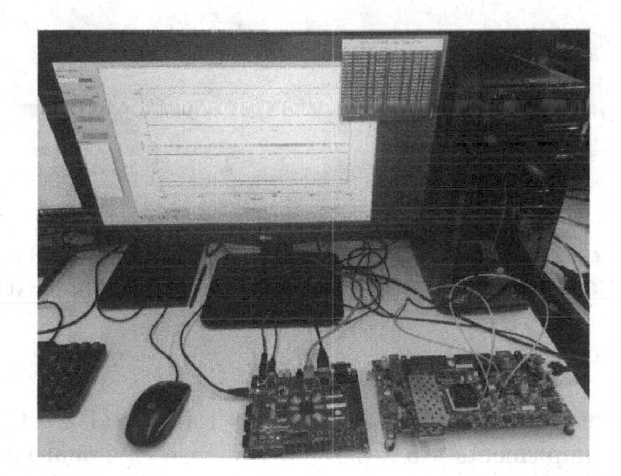

Fig. 6. HEENS architecture - hardware implementation

Figure 7, shows the raster plot for the standalone configuration, where the MC emulates 496 neurons distributed on 10×16 PE array with 5 virtual layers. Note, that, the different tone colors in the raster plot represent different virtualization layers. On the other hand, Fig. 8 illustrates the neural activity coming

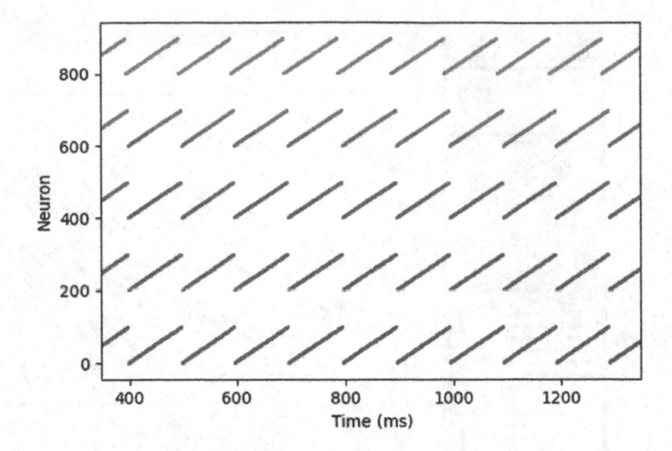

Fig. 7. Raster plot for topology implemented in the MC. 5 virtual layers are shown represented with different levels of tonality, each layer with 32 neurons.

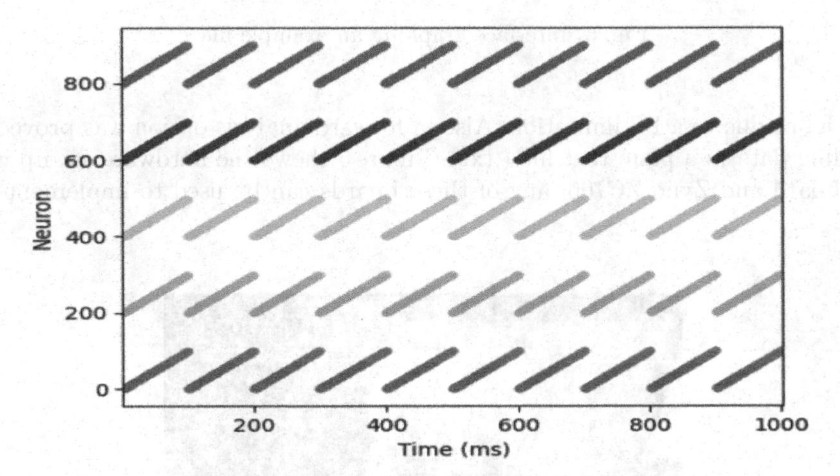

Fig. 8. Raster plot for topology implemented in 1 MC and 4 NC. Red color corresponds to MC, green to NC1, light blue to NC2, purple to NC3, and pink to NC4. (Color figure online)

from the 496 neurons network distributed in 1 MC and 4 NC configuration, where each chip implements a 5×5 PE array with 8 virtual layers. Note that each color in the raster plot represents a different chip ID. Besides, the interface flexibility allows changing the dot size that represents spikes, the scale, and graph zoom. Also, the start, pause, and resume buttons facilitate the analysis of neural activity. In addition, it allows us to easily distinguish the synapses topology as well as to detect not mapped spikes.

6 Conclusions

The interface developed permits the identification of Chip ID and 8 virtualization layers. Also, it allows monitoring all the neurons of the ring whether they are mapped or not. Finally, the spike activity can be easily represented and its evolution can be observed according to the neural algorithm computed.

From the results, it can be seen that the proposed interface allows monitoring real-time data and analyzing both online and historical data. However, real-time execution is currently limited by network throughput, which causes bottlenecks to occur in highly interconnected networks with a high rate of spike generation, increasing latency in the execution phases.

Future work considers improving port transmission speed, by evaluating Ethernet communication performance and optimizing packet encoding, to include new functionality that supports monitoring of analog neural or synaptic parameters such as membrane voltage, synaptic weights, or others.

Acknowledgements. Work supported in part under project RTI2018-099766-B-I00 and PID2021-123535OB-I00 funded by MCIN/AEI/10.13039/501100011033 and by ERDF A way of making Europe.

References

1. Frenkel, C., Bol, D., Indiveri, G.: Bottom-up and top-down neural processing systems design: neuromorphic intelligence as the convergence of natural and artificial intelligence (2021). arXiv preprint arXiv:2106.01288
2. Camuñas-Mesa, L.A., Linares-Barranco, B., Serrano-Gotarredona, T.: Neuromorphic spiking neural networks and their memristor-CMOS hardware implementations. Materials **12**(17), 2745 (2019). https://doi.org/10.3390/ma12172745
3. Guo, W., Fouda, M.E., Eltawil, A.M., Salama, K.N.: Neural coding in spiking neural networks: a comparative study for robust neuromorphic systems. Front. Neurosci. **15**(March), 1–21 (2021). https://doi.org/10.3389/fnins.2021.638474
4. Spilger, P., et al.: hxtorch: PyTorch for BrainScaleS-2: Perceptrons on analog neuromorphic hardware. Commun. Comput. Inf. Sci. **1325**, 189–200 (2020). https://doi.org/10.1007/978-3-030-66770-214
5. Benjamin, B.V., et al.: Neurogrid: a mixed-analog-digital multichip system for large-scale neural simulations. Proc. IEEE **102**(5), 699–716 (2014). https://doi.org/10.1109/JPROC.2014.2313565
6. Debole, M.V., et al.: TrueNorth: accelerating from zero to 64 million neurons in 10 years. Computer **52**(5), 20–29 (2019). https://doi.org/10.1109/MC.2019.2903009
7. Furber, S.B., Galluppi, F., Temple, S., Plana, L.A.: The SpiNNaker project. Proc. IEEE **102**(5), 652–665 (2014). https://doi.org/10.1109/JPROC.2014.2304638
8. Orchard, G., et al.: Efficient neuromorphic signal processing with loihi 2. In: IEEE Workshop on Signal Processing Systems, SiPS: Design and Implementation, 2021-October, no. 1, pp. 254–259 (2021). https://doi.org/10.1109/SiPS52927.2021.00053
9. Zapata, M., Vallejo-Mancero, B., Remache-Vinueza, B., Madrenas, J.: Monitoring implementation for spiking neural networks architecture on Zynq-7000 all programmable SoCs. In: Russo, D., Ahram, T., Karwowski, W., Di Bucchianico, G., Taiar, R. (eds.) IHSI 2021. AISC, vol. 1322, pp. 489–495. Springer, Cham (2021). https://doi.org/10.1007/978-3-030-68017-6_73

10. Zapata, M., Jadan, J., Madrenas, J.: Efficient configuration for a scalable spiking neural network platform by means of a synchronous address event representation bus. In: 2018 NASA/ESA Conference on Adaptive Hardware and Systems (AHS), 2018, pp. 241–248 (2018). https://doi.org/10.1109/AHS.2018.8541463
11. Oltra, J.A., Madrenas, J., Zapata, M., et al.: Hardware-software co-design for efficient and scalable real-time emulation of SNNs on the edge. In: IEEE International Symposium on Circuits and Systems (ISCAS), 2021, pp. 1–5 (2021). https://doi.org/10.1109/ISCAS51556.2021.9401615

Correction to: Artificial Life and Evolutionary Computation

Claudio De Stefano, Francesco Fontanella, and Leonardo Vanneschi

Correction to:
C. De Stefano et al. (Eds.): *Artificial Life and Evolutionary Computation*, CCIS 1780, https://doi.org/10.1007/978-3-031-31183-3

The following chapters were originally published electronically on the publisher's internet portal without open access:

"Computational Investigation of the Clustering of Droplets in Widening Pipe Geometries", written by Hans-Georg Matuttis, Johannes Josef Schneider, Jin Li, David Anthony Barrow, Alessia Faggian, Aitor Patiño Diaz, Silvia Holler, Federica Casiraghi, Lorena Cebolla Sanahuja, Martin Michael Hanczyc, Mathias Sebastian Weyland, Dandolo Flumini, Peter Eggenberger Hotz, Pantelitsa Dimitriou, William David Jamieson, Oliver Castell, Rudolf Marcel Füchslin;

"Network Creation During Agglomeration Processes of Polydisperse and Monodisperse Systems of Droplets", written by Johannes Josef Schneider, Alessia Faggian, Aitor Patiño Diaz, Jin Li, Silvia Holler, Federica Casiraghi, Lorena Cebolla Sanahuja, Hans-Georg Matuttis, Martin Michael Hanczyc, David Anthony Barrow, Mathias Sebastian Weyland, Dandolo Flumini, Peter Eggenberger Hotz, Pantelitsa Dimitriou, William David Jamieson, Oliver Castell, Patrik Eschle, Rudolf Marcel Füchslin;

"Artificial Chemistry Performed in an Agglomeration of Droplets with Restricted Molecule Transfer", written by Johannes Josef Schneider, Alessia Faggian, William David Jamieson, Mathias Sebastian Weyland, Jin Li, Oliver Castell, Hans-Georg Matuttis, David Anthony Barrow, Aitor Patiño Diaz, Lorena Cebolla Sanahuja, Silvia Holler, Federica Casiraghi, Martin Michael Hanczyc, Dandolo Flumini, Peter Eggenberger Hotz, Rudolf Marcel Füchslin.

With the authors' decision to opt for Open Choice the copyright of the chapters changed on 19 September 2023 to © Authors, 2023 and the chapters are forthwith

The updated versions of these chapters can be found at
https://doi.org/10.1007/978-3-031-31183-3_7
https://doi.org/10.1007/978-3-031-31183-3_8
https://doi.org/10.1007/978-3-031-31183-3_9

C. De Stefano et al. (Eds.): WIVACE 2022, CCIS 1780, pp. C1–C2, 2023.
https://doi.org/10.1007/978-3-031-31183-3_25

Funded by: the European Union's Horizon 2020 program "Artificial Cells with Distributed Cores to Decipher Protein Function" (ACDC), Grant Number: 824060.

Author Index

C. De Stefano et al. (Eds.): WIVACE 2022, CCIS 1780, pp. 303–304, 2023.
https://doi.org/10.1007/978-3-031-31183-3

Printed in the United States
by Baker & Taylor Publisher Services